Principles of modern technology

# Principles of modern technology

Adrian C. Melissinos
*Department of Physics*
*University of Rochester*

CAMBRIDGE UNIVERSITY PRESS
*Cambridge*
*New York Port Chester Melbourne Sydney*

Published by the Press Syndicate of the University of Cambridge
The Pitt Building, Trumpington Street, Cambridge CB2 1RP
40 West 20th Street, New York, NY 10011, USA
10 Stamford Road, Oakleigh, Melbourne 3166, Australia



First published 1990

Printed in Great Britain at the University Press, Cambridge

*British Library cataloguing in publication data*
Melissinos, Adrian C.
Principles of modern technology.
1. Technology
I. Title
600

*Library of Congress cataloguing in publication data available*

ISBN 0 521 35249 5 hard covers
ISBN 0 521 38965 8 paperback

MP

For
John and Andrew

# Contents

# Preface

Ours is the age of technology, rivaling the industrial revolution in its impact on the course of civilization. Whether the great achievements of technology, and our dependence on them, have improved our lot, or lead inexorably to a 'strange new world' we shall not debate here. Instead we focus on the physical laws that make technology possible in the first place. Our aim is to understand and explain modern technology, as distinct from describing it.

Even when the principles underlying a technical process or device are well understood, a great deal of engineering effort and a long manufacturing infrastructure are needed to translate them into practice. In turn, the technical skills that are developed lead to new possibilities in basic research and to new applications. For instance, the laser could have been easily built at the turn of the century; yet it was a long road starting with the development of radar and followed by the invention of the maser that led to the proposal for the laser. The use of computers in so many manufacturing areas and research fields is another example of the interplay between technology and basic science.

Because of the complexity of modern devices and of the rapid advances in all scientific fields, the need for specialization is acute. Thus, often, science students are only vaguely aware of the applications of the principles they have learned, whereas engineering students are too involved to appreciate the power of the physical law. Such disparity may carry over even into one's professional career, as modern life leaves little time for reflection. However, unless we understand the principles on which our technological developments are based, we remain simple users and are destined to be enslaved by our own inventions. It is therefore important that the common citizen be familiar with the principles of modern technology and we believe that this is possible with only modest scientific background but sufficient determination.

This book evolved from a course on the 'Physics of Modern Technology' that has been taught at the University of Rochester over the past 15 years. It was initiated by the late Elliott Montroll, a distinguished theoretical physicist of broad interests and intense intellectual curiosity. Elliott was a fascinating raconteur of scientific history and a wonderful colleague and friend. The course is directed to juniors and seniors in science or engineering, the only prerequisite being an introductory sequence in principles of flight in the atmosphere, including elements of fluid dynamics, and end with a description of the NASA shuttle. The next, and last chapter is devoted to extra-terrestrial travel. We discuss the dynamics of interplanetary missions and in particular the journey of the Voyager-2 craft. We conclude by assessing the possibilities of travel beyond the solar system and leave the reader to ponder where the future will lead us to.

The book should be useful as a general reference but also as a text. Typically, five chapters can be covered in one academic term. These can be chosen with some flexibility since the major parts are fairly independent. We have also included a few simple exercises at the end of each chapter, for the entertainment of the reader. Some material of more mathematical nature is presented in the appendices; it refers to Chapters 3 and 7 and is *not* needed for following the main text. Finally, a word of caution about units: they are mixed. To the extent possible we have tried to use MKS units, but in many cases, replacing the traditional engineering units is counter productive. To provide references to original work was neither practical nor possible. We have however included a list of books and review articles for further reading; many of these were used as sources for the course and others expand on the material presented.

I am indebted to many colleagues for their help and suggestions for the course and for their comments on various stages of the manuscript. In particular Drs R. Forward, S. Kerridge, M. Shlesinger and C. S. Wu provided me with original and published material for the course. Drs M. Bocko, S. Craxton, R. Forward, E. Jones, L. Mandel, M. Migliuolo, B. Moskowitz, R. Polling, J. Rogers and Y. Semertzidis read and commented on parts of the draft and their input was invaluable. Ms Constance Jones and Ms Judith Mack typed the many versions of the manuscript skillfully, efficiently and cheerfully and I wish to thank them sincerely. The artwork is due to Mr Roman Kucil. Finally I thank the students who took the course over the years and provided the enthusiasm and the rationale for completing the manuscript. Last but not least, thanks to my wife Joyce for her support and understanding of 'book writing'.

On a more personal note, the study of the physics of modern technology engenders a sense of admiration, but also of affection, for the men and women who had the vision, the perseverance and the good fortune to understand the physical law and use it correctly in their intriguing applications. What a wonderful adventure it must have been, an adventure which still goes on and in which, hopefully, the reader will participate in his own way.

A. C. Melissinos
Rochester, New York

# Acknowledgements

We wish to thank the following organizations for permission to use their original material. Addison-Wesley Inc. for use of Fig. 1.32; J. Wiley and Sons for use of Figs. 5.6, 6.2, 6.4(a), and 7.21; American Association of Physics Teachers and Professor T. D. Rossing for use of Fig. 2.32; McGraw-Hill Inc. for use of Figs. 6.4(b), 7.10, and 7.16; Gordon and Breach Publishers for use of Fig. 3.20; The Dallas Times Herald for use of Fig. 6.6; Itek Corporation for use of Fig. 6.12(a); U. S. Department of the Air Force for use of Fig. 6.12(b); the American Institute of Physics for use of Table 6.1; the U. S. National Aeronautics and Space Administration for use of Figs. 7.25 and 7.26; Bantam–Doubleday–Dell for use of Fig. 7.14; Jet Propulsion Laboratory of the California Institute of Technology for use of Figs. 8.13, 8.14 and 8.15.

# PART A

# MICROELECTRONICS AND COMPUTERS

Microelectronics are found today at the heart of almost every device or machine. Be it an automobile, a cash register or just a digital watch it is controlled by electronic circuits built on small semiconductor chips. While the complexity of the functions performed by these devices has increased by several orders of magnitude their size is continuously decreasing. It is this remarkable achievement that has made possible the development of powerful processors and computers and has even raised the possibility of achieving artificial intelligence.

The basic building block of all microcircuits is the transistor, invented in 1948 by John Bardeen, Walter Brattain and William Shockley at Bell Telephone Laboratories. The first chapter is devoted to a discussion of the transistor beginning with a brief review of the structure of semiconductors and of the motion of charge carriers across junctions. We discuss the $p$–$n$ junction and bipolar as well as field-effect transistors. We then consider modern techniques used in very large scale integration (VLSI) of circuit elements as exemplified by Metal-Oxide-Silicon (MOS) devices.

In the second chapter we take a broader look at how a processor, or computer, is organized and how it can be built out of individual logical circuit elements or gates. We review binary algebra and consider elementary circuits and the representation of data and of instructions; we also discuss the principles of mass data storage on magnetic devices. Finally we examine the architecture of a typical computer and analyze the sequence of operations in executing a particular task.

# 1

# THE TRANSISTOR

## 1.1 Intrinsic semiconductors

It is well known that certain materials conduct electricity with little resistance whereas others are good insulators. There also exist materials whose resistivity is between that of good conductors and insulators, and is strongly dependent on temperature; these materials are called *semiconductors*. Silicon (Si), germanium (Ge) and compounds such as gallium arsenide (GaAs) are semiconductors, silicon being by far the most widely used material. Solids, in general, are crystalline and their electrical properties are determined by the atomic structure of the overall crystal. This can be understood by analogy to the energy levels of a free atom.

A free atom, for instance the hydrogen atom, exhibits discrete energy levels which can be exactly calculated. A schematic representation of such an energy diagram is shown in Fig. 1.1(*a*). If two hydrogen atoms are coupled, as in the hydrogen molecule, the number of energy levels doubles as shown in part (*b*) of the figure. If the number of atoms that are coupled to each other is very large – as is the case for a crystal – the energy levels coalesce into *energy bands* as in Fig. 1.1(*c*). The electrons in the crystal can only have energies lying in these bands.

When an atom is not excited the electrons occupy the lowest possible energy levels. In accordance with the Pauli principle only two electrons (one with spin projection up and the other down) can be found at any one particular energy level. Thus the levels – or states – become progressively filled from the bottom. The same holds true in the crystal. The electrons progressively fill the energy levels within a given band, and only when the band is completely filled do they begin to populate the next band. The energies of the electrons are typically few *electron-Volts* (eV).

In an insulator the occupied energy bands are completely filled. As a result the electrons cannot move through the crystal. This is because motion implies slightly increased energy for the electrons but the next available energy level is in the conduction band which is far removed from the valence band. Thus the electron must acquire enough energy to overcome the *energy gap* between the *valence band* and the *conduction band* as shown in Fig. 1.2(*a*). In a conductor the valence and conduction bands overlap and the outermost electron of the atom is free to move through the lattice (Fig. 1.2(*c*)). In a semiconductor the energy gap is much smaller than for insulators and due to thermal motion electrons have a finite probability of finding themselves in the conduction band.

Fig. 1.1. Energy levels of an atomic system: (*a*) single atom, (*b*) two coupled atoms, (*c*) in a many-atom system the energy levels coalesce into energy 'bands'.

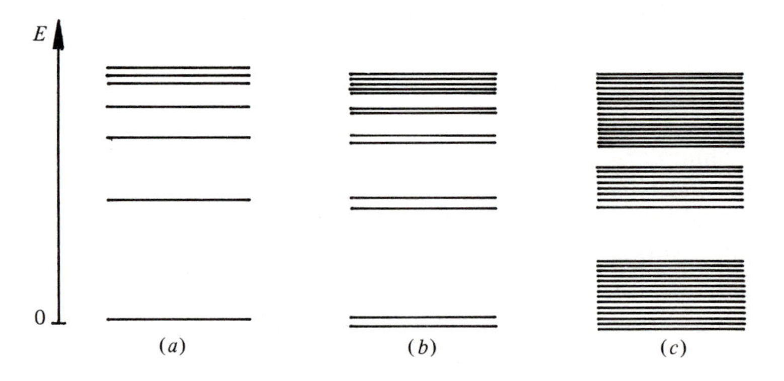


Fig. 1.2. Energy band structure for: (*a*) an insulator such as $SiO_2$, (*b*) a semiconductor, such as Ge, (*c*) a good conductor such as Al.

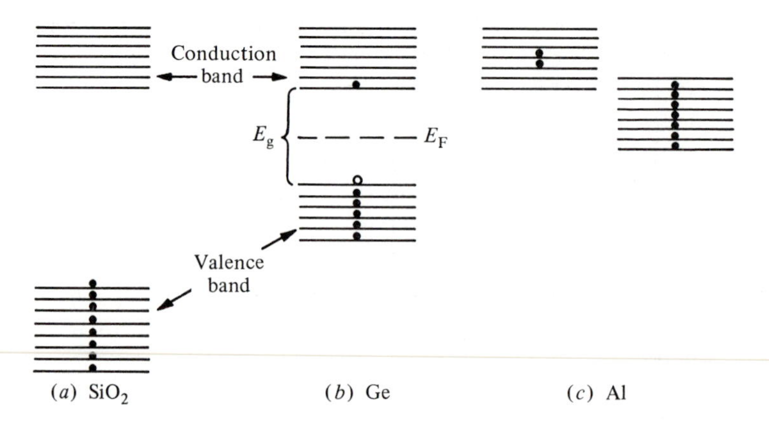

Furthermore when an electron makes a transition from the valence to the conduction band it leaves a vacancy in the valence band. This vacancy can move through the lattice (just as a bubble 'moves' through a liquid) and contribute to the flow of current; we speak of transport of electric charge by the motion of *holes* (Fig. 1.2(*b*)).

To obtain a feeling for the occupancy of the energy levels in a solid we can consider the following simple model for a conductor. We assume that one electron in each atom is so loosely bound that it is practically free inside the crystal. This is the case for copper which has $Z = 29$, and thus every atom has 29 electrons. Of these, 28 electrons completely fill the $n = 1$ (2 electrons), $n = 2$ (8 electrons) and $n = 3$ (18 electrons) shells, leaving one electron outside the closed shells. Such an electron is loosely bound to the atom and in fact it occupies a level in the conduction band; thus it can move freely through the crystal. It is simple to calculate the density of free electrons in copper. We have $Z = 29$, $A \simeq 63$, $\rho = 8.9\ \text{g/cm}^3$ and assume one free electron per atom; then

$$n_e = \frac{N_0}{A}\left(\frac{\text{atoms}}{\text{g}}\right) \times \rho\left(\frac{\text{g}}{\text{cm}^3}\right) = \frac{6 \times 10^{23}}{63} 8.9 = 8.5 \times 10^{22} \frac{\text{electrons}}{\text{cm}^3} \tag{1.1}$$

where $N_0 = 6 \times 10^{23}$ is Avogadro's number.

The free electrons in a metal can be described approximately as particles confined within a cubic box but with no other forces acting on them. This situation is depicted for one dimension in Fig. 1.3 and we speak of a 'potential well' of length $2L$. In this case the solution of Schrödinger's equation leads to wave functions of the form

$$\psi_n(x) = (1/\sqrt{L})\cos(k_n x) \qquad \text{or} \qquad (1/\sqrt{L})\sin(k_n x)$$

Fig. 1.3. The wave function for the lowest and next to lowest energy states of a particle confined to the region $-L < x < L$ by an infinitely high potential.

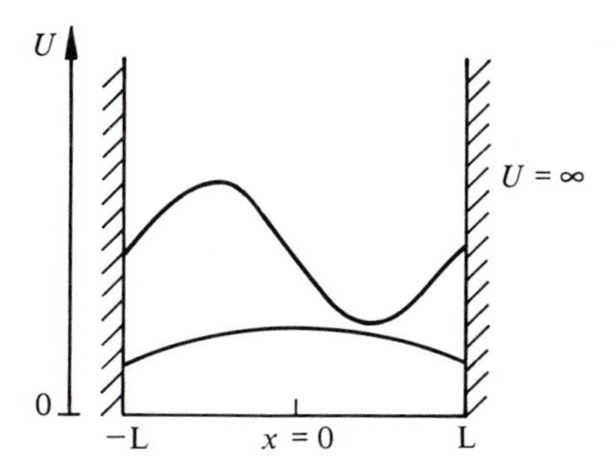

where the wave number $k_n$ can take only the discrete values

$$k_n = n\frac{\pi}{2L} \qquad n = 1, 2, 3, \ldots \tag{1.2}$$

so as to satisfy the boundary conditions $\psi(-L)=\psi(L)=0$. Thus the allowed energies of the particles in the potential well are quantized and given by

$$E_n = \frac{p^2}{2m} = \frac{\hbar^2 k_n^2}{2m} = \frac{\hbar^2\pi^2}{8mL^2}n^2 \tag{1.3}$$

If we generalize to three dimensions, we must use three quantum numbers, $n_x$, $n_y$ and $n_z$ and the energy is given by

$$E_n = \frac{\hbar^2\pi^2}{8mL^2}(n_x^2 + n_y^2 + n_z^2) \qquad n_x, n_y, n_z = 1, 2, 3, \ldots \tag{1.3'}$$

Every particular combination of $n_x, n_y, n_z$ represents a different energy level and only two electrons can occupy it. Note that several energy levels (different combinations of $n_x, n_y, n_z$) can have the same energy; we say that these levels are degenerate.

We can use Eq. (1.3′) to calculate the energy of the highest filled level given the density of free electrons $n_e$ in the crystal. This level is called the *Fermi level* and its energy is the Fermi energy for the system. It is given by

$$E_F = \frac{\hbar^2}{2m}(3\pi^2 n_e)^{2/3} \tag{1.4}$$

where $m$ is the mass of the electron. For Cu we use the result of Eq. (1.1) and

$$\hbar c = 2 \times 10^{-5} \text{ eV-cm}$$

$$mc^2 = 0.5 \times 10^6 \text{ eV}$$

to find $E_F = 7.1$ eV, in good agreement with observation. To see how Eq. (1.4) is derived we must count the number of $(n_x, n_y, n_z)$ combinations available when the maximal value of $(n_x^2 + n_y^2 + n_z^2)$ is specified. In Fig. 1.4 every combination of $(n_x, n_y, n_z)$ is indicated by a dot in 3-dimensional space. When $n_x, n_y, n_z$ are large, a given value of $(n_x^2 + n_y^2 + n_z^2)^{1/2} = \text{constant}$ defines the surface of a sphere in this space; all levels on the surface of the sphere have the same energy. The number of levels inside the sphere equals its volume, because the dots are spaced one unit apart from one another. Since $n_x, n_y, n_z$ must be positive the number of combinations $N_c$ is given by the volume of one octant

$$N_c = \tfrac{1}{8}(\tfrac{4}{3}\pi R^3) = \frac{\pi}{6}[(n_x^2 + n_y^2 + n_z^2)_{\max}]^{3/2}$$

Because of the Pauli principle the number of electrons occupying the $N_c$

levels is $N_e = 2N_c$. Thus the energy of the highest occupied level is

$$E_{\max} = \frac{\hbar^2\pi^2}{8mL^2}(n_x^2 + n_y^2 + n_z^2)_{\max}$$

$$= \frac{\hbar^2\pi^2}{8mL^2}\left(\frac{3}{\pi}N_e\right)^{2/3} = \frac{\hbar^2}{2m}\left[3\pi^2\frac{N_e}{8L^3}\right]^{2/3} \qquad (1.4')$$

Note that the $N_e$ electrons are confined in a volume of size $V = (2L)^3$ and therefore in Eq. (1.4′) $(N_e/8L^3) = n_e$ is the free electron density establishing the result of Eq. (1.4).

Let us now return to the free electron model. In the absence of excitations, that is at very low temperature, only the levels below the Fermi energy, $E_F$ will be occupied. Let $f(E)$ indicate the probability that a level at energy $E$ is occupied; clearly $f(E)$ is bounded between 0 and 1. If we plot $f(E)$ as a function of $E$ for $T = 0$ it must have the square form indicated by curve $A$ in Fig. 1.5. As the temperature increases some of the levels above $E_F$ will become occasionally occupied, and correspondingly some levels below $E_F$ will be empty. The probability of occupancy, $f(E)$ for a finite temperature $T_1 \neq 0$ is indicated by curve $B$ in Fig. 1.5. The

Fig. 1.4. Counting the number of states labeled by the integers $n_x$, $n_y$ and $n_z$ such that $(n_x^2 + n_y^2 + n_z^2) \leqslant R^2$. Each state is represented by a dot.

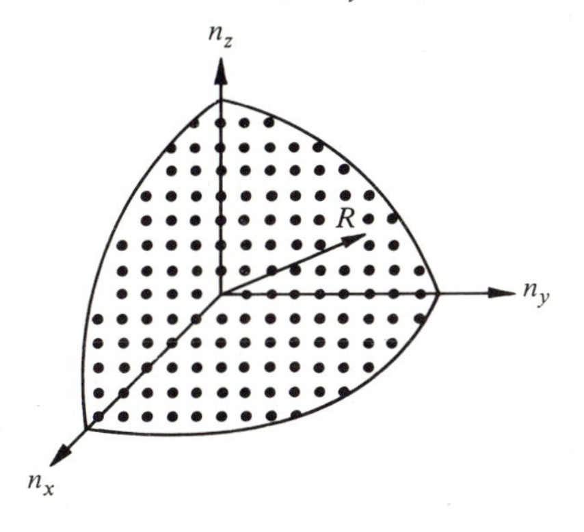


Fig. 1.5. The Fermi distribution function for zero temperature ($A$) and for finite temperature ($B$); $E_F$ is the Fermi energy.

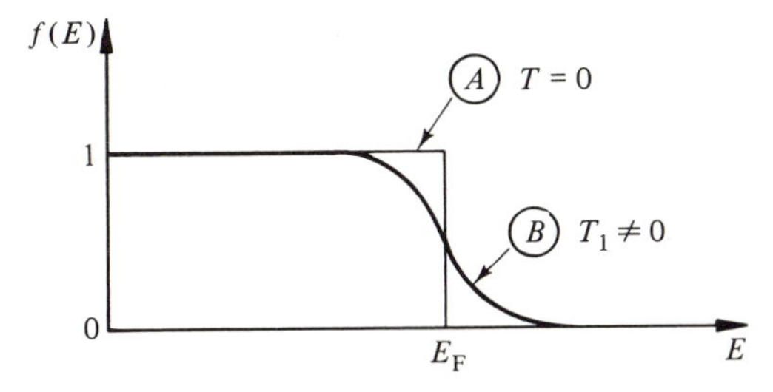

function $f(E)$ is known as the *Fermi function* and is given by

$$f(E)=\frac{1}{e^{(E-E_F)/kT}+1} \tag{1.5}$$

In the limit $T \to 0$ Eq. (1.5) reduces to $f(E)=1$ if $E<E_F$ or to $f(E)=0$ if $E>E_F$ in agreement with curve $A$ of Fig. 1.5.

For finite $T$, consider an energy level $E_k$ lying above $E_F$; we define $\varepsilon=(E_k-E_F)$. As long as $\varepsilon \gtrsim 3kT$ Eq. (1.5) can be approximated by

$$f(\varepsilon)\simeq e^{-\varepsilon/kT} \tag{1.6}$$

For a level $E_j$ lying below $E_F$ we define $\varepsilon'=(E_F-E_j)$. We are now interested in the probability that the level $E_j$ is empty, namely in $f'(\varepsilon')=1-f(\varepsilon')$. As long as $\varepsilon' \gtrsim 3kT$ a valid approximation to Eq. (1.5) is

$$1-f(\varepsilon')=1-\frac{1}{e^{-\varepsilon'/kT}+1}=e^{-\varepsilon'/kT} \tag{1.6'}$$

Eqs. (1.6) show that at finite temperature there are as many occupied states above the Fermi level as there are empty states below it. This result can serve as a rigorous definition of the Fermi level. Finally we note that the expansions of Eqs. (1.6, 6′) coincide with the classical Boltzmann distribution.

To get a better feeling for the implications of the Fermi function on the distribution of carriers in a semiconductor we first calculate $kT$ at room temperature. Boltzmann's constant

$$k=1.38\times10^{-23}\ \mathrm{J/K}$$

and if we take

$$T=300\ \mathrm{K}$$

$$kT=4.1\times10^{-21}\ \mathrm{J}=0.026\ \mathrm{eV}$$

The energy gap for an insulator is of order $\Delta E\sim 5\ \mathrm{eV}$ whereas for semiconductors it is $E_g\sim 1\ \mathrm{eV}$. Thus for semiconductors at room temperature a small fraction of the electrons in the valence band can be thermally excited into the conduction band.

For a pure semiconductor we designate the number density of (intrinsic) electrons in the conduction band by $n_i$. For an intrinsic semiconductor the density of holes will also equal $n_i$ and therefore the Fermi level will lie in the middle of the energy gap as shown in Fig. 1.2($b$). The *intrinsic carrier density* is then given by the probability of occupancy $f(\varepsilon)$ multiplied by $N_s$ the number of available states per unit volume. Using Eq. (1.6) we find

$$n_i=N_s e^{-E_g/2kT} \tag{1.7}$$

($N_s$ is an effective density of states near the band edge and for silicon it is of order $\sim 10^{19}\,\text{cm}^{-3}$). In general $n_i$ is much smaller than the free electron density in a good conductor. For instance, for silicon where $E_g = 1.1$ eV, at room temperature $n_i \simeq 10^{10}\,\text{cm}^{-3}$, whereas for germanium ($E_g = 0.7$ eV), $n_i \simeq 10^{13}\,\text{cm}^{-3}$. This should be compared to the free electron density in copper which we calculated to be $n_f \sim 10^{23}\,\text{cm}^{-3}$. Of course the crystal as a whole remains electrically neutral, but if an electric field is applied the carriers will be set in motion and this will lead to the transport of charge. It is evident from Eq. (1.7) that the conductivity of a pure semiconductor will be highly temperature dependent.

## 1.2 Doped semiconductors

We saw in the previous section that the intrinsic carrier densities are quite small. Thus, unless a semiconductor is free of impurities to a high degree, the phenomena associated with the motion of the intrinsic carriers will not be manifest. On the other hand, by introducing a particular impurity into the semiconductor one can greatly enhance the number of carriers of one or of the other kind (i.e. of electrons or of holes). The great technical advances in selectively and accurately controlling the concentration of impurities in silicon have made possible the development of microelectronics. We speak of doped semiconductors.

To understand the effect of doping we note that the electronic structure of Si or Ge is such as to have four electrons outside closed shells; they are elements of chemicals valence 4.

| | | | Filled shells | Valence |
|---|---|---|---|---|
| Si | $Z = 14$ | $A \sim 28$ | $(n = 1, n = 2)_{10}$ | $(3s)_2\ (3p)_2$ |
| Ge | $Z = 32$ | $A \sim 72$ | $(n = 1, n = 2, n = 3)_{28}$ | $(4s)_2\ (4p)_2$ |

If one examines the periodic table in the vicinity of Si and Ge, one finds the valence 3 elements boron (B, $Z = 5$), aluminum (Al, $Z = 13$) or indium (In, $Z = 49$). On the other side are valence 5 elements such as phosphorus (P, $Z = 15$), arsenic (As, $Z = 33$) or antimony (Sb, $Z = 51$). What will happen if impurities from these elements are introduced into pure silicon?

If valence 5 elements are introduced into the silicon lattice the extra electron will be loosely bound and can be easily excited into the conduction band. We say that these elements are *donor* impurities. If valence 3 elements are introduced they will have an affinity for attracting an electron from the lattice, creating a vacancy or hole in the valence band. We say that valence 3 elements are *acceptor* impurities.

Because of their different electronic structure as compared to that of the crystal lattice, the donor levels are situated just below the conduction band, as shown in Fig. 1.6(*a*). The acceptor levels are instead located slightly above the valence band (Fig. 1.6(*b*)). This energy difference is so small that at room temperature the impurity levels are almost completely ionized. Thus in the case of donor impurities the charge carriers are electrons and we speak of an *n-type* semiconductor whereas for acceptor impurities the carriers are the holes and we speak of a *p-type* semiconductor. Recall that the crystal is always electrically neutral and that the charge of the carriers is compensated by the (opposite) charge of the ionized impurity atoms, the ions however, remain at fixed positions in the lattice.

In the presence of impurities the position of the Fermi level is determined by the concentration of the impurities and moves toward the conduction band if the dominant free carriers are electrons, toward the valence band if the dominant free carriers are holes. This is sketched in Fig. 1.6. The position of the donor level is indicated by the plus signs in Fig. 1.6(*a*), of the acceptor level by the minus signs in (*b*) of the figure.

As an example we consider an *n*-type semiconductor, and as usual, designate the (extrinsic) conduction electron density by $n$, and the hole density by $p$. Then according to Eqs. (1.6, 6′)

$$\left.\begin{aligned} n &= N_c e^{-(E_c - E_F)/kT} \\ p &= N_v e^{-(E_F - E_v)/kT} \end{aligned}\right\} \tag{1.8}$$

Here we introduced a new concept, the effective density of states $N$. This is the number of available energy states per unit volume, the subscripts c and v referring to the condition and valence band correspondingly. In general $N_c$ and $N_v$ need not be equal to one another.

Similar relations hold for the intrinsic carriers except that we designate

Fig. 1.6. Energy band diagram for doped semiconductors. Dots represent electrons and open circles holes: (*a*) for an *n*-type semiconductor (note the position of the donor level), (*b*) for a *p*-type semiconductor (note the position of the acceptor level).

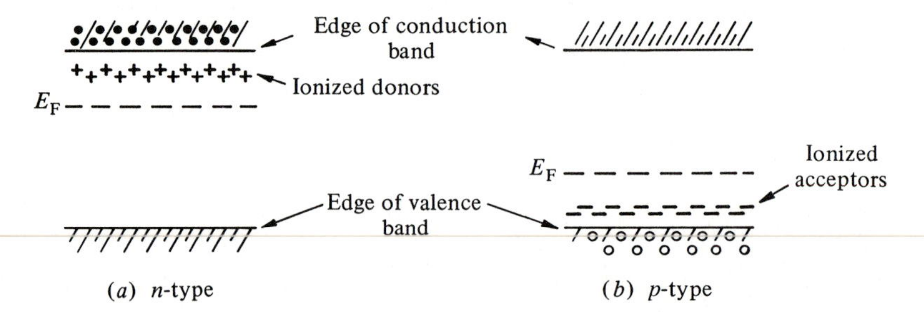

the corresponding Fermi level by $E_i$, and $n_i$ must equal $p_i$. Thus

$$n_i = N_c e^{-(E_c - E_i)/kT} = p_i = N_v e^{-(E_i - E_v)/kT} \tag{1.9}$$

This relationship can be solved to yield the exact value of $E_i$

$$E_i = \tfrac{1}{2}(E_c + E_v) + \tfrac{1}{2}kT \ln(N_v/N_c) \tag{1.9'}$$

as well as a convenient expression for $n_i$

$$n_i = (N_v N_c)^{1/2} e^{-(E_c - E_v)/2kT} \tag{1.9''}$$

Finally we can multiply the two Eqs. (1.8) with one another

$$np = N_c N_v e^{-(E_c - E_v)/kT}$$

and by comparing with Eq. (1.9″) obtain the very important relation

$$np = n_i^2 \tag{1.10}$$

The product of the electron and hole densities is independent of the doping and depends on the intrinsic properties of the semiconductor and the temperature. This is true under equilibrium conditions and provided the intrinsic carriers are not highly excited.

In a doped semiconductor we have majority and minority carriers. For instance in an $n$-type semiconductor the electrons are the *majority* carriers and the holes the *minority* carriers; the opposite is of course true for $p$-type semiconductors. The density of ionized donors and acceptors is designated by $N_D$ and $N_A$ respectively. Then if the electrons are the majority carriers it holds

$$N_D \gg N_A \qquad \text{and} \qquad n \sim N_D$$

We can obtain more accurate relations by taking into account the electrical neutrality of the crystal. The charge density $\rho$ must equal zero and therefore

$$\rho = q(p - n + N_D - N_A) = 0 \tag{1.11}$$

Solving Eq. (1.11) for $p$ and inserting the result in Eq. (1.10) we obtain a quadratic equation for $n$, whose solution is

$$n = \tfrac{1}{2}(N_D - N_A) \pm \tfrac{1}{2}[(N_D - N_A)^2 + 4n_i^2]^{1/2}$$

For an $n$-type semiconductor, where $(N_D - N_A) > 0$ we must keep the solution with the positive radical. And if $(N_D - N_A) \gg n_i$ we have

$$\begin{aligned} n_n &= (N_D - N_A) \sim N_D \\ p_n &= \frac{n_i^2}{(N_D - N_A)} \sim \frac{n_i^2}{N_D} \end{aligned} \tag{1.12}$$

where the subscript indicates the type of semiconductor. For instance, $n_n$ or $p_p$ are majority carrier concentrations. Similar relations are valid for $p$-type semiconductors.

Thus we see that by the controlled introduction of impurities we can create materials with a particular type of majority carriers. It is the junction of two or more such materials that makes possible the control and amplification of electric current by solid state devices.

## 1.3 Charge transport in solids

One is familiar with the notion of an electric current 'flowing' through a wire. What we are referring to is the transport of electric charges through the wire, and this in turn is a consequence of the motion of the carriers in the wire. In a good conductor the carriers are electrons, while in a gas discharge or in a liquid the carriers are both electrons and positive ions. In a semiconductor the carriers are electrons, or holes, or both, depending on the material. The current at a point $x$ along the conductor is defined as the amount of charge crossing that point in unit time $I = \Delta Q/\Delta t$. It is more convenient to use the current density $\mathbf{J}$ which is the amount of charge crossing unit area (normal to the direction of $\mathbf{J}$) per unit time. By definition then

$$\mathbf{J} = qn\mathbf{v}_d \tag{1.13}$$

Here $q$ is the charge of the electron (carrier), $n$ is the carrier density, and $\mathbf{v}_d$ is the *drift velocity* of the carriers.

The carriers in a solid are in continuous motion because of their thermal energy. This motion is completely random as the carriers scatter from the lattice and it does not contribute to *net* transport of charge. Thus to transport charge a drift velocity must be superimposed on the random motion. This can be achieved by applying an external electric field. The motion is then modified as shown schematically in Figs. 1.7(*a*), (*b*). (We have assumed that the carriers are electrons so their

Fig. 1.7. Idealized motion of free electrons in a metal: (*a*) in the absence of an external electric field, (*b*) in the presence of an external electric field a net drift current is established.

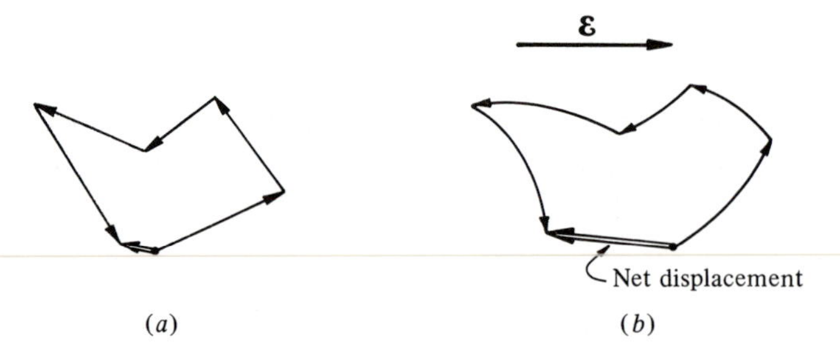

motion is opposite to the direction of $\mathscr{E}$.) Another cause for net carrier motion is the presence of density gradients. The carriers will then move so as to equalize the density and we speak of diffusion. Finally, carriers can be lost by recombination with impurities, or conversely, they can be created by photo-ionization or thermal excitation.

We first examine the motion of carriers under the influence of an electric field $\mathscr{E}$. The acceleration of a charged particle will be $\mathbf{a} = \mathbf{F}/m = q\mathscr{E}/m^*$ where we have replaced the mass, $m$, of the particle by an *effective mass* $m^*$ because the carriers do not move in free space but in the lattice. If the time between collisions is $t_{\text{coll}}$ then the average or drift velocity in the direction of the electric field will be

$$\mathbf{v}_{\text{d}} = \tfrac{1}{2}\mathbf{a}t_{\text{coll}} = q\,\frac{t_{\text{coll}}}{2m^*}\,\mathscr{E} \tag{1.14}$$

Namely, the drift velocity is proportional to the electric field $\mathscr{E}$. This is a very general result and the proportionality coefficient is called the *mobility* $\mu$. Thus

$$\mathbf{v}_{\text{d}} = \mu\mathscr{E} \tag{1.14$'$}$$

From Eqs. (1.14, 14′) we can express the current density as

$$\mathbf{J} = qn\mu\mathscr{E} \tag{1.15}$$

This result is equivalent to Ohm's law which states that the current density is proportional to the electric field and is related to it by the *conductivity* $\sigma$

$$\mathbf{J} = \sigma\mathscr{E} \tag{1.15$'$}$$

Thus

$$\sigma = qn\mu \tag{1.16}$$

Conductivity has dimensions of ($\text{ohm}^{-1}\,\text{m}^{-1}$) and has been recently defined as the 'siemens'. When both types of carriers contribute to the transport of charge, Eq. (1.16) must be modified to read

$$\sigma = q(n\mu_- + p\mu_+) \tag{1.16$'$}$$

The inverse of the conductivity is the *resistivity*, $\rho$, and the resistance of a conductor of cross sectional area $A$ and length $L$ is given by

$$R = \rho\,\frac{L}{A} = \frac{1}{\sigma}\frac{L}{A} \tag{1.16$''$}$$

We can evaluate the mobility if we knew the time between collisions $t_{\text{coll}}$. Instead, it is convenient to introduce the mean free path (m.f.p.), $l$, between collisions. Then, $t_{\text{coll}} = l/v_{\text{rms}}$ where $v_{\text{rms}}$ is the velocity due to thermal motion. We can write

$$\tfrac{1}{2}m^*(v_{\text{rms}})^2 = \tfrac{3}{2}kT$$

and therefore $v_{\text{rms}} = (3kT/m^*)^{1/2}$, so that from Eq. (1.14)

$$\mu = \frac{ql}{2(3kTm^*)^{1/2}}$$

Thus the mobility is a property of the crystal and depends on the temperature. As the electric field is increased the drift velocity increases and reaches a saturation value $v_s$. Typical values for negative carriers in silicon are

$$\mu \sim 10^2\ \text{cm}^2/\text{V-s} \qquad v_s \sim 10^7\ \text{cm/s}$$

In general, the mobility of the positive carriers is much smaller than that of the negative carriers.

When density gradients are present in the solid, the carriers will *diffuse* from regions of high concentration to those of lower concentration. The flux of carriers is proportional to the density gradient. In one dimension we have

$$\mathscr{F}_x = -D\frac{\mathrm{d}n}{\mathrm{d}x} \tag{1.17}$$

where $D$ is the diffusion coefficient. We can expect that the diffusion coefficient is related to the mobility of the carriers, and this relationship was first established by Einstein. One finds that

$$D = \frac{kT}{q}\mu \tag{1.17$'$}$$

Therefore the current due to diffusion is given by

$$J_x = -\mu kT\left(\frac{\mathrm{d}n}{\mathrm{d}x}\right) \tag{1.17$''$}$$

a result that should be compared to Eq. (1.15).

In addition to the drift and diffusion currents, carriers may be being lost due to recombination. Recombination often takes place at traps, which are locations in the crystal where a hole is trapped near the conduction band. In many semiconductors, under the influence of light or other radiation, an electron can become excited from the valence to the conduction band, increasing the density of carriers; thus a photocurrent can flow through the circuit if it is suitably biased.

## 1.4 The *p–n* junction

So far we have considered current flow in semiconductors which were uniformly doped to make *n*-type or *p*-type material. If two such semiconductor materials of different type are joined the current flow through the junction depends on the polarity of the external bias. The technology for making *p–n* junctions is an important development which we discuss later. For the analysis of the junction it is sufficient to use a one-dimensional approximation as shown in Fig. 1.8. We assume that for the *p*-type material (to the left of the junction) the ionized acceptor density is $N_A$, while for the *n*-type material the ionized donor density is $N_D$ and that $N_A > N_D$; this is shown in Fig. 1.8(*a*). In the idealized case the distribution of positive carriers follows the impurity distribution and would be as in Fig. 1.8(*b*), where the numbers in brackets are typical

Fig. 1.8. Carrier density distribution in the immediate vicinity of a *p–n* semiconductor junction: (*a*) donor and acceptor densities define a step junction, (*b*) positive and (*c*) negative carrier densities for the idealized case, (*d*) and (*e*) represent the realistic equilibrium distribution of carrier densities. Values in brackets are typical concentrations per cm$^3$.

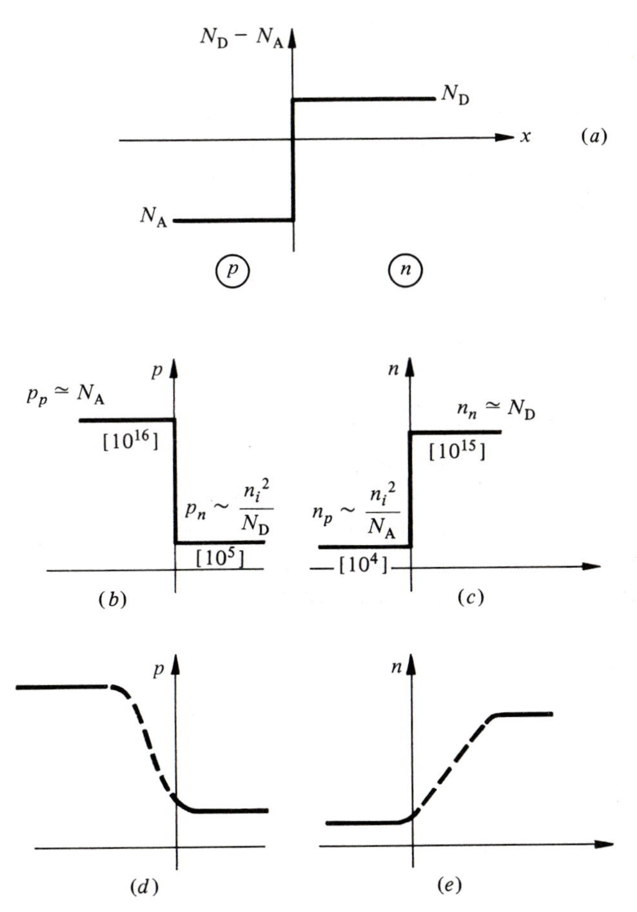

values. To the left of the junction the holes are the majority carriers while they are minority carriers to the right of the junction; the converse is true for the negative carriers as indicated in Fig. 1.8(*c*). For silicon at room temperature $n_i = 10^{10}$, so we assume that $np = n_i^2 = 10^{20}$.

The idealized distributions shown in Figs. 1.8(*b*), (*c*) are modified in practice because the majority carriers diffuse across the junction. As the holes move into the *n*-type material they very quickly recombine with the free electrons and this results in a *reduction* of the majority carriers to the right of the junction; similarly, as the electrons diffuse into the *p*-type material recombination takes place reducing the majority carriers to the left of the junction. Thus the carrier distribution has the form shown in (*d*), (*e*) of Fig. 1.8; a finite *depletion* zone is created in the vicinity of the junction.

Following the above discussion we sketch our model junction as in Fig. 1.9(*a*) where we indicate the holes by open circles and the electrons by dots; the junction is at $x = 0$ and the boundaries of the depletion region are labeled by $-x_p$ and $x_n$. For $x < 0$ there exists an excess of ionized acceptors, that is an excess of negative charge. For $x > 0$ there exists an excess of ionized donors, that is an excess of positive charge. Thus the

Fig. 1.9. The electrostatic parameters in the vicinity of a junction: (*a*) definition of the depletion region, (*b*) electric charge density, (*c*) electric field, (*d*) electrical potential.

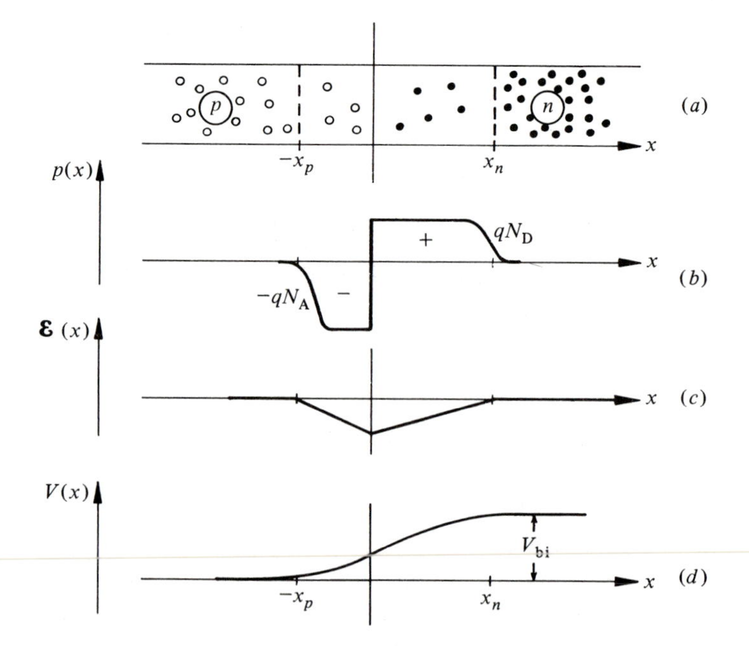

charge density $\rho(x)$ is distributed as shown in Fig. 1.9(*b*). A non-zero charge distribution gives rise to an electric field, which in its simplest form, is directed from the positive to the negative charge. Thus the field is negative and as shown in Fig. 1.9(*c*). Finally by integrating the electric field we can find the potential in the vicinity of the junction; this is indicated in (*d*) of the figure.

Clearly, the electric field 'pushes' the electrons towards positive $x$, and the holes toward negative $x$; that is, *against* the direction in which the carriers tend to diffuse. The electric field can be calculated by integrating Gauss' law

$$\mathscr{E} = \frac{1}{K_s \varepsilon_0} \int_{-\infty}^{x} \rho(x)\,\mathrm{d}x \tag{1.18}$$

Here $K_s$ is the dielectric constant of silicon; $K_s \simeq 11.8$. In our example peak field is reached at $x=0$ and $\mathscr{E}(x=0)$ is negative. Similarly, the potential $V(x)$ is given by integrating the electric field

$$V(x) = -\int_{-\infty}^{x} \mathscr{E}(x)\,\mathrm{d}x \tag{1.18$'$}$$

The difference in potential across the junction is designated by $V_{bi}$ (where bi stands for 'built-in') as shown in Fig. 1.9(*d*).

We can evaluate $V_{bi}$ by noting that under equilibrium conditions both the electron current and the hole current across the junction must be zero. The total current is the sum of the drift current $J_{dr}$ and the diffusion current $J_D$. Looking at the current we have, using Eqs. (1.15, 1.17″)

$$J_{dr/n} + J_{D/n} = qn\mu\mathscr{E} - \mu kT\frac{\mathrm{d}n}{\mathrm{d}x} = 0$$

or

$$V_{bi} = \int_{-\infty}^{+\infty} \mathscr{E}\,\mathrm{d}x = \frac{kT}{q}\int_{-\infty}^{+\infty}\frac{\mathrm{d}n}{n} = \frac{kT}{q}\ln\left(\frac{n_{\infty}}{n_{-\infty}}\right)$$

The electron densities at large positive $x$ and large negative $x$ can be taken as $n_{\infty} = N_D$ and $n_{-\infty} = n_i^2/N_A$ so that

$$V_{bi} = \frac{kT}{q}\ln\left[\frac{N_D N_A}{n_i^2}\right] \tag{1.18$''$}$$

For the densities used in Fig. 1.8 we find $V_{bi} = 0.65$ eV which is typical of most commercial junctions. The typical thickness of the depletion region is of the order of 1 micron ($10^{-6}$ m) or less.

A most convenient way for looking at the potentials and the carrier motion at a junction is to consider the energy band diagram. This is shown in Fig. 1.10. As we recall the position of the Fermi level with respect to the edge of the valence or the conduction band is different for $p$-type and $n$-type materials (see Fig. 1.6). However when the two materials are joined, the Fermi levels *must be* at the *same* energy when the system is in *equilibrium*; otherwise there would be flow of charge until the Fermi levels equalized.* Thus the band diagram takes the form shown in Fig. 1.10, the relative displacement of the bands being given by $qV_{bi}$; (here negative potential is toward the top of the page in contrast to Fig. 1.9). The importance of this diagram is that the electrons must gain energy to move upwards, thus their motion from the $n$-region to the $p$-region is impeded. Similarly the holes must gain energy to move downwards (uphill for the holes is down), thus their motion from the $p$-region to the $n$-region is impeded. The depletion region is characterized by the sloping part of the band edges and in fact the electric field is proportional to that slope.

When an external voltage is applied between the ends of the $p$-type and $n$-type material, the potential difference will appear across the junction and modify the energy diagram by *displacing* the relative position of the Fermi levels. We say that the junction is *biased.* There are two possibilities: if a positive voltage is applied to the $n$-type material the potential difference across the junction will increase and there can be no current flow across the junction as shown in Fig. 1.11($a$). The junction is *reverse* biased. If

Fig. 1.10. Energy band diagram for a $p$–$n$ junction; at thermal equilibrium the Fermi level must be at the same energy in both parts of the junction.

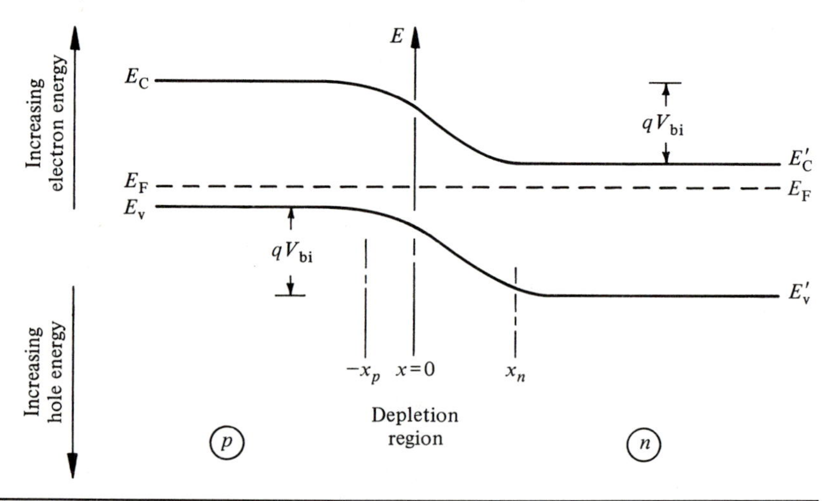


* As an analogy one can think of two containers of fluid which are filled to different heights; when the containers are put in communication the water will flow until the levels equalize.

the negative voltage is applied to the *n*-type material the potential across the junction is reduced and there is flow of electrons toward the left end of holes toward the right. When the carriers cross the junction they recombine but the flow is sustained because the potential drives the majority carriers on both sides toward the junction. In this configuration the diode is *forward biased*, as shown in Fig. 1.11(*b*). In the case of a forward biased junction there are enough carriers lying sufficiently high in the conduction band to have energy$E' > E_C$; these carriers drift across the junction under the influence of the external potential. As can be deduced from Eq. (1.18″) for a typical *n–p* junction a bias of 0.5–1.0 volt is sufficient to reach saturation. We remind the reader that the current is carried by the majority carriers in each part of the bulk material, i.e. by holes in the *p*-region and by electrons in the *n*-region.

The simple *p–n* junction such as described here forms a very useful device widely used in electrical circuits. It is referred to as a *diode* and represented by the symbol shown in Fig. 1.12(*a*). For positive voltage the junction is forward biased and the current flow grows exponentially until it reaches saturation. The current v. voltage (*I–V*) characteristic of a typical diode is shown in Fig. 1.12(*b*); the *non-linear* nature of the diode is clearly exhibited. The *I–V* curve can be described analytically by an

Fig. 1.11. Energy band diagram for a biased *p–n* junction: (*a*) reverse bias, (*b*) forward bias. The physical connection to the voltage source is indicated in the sketches.

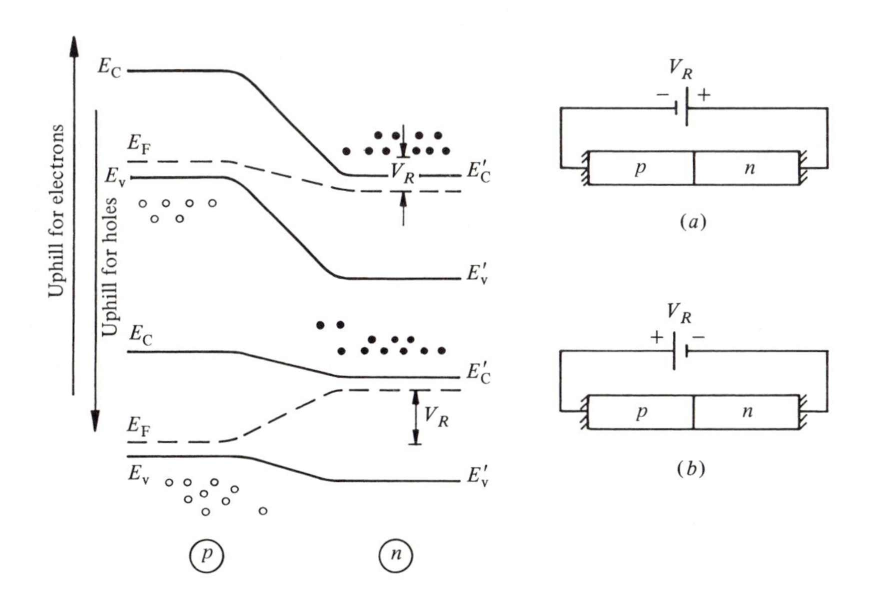

Fig. 1.12. A $p$–$n$ junction forms a diode: ($a$) circuit symbol, ($b$) $I$–$V$ characteristic.

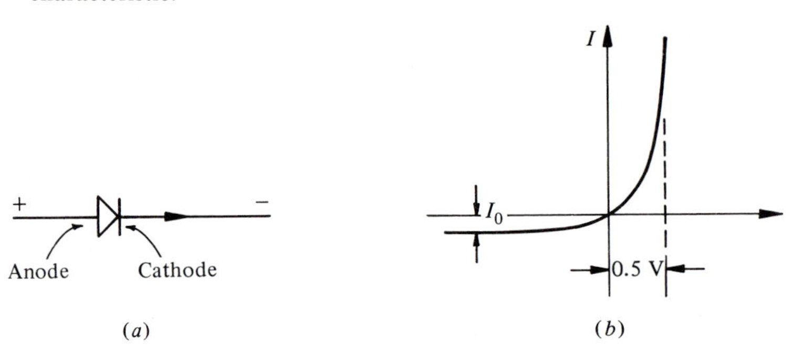


equation of the form

$$I = I_0(e^{qV/kT} - 1)$$

where $V$ is the biasing voltage.

## 1,5 The junction transistor

The junction transistor consists of two $p$–$n$ junctions connected back to back with the common region between the two junctions made very thin. A model of the $n$–$p$–$n$ transistor is shown in Fig. 1.13. Note that one junction is forward biased at a relatively low voltage, whereas the other junction is reverse biased at a considerable voltage. The three distinct regions of differently doped material are labeled *emitter*, *base* and *collector* respectively.

According to the biasing shown in the figure, electrons will flow from the emitter into the base; one would expect a large positive current $I_B$

Fig. 1.13. A $n$–$p$–$n$ transistor consists of two back to back diode junctions. The biasing scheme and current flow are indicated.

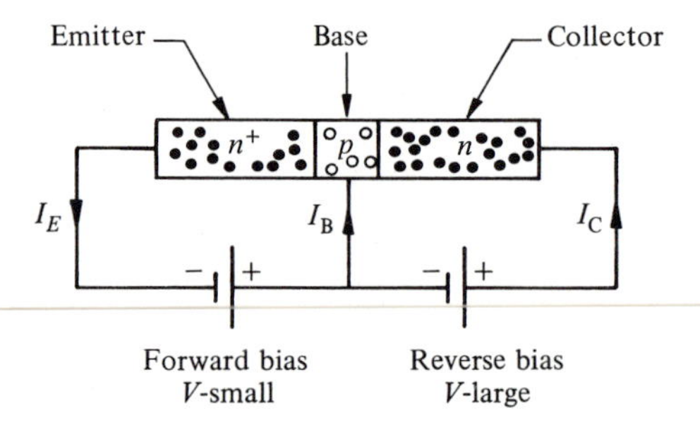

from the base to the emitter. If however the base is thin enough, the electrons injected into the base will reach the base–collector junction before recombining or diffusing in the base. Once electrons cross the base–collector junction they can move freely in the collector since they are majority carriers. Furthermore, the base–collector voltage difference is large so that the electrons gain much more energy than they lost in overcoming the voltage difference between the collector and base. Such a system can provide power amplification. Thus $I_B$ is a small current, while $I_E$ and $I_C$ are much larger when the transistor is in the conducting state.

The energy band diagram for the *n–p–n* transistor is shown in Fig. 1.14. The majority carriers are electrons and therefore once they reach the collector they fall through the potential hill. The *small* voltage between base and emitter can be used to control the flow of current across the much *larger* base–collector voltage. For the device to operate in this fashion the electrons injected from the emitter must traverse the junction without attenuation. In a good transistor $\sim$0.95 to 0.99 of the injected carriers traverse the base. Typical widths for the base are of order $W_B \sim 1$–$5\ \mu$m. The emitter is heavily doped with donors so as to be able to provide the necessary current even with small base emitter bias. Junction transistors are referred to also as 'bipolar' transistors to distinguish them from field effect devices.

The symbols for a transistor are shown in Fig. 1.15. The arrows indicate the direction of *positive* current flow so that on the left of the figure we recognize an *n–p–n* transistor (as in Fig. 1.13) and a *p–n–p* transistor on

Fig. 1.14. Energy band diagram for a biased *n–p–n* transistor. For electrons, positive energy is toward the top of the page (uphill); thus electrons flow in the direction indicated.

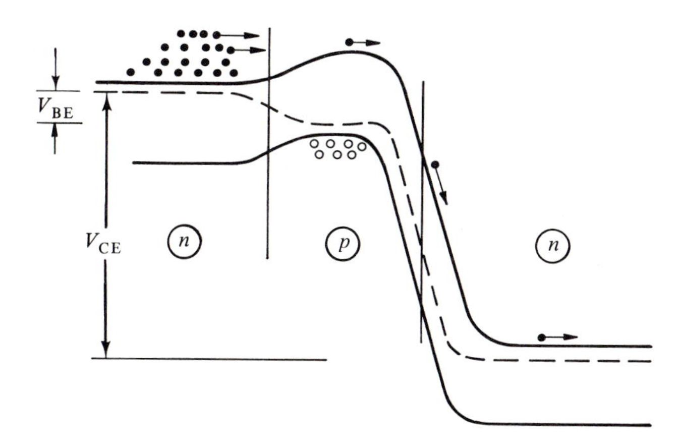

the right. Typical values are $V_{BE} \sim 0.2$ V whereas $V_{CE} \sim 5$–$10$ V. The performance of a transistor can be characterized by the *current transfer ratio* $\alpha$, which is defined as

$$\alpha = \frac{\Delta I_C}{\Delta I_E} \tag{1.19}$$

Here we use $\Delta$ to indicate changes in current rather than steady state currents that may result from any particular biasing arrangement. Clearly $\alpha < 1$, but for a good transistor $\alpha$ must be close to one.

To calculate the *current gain* of a transistor, we recall that the emitter current is the sum of the collector and base currents.

$$\Delta I_E = \Delta I_C + \Delta I_B \tag{1.19'}$$

The current gain $\beta$ is defined by

$$\beta = \Delta I_C / \Delta I_B \tag{1.20}$$

Using this definition and that of Eq. (1.19) in Eq. (1.19′) we obtain

$$\frac{1}{\alpha}\Delta I_C = \beta \Delta I_B + \Delta I_B = (1+\beta)\Delta I_B$$

or

$$\beta = \alpha(1+\beta), \qquad \beta = \frac{\alpha}{1-\alpha} \tag{1.20'}$$

For a typical value of $\alpha \simeq 0.98$ one finds $\beta \simeq 49$. Namely, we can use a small current into the base of the transistor to control the flow of a much larger current from the collector to the emitter. Thus a transistor is a device that controls current flow.

The transistor is a three-terminal device and thus there are more biasing possibilities than for a diode. There are three basic biasing configurations which can be classified as: common or *grounded base*; common emitter;

Fig. 1.15. Circuit symbol for a junction transistor; the current flow and voltage definitions are indicated: (*a*) *n–p–n*, (*b*) *p–n–p*.

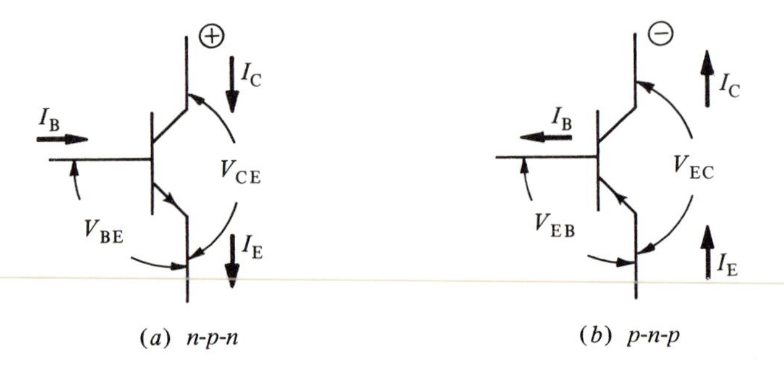

and common collector, also referred to as *emitter follower*. These configurations are shown in Figs. 1.16 to 1.18 and serve different functions. For instance the *grounded base* configuration shown in Fig. 1.16 leads to *voltage gain*. The base–emitter resistance can be taken to be of order $R_{BE} \sim 100\ \Omega$. If the load resistor is $R_L = 10^4\ \Omega$ then the voltage gain is of order $R_L/R_{BE} \sim 100$ as shown below. The base–emitter voltage $\Delta V_{BE}$ will in general be small ($\sim 10^{-3}$ V) and

$$\Delta I_E = \frac{\Delta V_{BE}}{R_{BE}} = \frac{1}{\alpha}\Delta I_C$$

Further

$$V_{CE} = V_{CC} - R_L I_C$$

but

$$\Delta V_{CE} = -R_L \Delta I_C = -\alpha R_L \frac{\Delta V_{BE}}{R_{BE}}$$

so that (note $\Delta V_{EB} = -\Delta V_{BE}$)

$$\frac{\Delta V_{CE}}{\Delta V_{EB}} = \alpha \frac{R_L}{R_{BE}} \sim \frac{R_L}{R_{BE}} \tag{1.21}$$

Note that the collector voltage $V_C$ maintains itself only slightly above the emitter voltage when the transistor is conducting; this is true in this configuration where the base is grounded.

The common or *grounded emitter* configuration shown in Fig. 1.17 is used to provide current gain. In this case the collector current is obtained directly from the definition of Eq. (1.20′). A small current flow into the base controls the current $\Delta I_C$ flowing through a particular device, such as the light bulb shown in the figure. The current gain is given by $\beta$, provided of course that the device is not saturated ($\beta \Delta I_B < V_{CC}/R_L$).

Fig. 1.16. Circuit diagram for grounded base operation.

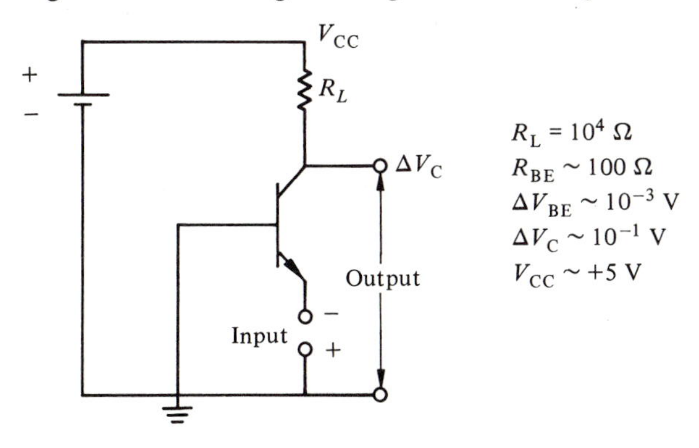

The third configuration is that of common collector, better known as the *emitter follower*, and is shown in Fig. 1.18. As the name indicates the emitter voltage follows the base voltage, but the current flowing through the emitter resistor $R_E$ is $\beta$ times larger than the current supplied to the base. Therefore a high impedance source can drive a load of low impedance; the emitter follower acts as an impedance transformer without voltage gain. In view of the condition $\Delta V_E = \Delta V_B$ we have

$$\Delta I_E = \frac{\Delta V_E}{R_E} = \frac{\Delta V_B}{R_E}$$

Furthermore

$$\Delta I_B = \frac{1}{1+\beta}\Delta I_E = \frac{\Delta V_B}{R_E(1+\beta)}$$

and since the impedance of the source (*input impedance*) is defined as

Fig. 1.17. Circuit diagram for grounded emitter operation.

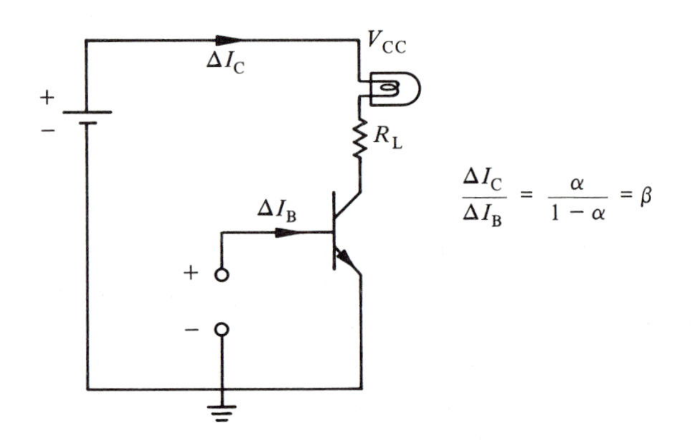


Fig. 1.18. Emitter follower circuit.

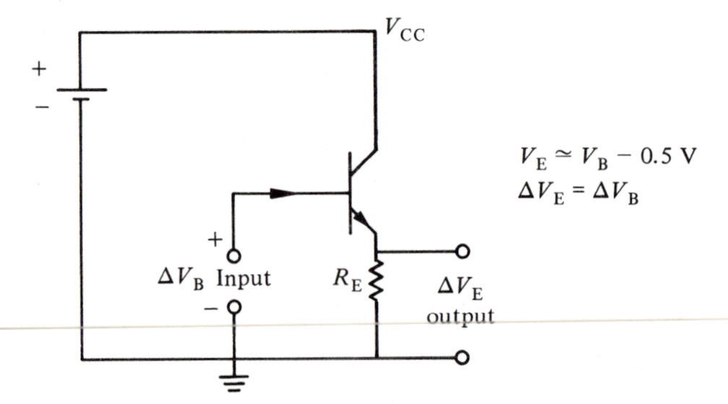

$r_{\text{in}} = \Delta V_{\text{B}}/\Delta I_{\text{B}}$, we have

$$R_{\text{E}} = \frac{r_{\text{in}}}{1+\beta} \tag{1.22}$$

where $R_{\text{E}}$ is the *output impedance*. Emitter followers are used to drive long lines, meters, or other devices that have low impedance.

The design of electrical circuits is a vast subject of great technical importance. The interested reader is referred to the excellent text *The Art of Electronics* by P. Horowitz and W. Hill, Cambridge University Press, N.Y., 1980.

## 1.6 Manufacture of transistors; the planar geometry

The manufacture of transistors depends first on the availability of pure silicon. The second requirement is the ability to introduce donors or acceptors in a highly controlled fashion. Finally it must be possible to make very thin junctions between materials of different types. This technology is by now highly developed and relies on manufacturing transistors and entire circuits on the surface of thin silicon wafers; we speak of planar geometry.

Ultrapure silicon is obtained by zone melting and other refining techniques. Single crystals of silicon are grown from a melt and the resulting ingots are typically 4 inches in diameter and 20 inches long. The silicon can be doped by adding the desired impurity to the melt. The ingot is then sliced into wafers approximately 0.5 mm thick and polished; the cut is usually along the (1, 0, 0) crystallographic plane. When good resistivity is desired the silicon is doped so as to result in *p*-type material. Germanium and gallium arsenide are also produced in wafers.

To make junctions one introduces the desired impurity, or silicon of opposite doping into the substrate (wafer). This can be done by alloying, by epitaxy, by diffusion, or by ion implantation. *Alloying* was one of the first techniques used in producing junctions; it is shown schematically in Fig. 1.19. A pellet of *p*-type silicon is placed on top of an *n*-type wafer.

Fig. 1.19. Formation of a junction by alloying: (*a*) initial configuration, (*b*) melting at high temperature, (*c*) the junction after re-crystallization.

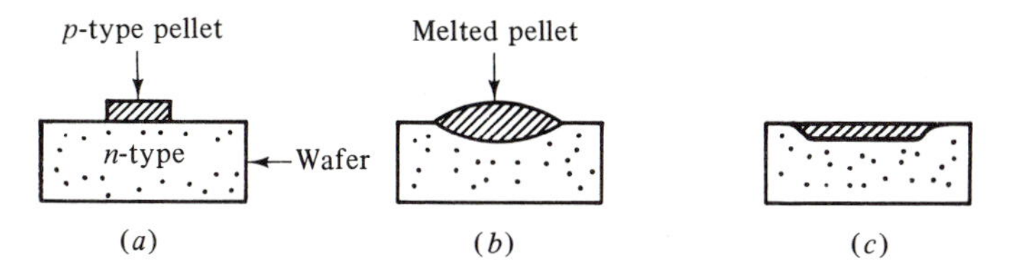

When this assembly is heated in an oven the pellet melts and penetrates into the wafer. As the system cools and the silicon recrystallizes, a structure such as shown in (*c*) of the figure is attained. Repeating this sequence, one can obtain a junction transistor. In this process the base thickness can be controlled reasonably well and several units can be manufactured simultaneously. Molecular beam epitaxy (MBE) is the preferred technique when very thin surface layers are to be deposited; it is becoming fairly widely used. Ion implantation allows the very exact positioning of impurity ions in the substrate and also the creation of very thin layers of donors or acceptors.

Diffusion is by far the most common manufacturing process and it led to a real breakthrough because it could be combined with photolithography to make devices of very small dimensions. This was facilitated by the discovery that silicon dioxide ($SiO_2$) was easily formed by simply exposing the pure silicon to an atmosphere of oxygen. $SiO_2$ is glass and even thin layers completely prevent the diffusion of impurities in or out of the substrate.

The manufacture of transistors in planar geometry by the diffusion process is outlined in the sketches of Fig. 1.20. One starts with *p*-type silicon and the first step is to produce a thin layer of $SiO_2$ on its surface; it suffices to heat the wafer in an oxygen-rich atmosphere. Next the wafer is coated with a layer of photoresist. This is a chemical whose properties are affected when it is exposed by ionizing radiation, i.e. visible light, U.V., X-rays, etc. The exposure takes place through a photographic mask which has been highly reduced from its original layout; the mask carries the pattern that is to be transferred to the wafer. Then the exposed resist is dissolved away by dipping the wafer in a solvent, leaving the unexposed areas intact. At that stage the assembly looks as in (*e*) of the figure.

Next the wafer is introduced in an etching solution, for instance HF, and the exposed $SiO_2$ is removed, exposing the silicon in the pattern inscribed by the mask. Impurities are then diffused into the wafer as indicated in (*g*). By controlling the temperature and length of the exposure one can obtain the desired concentration and depth in the diffused layer to great accuracy. When the diffusion process is completed, the remaining resist coat is washed away using a stronger solvent. Finally a new oxide layer is established over the wafer to provide electrical isolation between the different parts of the circuit and to protect the silicon. If a second layer of diffusion is desired, the process is repeated. Eventually, metallic contacts are placed at the appropriate locations to provide the electrical connections for the circuit. This process involves quite a lot of toxic chemicals and the workers must be adequately protected.

The physical arrangement of the *n*- and *p*-type silicon regions in a junction transistor manufactured by diffusion technology is shown in Fig. 1.21. In general one starts with a *p*-type substrate into which an *n*-type 'island' is diffused. This provides electrical isolation and long-term stability. For an *n–p–n* type transistor (see Fig. 1.21(*a*)) a *p*-type island is established to provide the base and then two highly doped *n*-type regions are introduced to serve as the base emitter and collector. The insulating $SiO_2$ layer and the electrical connections are also indicated. In this case the majority carriers are electrons and they flow along the arrow (the

Fig. 1.20. Steps involved in the production of a transistor: (*a*) *p*-type silicon is used for the substrate, (*b*) a $SiO_2$ layer is formed, (*c*) the wafer is coated with photoresist, (*d*) the photoresist is exposed through a mask, (*e*) exposed photoresist is washed away, (*f*) in the exposed area the oxide layer is chemically removed, (*g*) *n*-type material is diffused into the substrate, (*h*) the remaining photoresist is removed. The process can now be repeated to introduce additional *n*-type or *p*-type regions.

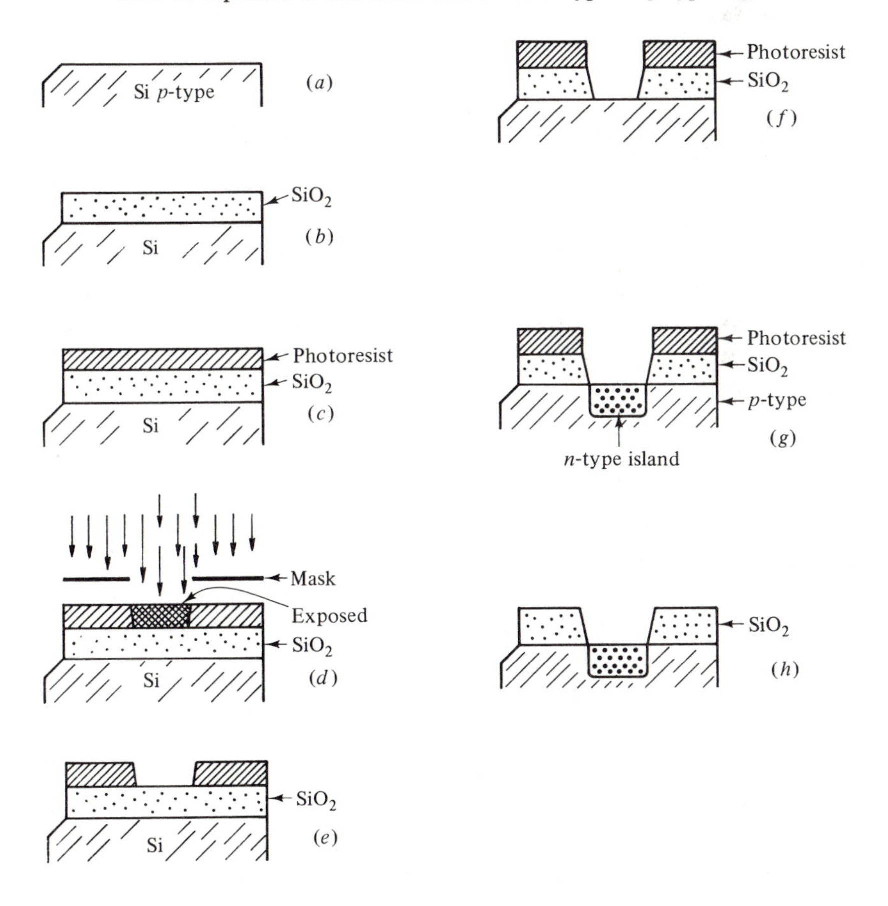

positive current is in the inverse direction). Note that the thin base region (*p*-type) is between the highly doped emitter region and the *n*-type island through which the carriers can flow to the collector.

For a *p–n–p* transistor the current flow geometry is different as indicated in Fig. 1.21(*b*). The emitter and collector islands are adjacent to one another and the holes have to travel across the base which is part of the *n*-type island. A highly doped *n*-type region around the base electrode provides the necessary minority carriers to control the emitter–collector current flow.

The semiconductor technology developed in the manufacture of junctions could be adapted to make circuit elements such as resistors or capacitors on the surface of a silicon chip. Examples are shown in Fig. 1.22. For instance, the thickness of the *n*-type region in Fig. 1.22(*a*) determines the resistance *R* between the electrodes at *A* and *B*. In (*b*) of the figure

Fig. 1.21. Typical layout of junction transistors manufactured by diffusion methods. Note the 'islands', location of the contacts and current flow: (*a*) *n–p–n*, (*b*) *p–n–p*.

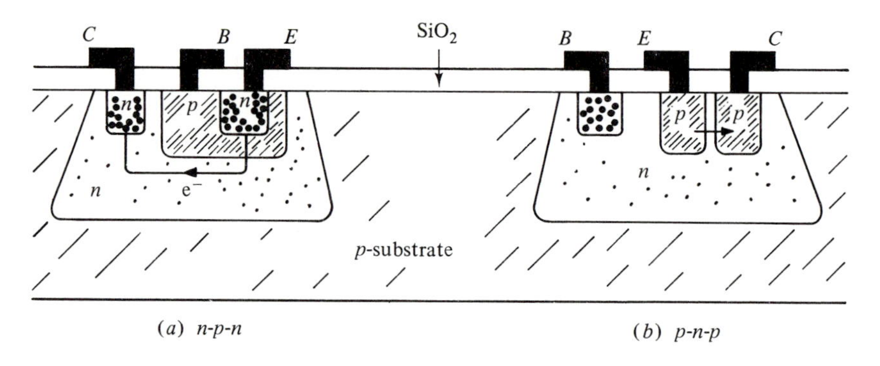


Fig. 1.22. Circuit elements can be realized with semiconductor materials: (*a*) resistor, (*b*) capacitor.

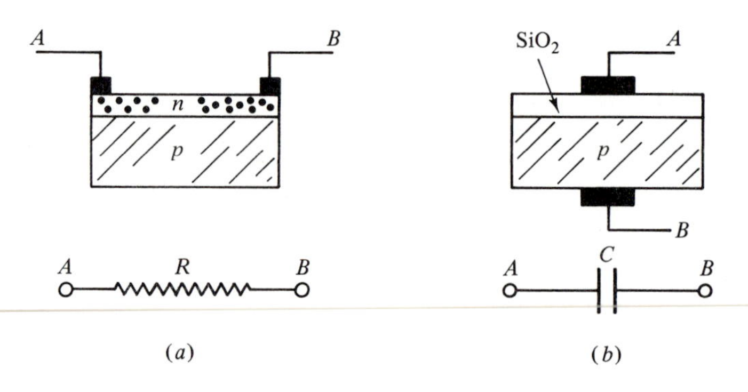

the capacitance $C$ between the two electrodes is controlled by the thickness of the dielectric layer. In this way it became possible to *integrate* the transistor with the discrete components needed to make a complete circuit.

An example of the evolution of a simple circuit from its discrete component form to its integrated form is shown in Fig. 1.23. The $n$-regions are indicated by the dots, the $p$-regions are shaded whereas the undoped silicon is left clear. By controlling the depth of the diffused layers and their area the performance of the circuit can be optimized. This type of integrated circuit is not very common any more, having been replaced by transistor–transistor logic. For high performance circuits, however, discrete components are still utilized.

## 1.7 The field effect transistor (FET)

In the junction transistor the current flow from the collector to the emitter is controlled by the current flowing into the base. In the field effect transistor the current flow is controlled by the electric field between

Fig. 1.23. Monolithic integrated circuit construction: (*a*) circuit diagram, (*b*) discrete elements and junctions, (*c*) combined elements, (*d*) wafer layout.

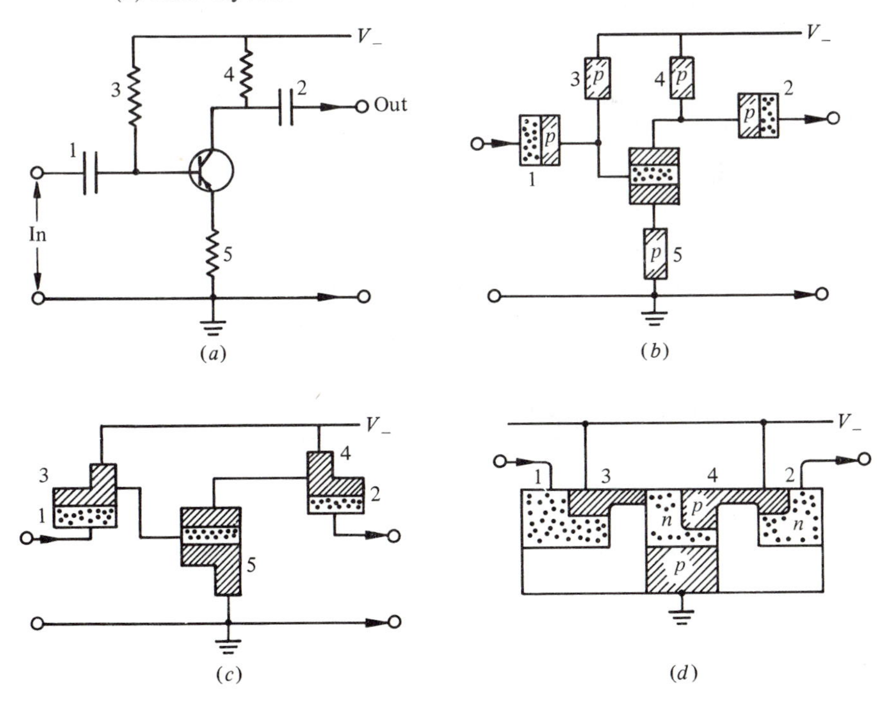

the gate and the other electrodes. In this sense the FET resembles a triode electronic vacuum tube and efforts to construct an FET go back to the late thirties, even though the first FET was not built until 1953. The great success of the FET is due to the advances in the technology which permitted the construction of very thin $SiO_2$ layers, the diffusion of *n*-type silicon into a substrate, and the application of metallic gates and contacts. Thus the technology is know as MOS for Metal-Oxide-Silicon.

As a first example we consider a small slab of *n*-type silicon to which *source* and *drain* contacts are made as shown in Fig. 1.24(*a*). Furthermore, two heavily doped *p*-regions are diffused at the other two sides of the slab; they form the gate of the device. With no voltages applied a depletion region will be created around the $p^+$-islands as shown in Fig. 1.24(*a*). If a positive voltage $V_D$ is applied to the drain, with the source grounded, the electrons in the *n*-type substrate will move from source to drain; the depletion region will widen because the electrons are being pulled toward the drain. As $V_D$ is increased, the current $I_{DS}$ increases and so does the depletion region until it reaches 'pinch off' as shown in Fig. 1.24(*b*); at that point $I_{DS}$ becomes independent of $V_D$. The $I_{DS}$ v. $V_D$ characteristics for $V_G = 0$ and $V_G < 0$ are shown in Fig. 1.24(*c*).

If the gate voltage $V_G$ is made negative, holes will be pulled toward the gate and electrons pushed away from it. Thus the depletion zone will widen and as a consequence the impedance to the flow of electrons from

Fig. 1.24. Representation of a junction field effect transistor where the source, and drain electrodes and the gate regions are shown: (*a*) no bias, in which case the depletion region is small, (*b*) biased to saturation, (*c*) the current–voltage characteristic for different gate voltages.

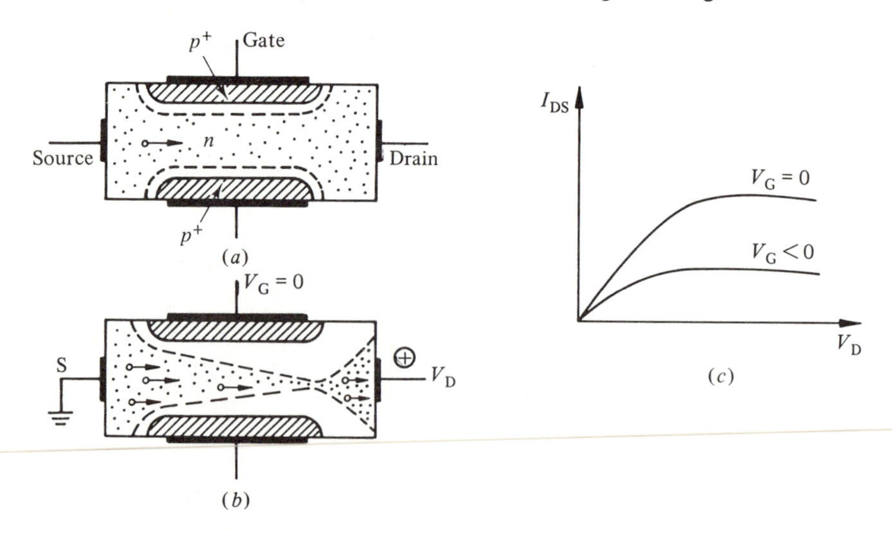

source to drain will increase. Furthermore, 'pinch-off' will occur at lower values of $V_D$ as shown by the $I_{DS}$–$V_D$ characteristic for $V_G < 0$. The region between source and drain where the carriers flow is called the *channel*, and the particular device we described is known as a J-FET.

The most commonly used FETs are manufactured by MOS technology and are referred to as MOSFETs; they lend themselves readily to very large scale integration (VLSI). There are two types of MOSFETs, the *enhancement* and *depletion* types. An enhancement FET is shown in Fig. 1.25. Starting with a *p*-type substrate two *n*-type islands are introduced to form the source and drain. Since the channel is *p*-type material it will not conduct when the gate is at ground voltage. If however a sufficiently positive voltage is applied to the gate the electric field between the gate and the source will draw negative carriers into the channel and electrons will flow from the source to the drain unimpeded as indicated in (*b*) of the figure.

The *depletion* FET is shown in Fig. 1.26. Here we want the channel to be conducting when no voltage is applied to the gate. To achieve this,

Fig. 1.25. Realization of an enhancement FET in MOS technology: (*a*) with no voltage on the gate the device is not conducting, (*b*) when $V_{GS}$ is positive there is current flow.

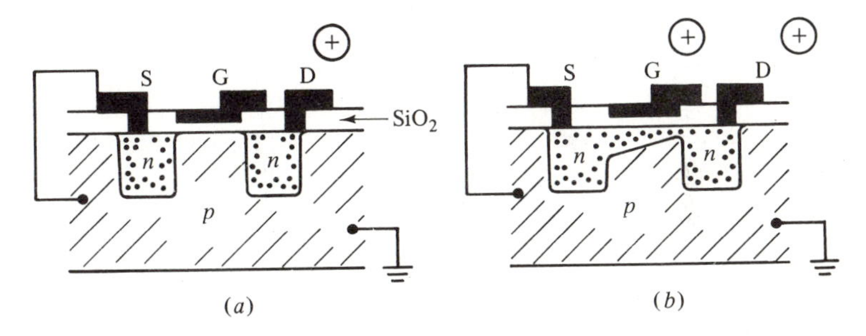


Fig. 1.26. Realization of a depletion FET in MOS technology: (*a*) with no voltage on the gate the device conducts, (*b*) negative $V_{GS}$ turns the current off.

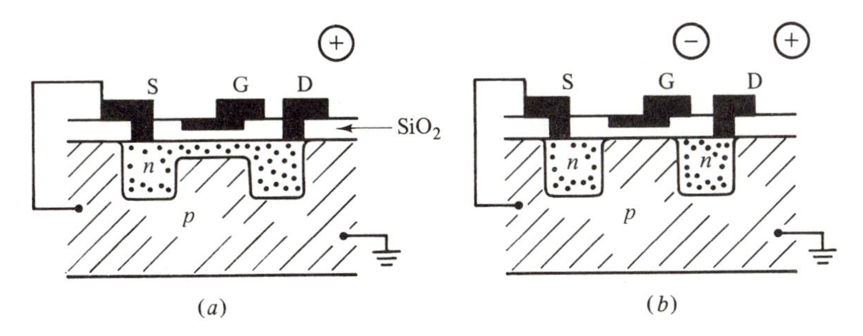

donors are introduced by ion implantation into a *very shallow* channel between source and drain as shown in (*a*) of the figure. If a negative voltage is applied at the gate the channel becomes narrower as the electrons are repelled from the gate and eventually conduction stops altogether as shown in Fig. 1.26(*b*). Thus the operation of the depletion FET resembles that of the J-FET.

The physical construction of an FET can be better understood with the help of the sketches of Fig. 1.27; note that the transverse dimensions are highly exaggerated. The silicon dioxide layer is extremely thin and its thickness $D$ is of order $D \simeq 0.02\ \mu\text{m} = 200$ Å; this is only a small fraction of a wavelength of visible light and of the order of a hundred atomic layers or less. On top of the $SiO_2$ the gate is grown out of polycrystalline silicon (poly); the lateral dimensions of the gate are length $L$ and width $W$, typically 5 $\mu$m each. The drain and source are formed by diffusing donors into the substrate creating the $n$-type regions. A plan view of the FET is shown in part (*b*) of the figure; note that the gate is placed transversely with respect to the two diffused regions. For a depletion type FET a very thin layer of donors is introduced into the channel by ion implantation.

The symbols used to indicate FETs are shown in Fig. 1.28. For each particular manufacturing technology a typical supply voltage is used. This is referred to as $V_{DD}$ and is of order of few volts, i.e. $V_{DD} \sim 5$ volts. An enhancement FET will conduct if the gate–source voltage exceeds a certain threshold $V_{thr}$; typically $V_{thr} \sim 0.2V_{DD}$. When the FET conducts the drain–source voltage, $V_D$ can be less than $V_{DD}$. The dependence of the drain–source current $I_{DS}$ on $V_G$ and $V_D$ is as shown in Fig. 1.29.

To become familiar with orders of magnitude we estimate the current

Fig. 1.27. FET construction in MOS planar geometry: (*a*) perspective view with the vertical dimensions greatly exaggerated, (*b*) plan view.

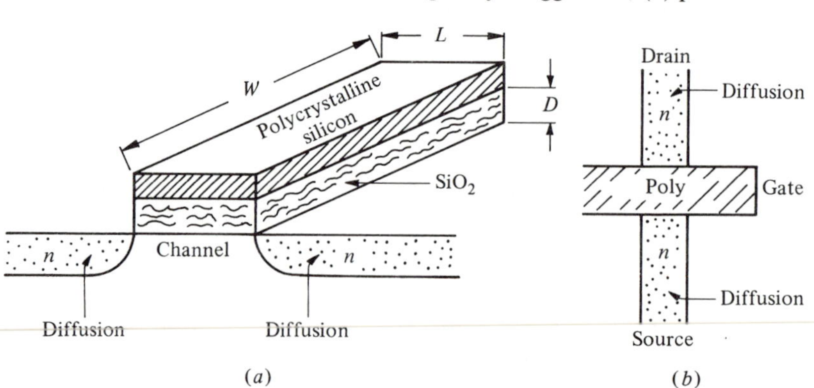

through an FET. We first calculate the capacitance of the gate

$$C_g = K_s \varepsilon_0 \frac{A}{D} = K_s \varepsilon_0 \frac{WL}{D} \tag{1.23}$$

where $K_s$ is the relative dielectric constant of $SiO_2$, $K_s \sim 11$. The charge in transit in the channel is

$$\Delta Q_c = C_g(V_G - V_{thr}) \tag{1.23'}$$

and the time of transit is given by

$$\Delta t = \frac{L}{v_{dr}} = \frac{L}{\mu|\mathscr{E}|} = \frac{L^2}{\mu V_D} \tag{1.23''}$$

Combining the above equations the source to drain current can be expressed as

$$I_{DS} = \frac{\Delta Q_c}{\Delta t} = \frac{K_s \varepsilon_0 W}{LD} \mu(V_G - V_{thr})V_D \tag{1.24}$$

Fig. 1.28. Circuit symbols for FETs: (*a*) enhancement FET, (*b*) depletion FET.

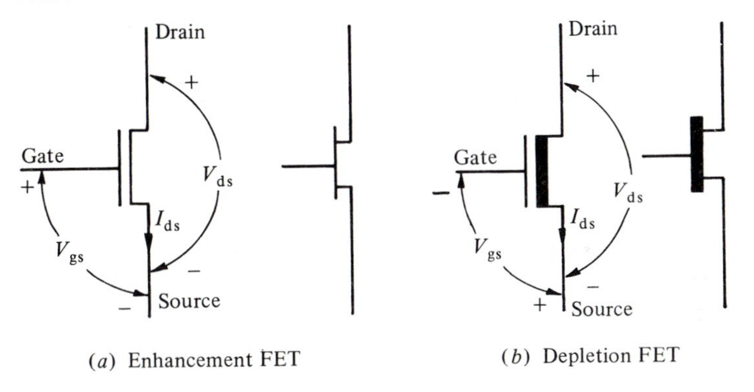


Fig. 1.29. FET current–voltage characteristics for different gate voltages: (*a*) enhancement FET, (*b*) depletion FET.

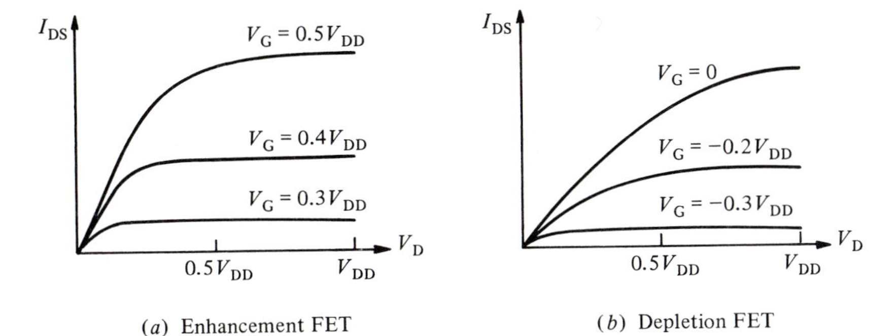

In the above equations $W$, $L$ and $D$ are the width, length and depth of the gate as defined in Fig. 1.27, and $\mu$ is the mobility. Eq. (1.24) is valid in the linear region where $I_{DS}$ is proportional to $V_D$. The saturation current is independent of $V_D$ and given to a good approximation by

$$I_{sat} = K_s \varepsilon_0 \frac{W}{D} \frac{\mu}{2L} (V_G - V_{thr})^2 \tag{1.25}$$

This quadratic dependence on $(V_G - V_{thr})$ can be seen in the characteristic curves of Fig. 1.29.

As a numerical example we choose $W = L = 6\ \mu\text{m}$, $D = 0.02\ \mu\text{m}$ and $(V_G - V_{thr}) = 1\text{ V}$; furthermore let $\mu = 1000\text{ cm}^2/\text{V-s}$ and we have that $K_s = 11$ and $\varepsilon_0 = 8.85 \times 10^{-12}$ coul/V-m. Thus

$$I_{sat} \simeq 10^{-10} \frac{10}{2 \times (2 \times 10^{-8})} \frac{\text{coul}}{\text{s}} = 0.025\text{ A}$$

which is typical for a high mobility FET.

## 1.8 Transistor–transistor logic

One of the simplest digital logic operations is that of inversion, and we show how it can be easily implemented using FETs. The circuit diagram for an inverter is shown in Fig. 1.30($a$). If we designate by $R_{DS}$ the drain to source impedance when the FET is in the conducting state we have the conditions

$$V_{in} < V_{thr}: I_{DS} = 0,\ V_{out} = V_{DD};$$
$$V_{in} > V_{thr}: I_{DS} = V_D/R_{DS},\ V_{out} \to 0$$

Fig. 1.30. Circuit diagram for an inverter using FETs: ($a$) with load resistor, ($b$) with depletion FET replacing the load.

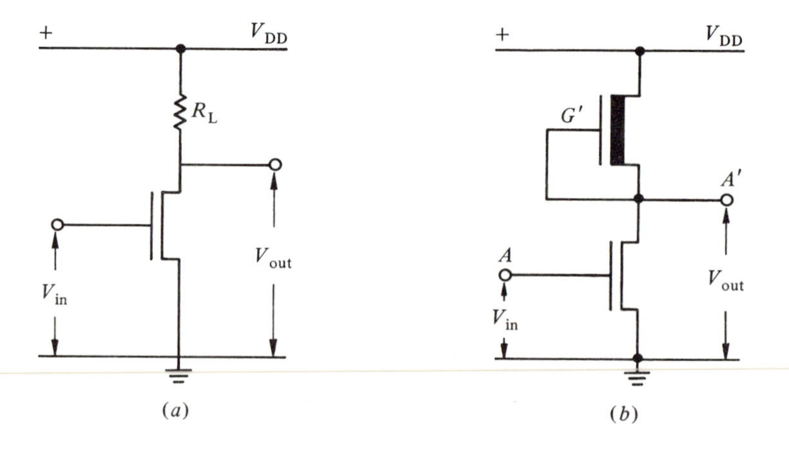

The result for $V_{in} > V_{thr}$ assumes that $R_L \gg R_{DS}$, because

$$V_{out} = V_{DD} - I_{DS}R_L = V_{DD} - V_{out}(R_L/R_{DS})$$

Thus

$$V_{out} = V_{DD}\frac{R_{DS}}{R_L + R_{DS}}$$

In MOS technology it is difficult to create a suitably high resistance $R_L$ to serve as the load for the inverter circuit such as the one shown in Fig. 1.30(*a*). This is because all paths are kept short to accommodate a high density of elements on the chip; short paths have low resistance. The solution is to use a depletion FET with its gate connected to its source as shown in (*b*) of the figure. Under these conditions the FET acts like a resistive load (see Fig. 1.29(*b*) for $V_G = 0$) and it is always in the conducting state. However, the current flowing through it is controlled by the state of the enhancement FET to which it is connected in series. The circuit is made up of two transistors coupled to one another, hence the designation as transistor–transistor-logic.

The physical realization of the inverter using an enhancement and a depletion FET is shown in Fig. 1.31. The gate of the depletion FET is wide so as to provide the necessary resistance. The $V_{DD}$, ground and output connections are made directly to the diffused *n*-type regions, whereas the input connection is made directly to the gate of the enhancement FET. Note the connection between the gate and source of the depletion FET.

The actual processes involved in the construction of such an inverter gate using *n*MOS technology can be chronicled with the help of Fig. 1.32

Fig. 1.31. Plan view of the MOS inverter using transistor–transistor-logic.

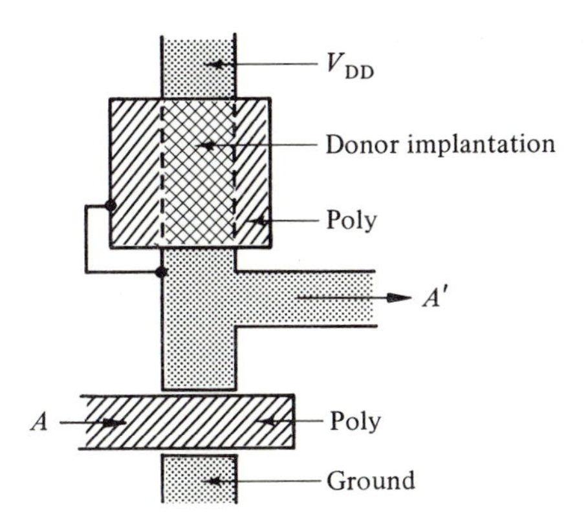

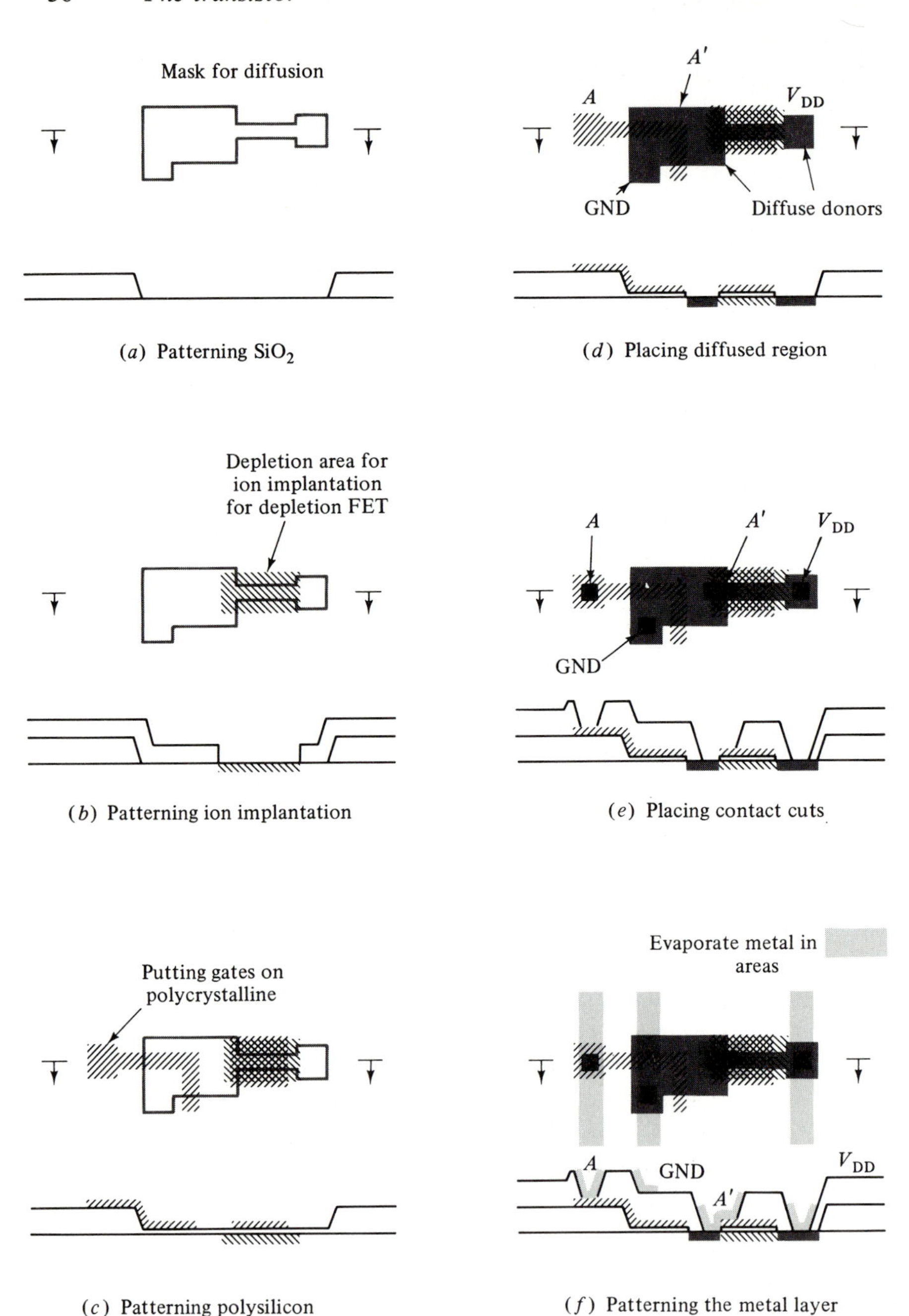


Fig. 1.32. Steps involved in the construction of the inverter shown in the previous figure. (From C. Mead and L. Conway, *Introduction to VLSI*, Addison-Wesley, Reading, MA, 1980.)

(from *Introduction to VLSI Systems* by C. Mead and L. Conway, Addison-Wesley, 1980). The transverse dimensions are, as usual, exaggerated to make them visible in the drawing. First a thin layer of $SiO_2$ is placed on the *p*-type substrate. Then a pattern as shown in (*a*) is etched away. Note the design in the plane of the wafer; the cross-section shown is taken at the registration marks. Next the $SiO_2$ is masked so that only the implantation region is exposed and the donors are implanted to form the channel of the depletion FET; this is shown in (*b*). The photoresist is removed, a thin $SiO_2$ layer is formed on the exposed silicon, and a new mask is used to form the polycrystalline gates for both FETs; this is shown in (*c*). The $SiO_2$ is etched away from the diffusion regions and the donors are introduced to form the *n*-type regions as indicated in (*d*). Our device is now finished but we still must make contacts to the ground, $V_{DD}$, $A$ and $A'$ points of the inverter. An insulating layer is placed over the wafer except in the region where the contacts will be made, as shown in (*e*), and finally the metallic contacts are evaporated onto the assembly. The gate to source connection for the depletion FET is made at this stage as well as the four connections of the circuit. Final steps in the manufacturing process are connection of the leads and placing the wafer in an appropriate package. The whole idea of large scale integration is to interconnect many gates on the same wafer.

## 1.9 Logic gates

The circuits that perform the simplest digital logic functions are quite generally referred to as *gates*. In a digital circuit a signal level (voltage) can be in one of two conditions: either high or low. The exact voltage levels and their tolerances depend on the technology used. One of the conditions corresponds to the *true*, or *asserted*, or '1' state while the other condition is the *false*, or *negated*, or '0' state.

A particular logic function is specified in terms of a *truth table* that gives the output state for all possible combinations of input states. For instance the truth table for the inverter (also referred to as NOT gate),

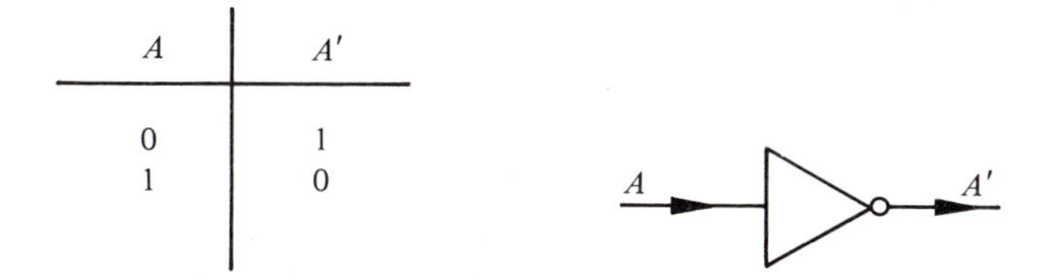

| $A$ | $A'$ |
| --- | --- |
| 0 | 1 |
| 1 | 0 |



Truth table and symbol for inverter

and its symbol are given above. The circuits of Figs. 1.30 or 1.31 perform exactly this function. In general we will use primes to indicate the *complement* of any variable: that is if $B$ is true, $B'$ is false and if $B$ is false then $B'$ is true. The open circle at the output of a gate indicates inversion.

In addition to inversion there are two other basic logic functions. These are the AND and OR functions. The output of an AND function is true if *all* inputs are true. The output of an OR function is true if *any one* of its inputs is true. From these basic conditions several variations can be generated. We shall discuss one particular variation, that of the NAND and of the NOR gate. These are AND and OR gates but with their outputs complemented.

The truth table and symbol for the NAND gate (Not AND) with two inputs are shown below. The circuit diagram and its physical implementation in the $n$MOSFET technology are shown in Fig. 1.33.

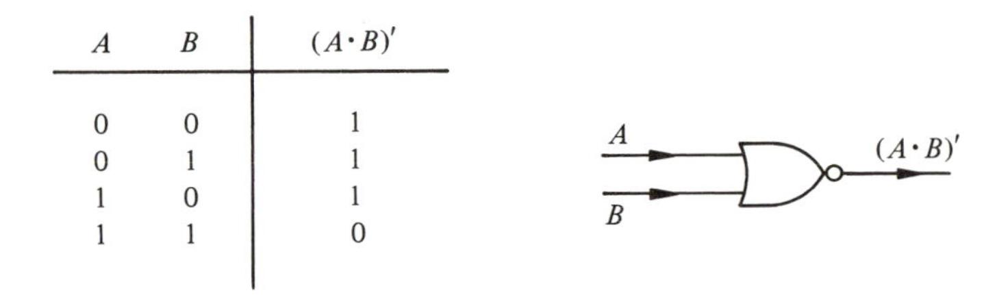

| $A$ | $B$ | $(A\cdot B)'$ |
|---|---|---|
| 0 | 0 | 1 |
| 0 | 1 | 1 |
| 1 | 0 | 1 |
| 1 | 1 | 0 |



Truth table and symbol for NAND gate

Fig. 1.33. Circuit diagram and plan view of a NAND gate in MOS technology.

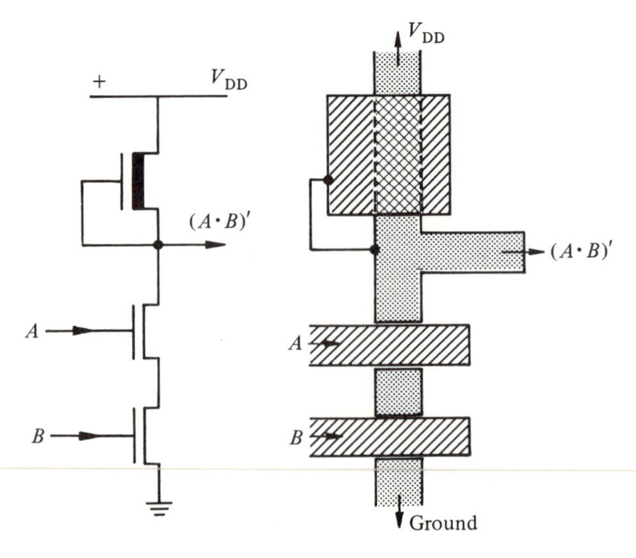

When input $A$ is high but not input $B$ the enhancement FET, $A$, conducts but no current will flow because FET, $B$, is off. The same is true when input $B$ is high but input $A$ is low. When both inputs are high, both FETs conduct and current flows from $V_{DD}$ to ground. This brings the output of the circuit (at the source of the depletion FET) to ground, namely to the low state. There is a limitation on how many inputs can be attached to a gate because of the increase of the impedance as many channels are connected in series, but four or five inputs can be easily used; we speak of a two-fold, three-fold, etc. NAND gate. The NOR gate (Not OR), for two inputs, has the truth table and symbol shown below. The circuit diagram and its physical implementation are shown in Fig. 1.34. Here if either input $A$ or input $B$ are high, current will flow from $V_{DD}$ to ground because one of the enhancement FETs conducts. Thus the output of the circuit will drop towards ground, that is it will go to the low state. One can OR several logic signals together, with a limit of five to ten inputs to a single gate.

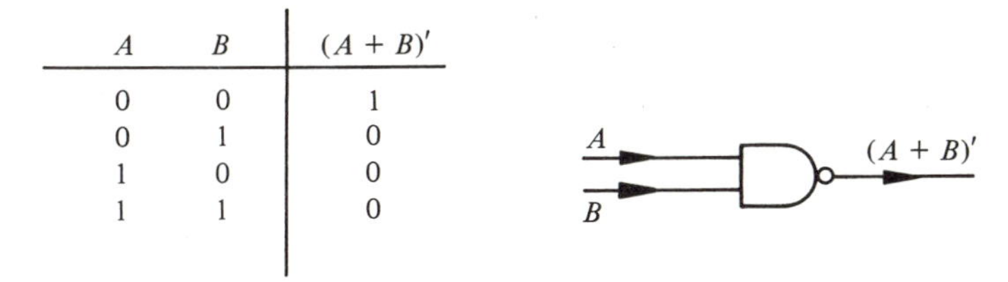

| $A$ | $B$ | $(A + B)'$ |
|---|---|---|
| 0 | 0 | 1 |
| 0 | 1 | 0 |
| 1 | 0 | 0 |
| 1 | 1 | 0 |



Truth table and symbol for NOR gate

Fig. 1.34. Circuit diagram and plan view of a NOR gate in MOS technology.

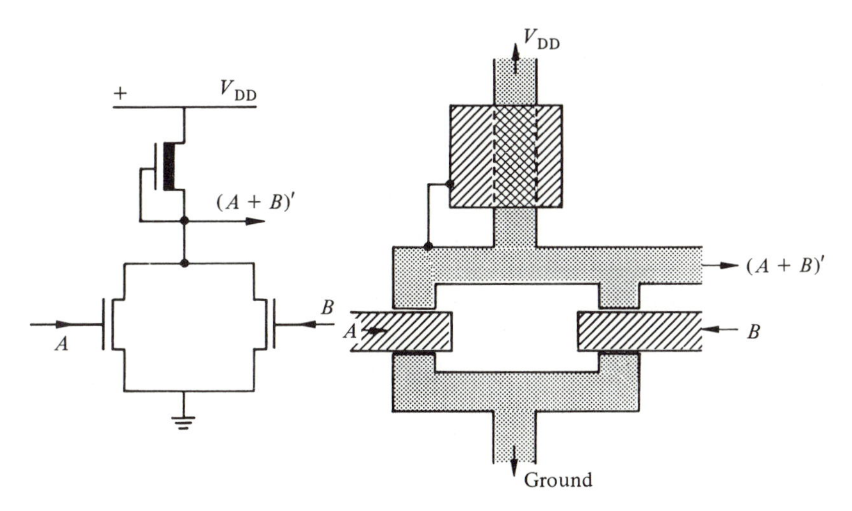

We will see in the next chapter how all logic circuits can be reduced to the basic functions that we introduced here and can therefore be built in terms of simple logic gates. We should also keep in mind that any particular signal is asserted (or negated) only for a finite time interval. Thus, for instance, an AND gate will not be asserted unless the two inputs are true at the same time. These time intervals can be very short, of order 100 ns or less, and this is why a digital circuit can perform many consecutive operations in one second, typically in excess of $10^6$.

## Exercises

### *Exercise 1.1*

(*a*) Look up the atomic mass number $A$, and the density $\rho$ of Si and Ge and find the number of atoms per $cm^3$.

(*b*) Assuming that the atoms are in a diamond structure (8 atoms/unit cell) find the lattice spacing.

(*c*) Find the resistivity of Ge at room temperature if it is doped with $10^{15}$ atoms/$cm^3$ of Sb. Assume a mobility of the donor's electrons of $\mu_e = 1200$ $cm^2$/V-s.

### *Exercise 1.2*

Calculate the forward saturation current in a silicon $n$–$p$ junction of area $10^{-4}$ $cm^2$. Let the impurity concentration be $1/10^6$ for both holes and electrons. Assume that the width of the depletion region is $W = 200$ $\mu$m and a forward bias of 0.5 V; assume some reasonable values for the mobility of electrons and holes.

### *Exercise 1.3*

Consider germanium doped with $10^{14}/cm^3$ atoms of arsenic.

(*a*) Find the conductivity assuming a reasonable value for the mobility of the impurities.

(*b*) The energy gap of germanium is $E_g = 0.67$ eV and the density of states at the edge of the conduction band can be taken as $N_c = 10^{19}/cm^3$. Estimate the intrinsic carrier density for germanium at room temperature.

(*c*) Use the result of (*b*) to find the density of holes in the doped sample.

***Exercise 1.4***

Make a plot of the Fermi–Dirac distribution at $T = -78°C$, room temperature, and at $T = 500°C$ when $E_F = 1$ eV.

***Exercise 1.5***

(*a*) Sketch an FET transistor in *p*MOS technology, that is, one using an *n*-type substrate. Estimate all dimensions including the depth of the diffusion layer.
(*b*) Give the type and density of carriers in each region.
(*c*) Give typical values for the voltages and currents through the device when in the 'on' state and when in the 'off' state.

# 2

# Digital electronics

Modern electronic devices operate in general, on digital principles. That is, signals are transmitted in numerical form such that the numbers are coded by binary digits. A binary digit has only two states: 'one' and 'zero', or 'high' and 'low' etc. The reason for relying almost exclusively on digital information is that binary data can be easily manipulated and can be reliably stored and retrieved. That this approach is practical and economically advantageous is due to the great advances in large scale integration and chip manufacture as already discussed. In this chapter we will consider digital systems and the representation and storage of binary data. We will conclude by discussing the architecture of a small 3-bit computer, which nevertheless, contains all the important features of large machines.

## 2.1 Elements of Boolean algebra

In digital logic circuits a variable can take only one of the two possible values: 1 or 0. The rules for operating with such variables were first discussed by the British mathematician George Boole (1815–64) and are now referred to by his name. Since in pure logic a statement is either true or false, Boolean algebra can be applied when manipulating logic statements as well. This material is conceptually simple yet it is most relevant to the understanding of complex logic circuits.

Boolean algebra contains three basic operations: *AND*, *OR* and *Complement*. The result of these operations can be best represented by a truth table as introduced in Section 1.9, where also the symbols for the corresponding circuits were given. We will use a + sign to indicate the *OR* function, the · sign to define the *AND* and a prime to represent the

*Complement* of any variable. As an example of a Boolean function of three variables we consider

$$F = x' + (y \cdot z) \tag{2.1}$$

The truth table for Eq. (2.1) and the corresponding logic circuit are shown in Fig. 2.1.

The rules of Boolean algebra are summarized in Table 2.1. Relations (1–8) seem obvious and define the basic arithmetic operations; relations (9–13) define the properties for operating on products and sums, and are similar to those found in ordinary algebra. However the remaining three relations are peculiar to Boolean algebra. Relations (15, 16) are known as De Morgan's theorems. Relation (14) is similar to what holds true for the inverse of a number $(A^{-1})^{-1} = A$ but is by no means equivalent; recall that $x' \cdot x = 0$ whereas $A^{-1}A = 1$; a corresponding equality in binary is

Table 2.1. *The rules of Boolean algebra*

| | | |
|---|---|---|
| *OR* | *AND* | |
| (1) $x + 0 = x$ | (2) $x \cdot 0 = 0$ | |
| (3) $x + 1 = 1$ | (4) $x \cdot 1 = x$ | |
| (4) $x + x = x$ | (6) $x \cdot x = x$ | |
| (7) $x + x' = 1$ | (8) $x \cdot x' = 0$ | |
| *Addition and multiplication* | | |
| (9) $x + y = y + x$ | (10) $x \cdot y = y \cdot x$ | commutative |
| (11) $x + (y + z) = (x + y) + z = x + y + z$ | | associative |
| (12) $x \cdot (y \cdot z) = (x \cdot y) \cdot z = x \cdot y \cdot z$ | | |
| (13) $x \cdot (y + z) = (x \cdot y) + (x \cdot z)$ | | distributive |
| *Complementation* | | |
| (14) $(x')' = x$ | | |
| (15) $(x + y)' = (x' \cdot y')$ | | De Morgan's |
| (16) $(x \cdot y)' = (x' + y')$ | | theorems |

Fig. 2.1. Truth table and circuit for the function $F = x' + (y \cdot z)$.

| $x$ | $y$ | $z$ | $F$ |
|---|---|---|---|
| 0 | 0 | 0 | 1 |
| 0 | 0 | 1 | 1 |
| 0 | 1 | 0 | 1 |
| 0 | 1 | 1 | 1 |
| 1 | 0 | 0 | 0 |
| 1 | 0 | 1 | 0 |
| 1 | 1 | 0 | 0 |
| 1 | 1 | 1 | 1 |

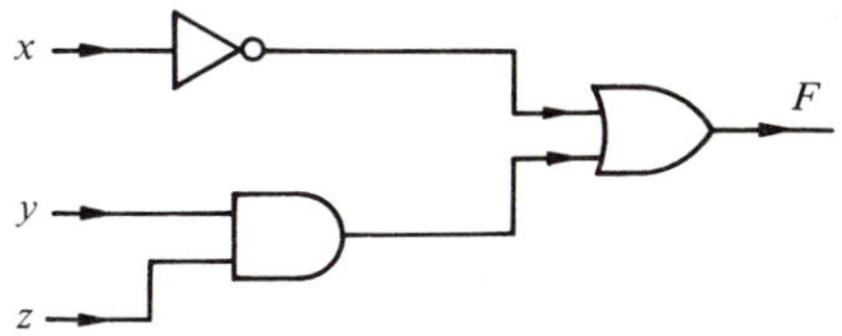

$x' + x = 1$. As a further illustration a relation such as

$$x + (y \cdot z) = (x + y) \cdot (x + z) \tag{2.2}$$

is true by definition for binary variables but is far from valid for ordinary algebra.

Boolean algebra can be used to simplify logical expressions or circuits. For instance the expression

$$F = [(A' \cdot B)' \cdot B]'$$

can be reduced using De Morgan's theorems as follows

$$F = [(A' \cdot B)' \cdot B]' = [(A + B') \cdot B]' = [A \cdot B + B' \cdot B]' = (A \cdot B)'$$

where by relation (16) the last result is also equivalent to $F = A' + B'$. The two equivalent circuits and the truth table are shown in Fig. 2.2. As further examples the reader should convince himself that

$$x' + y' \neq (x + y)'$$

Fig. 2.2. Two equivalent digital circuits.

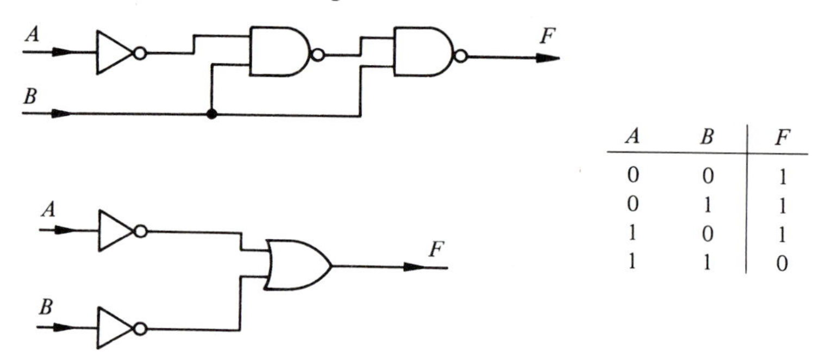


| $A$ | $B$ | $F$ |
|---|---|---|
| 0 | 0 | 1 |
| 0 | 1 | 1 |
| 1 | 0 | 1 |
| 1 | 1 | 0 |

Fig. 2.3. Simple circuit with two inputs and two outputs.

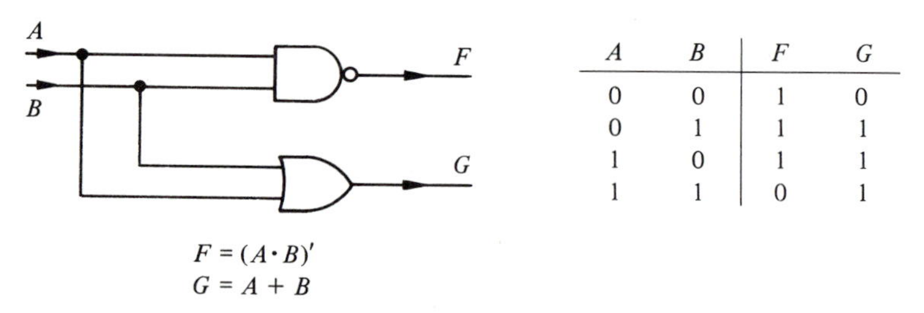


| $A$ | $B$ | $F$ | $G$ |
|---|---|---|---|
| 0 | 0 | 1 | 0 |
| 0 | 1 | 1 | 1 |
| 1 | 0 | 1 | 1 |
| 1 | 1 | 0 | 1 |

Fig. 2.4. Symbol for a combinatorial circuit.

and establish the equivalence of the two relations shown below:

(a) $F = [(A \cdot B')' \cdot B']' = A + B$

(b) $G = (A' + B') \cdot (C' + D') \cdot (B' + D) = B' \cdot D' + B' \cdot C' + (A' \cdot C') \cdot D$

It suffices to show that the two sides of the equation obey the same truth table, or one can use Boolean algebra techniques as in the previous example.

A number of digital inputs can be combined in different ways to give more than one output. For instance in Fig. 2.3 we show a circuit with 2 inputs and 2 output functions. In general a *combinatorial circuit* can have $n$ inputs and $m$ outputs. It is specified by a truth table with $2^n$ input rows and $m$ output columns. Symbolically it is designated by a box with $n$ input and $m$ output lines as in Fig. 2.4.

## 2.2 Arithmetic and logic operations

By a suitable arrangement of gates it is possible to perform arithmetic operations on two operands, $x$ and $y$. We first consider *addition* of two binary digits (bits) $A$ and $B$; the *sum* bit is designated by $S$ and the *carry* bit by $C$, and they have their usual meaning; they are defined as follows

$$S = (A + B) \cdot (A \cdot B)' \tag{2.3}$$

$$C = A \cdot B \tag{2.3'}$$

A circuit that performs these functions is called a *half-adder*, a straightforward realization being shown in the top of Fig. 2.5.

Fig. 2.5. Two equivalent circuits for the half-adder.

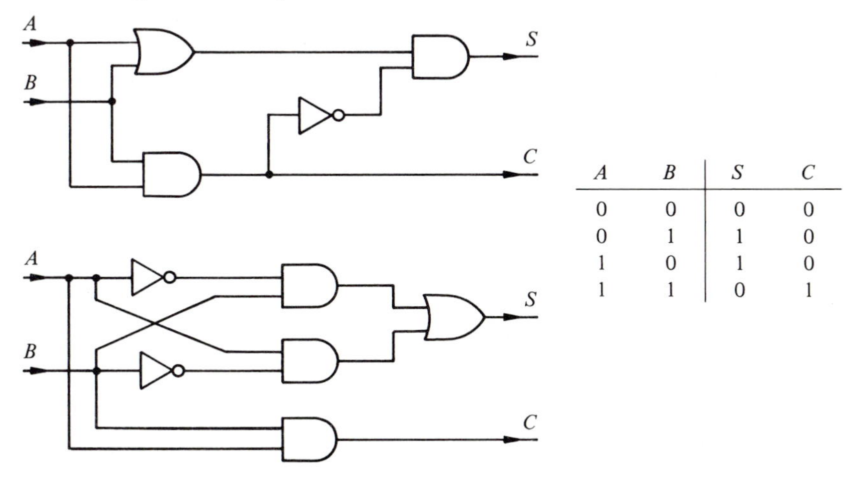


| $A$ | $B$ | $S$ | $C$ |
|---|---|---|---|
| 0 | 0 | 0 | 0 |
| 0 | 1 | 1 | 0 |
| 1 | 0 | 1 | 0 |
| 1 | 1 | 0 | 1 |

The sum bit can also be expressed by the relation

$$S = (A + B)\cdot(A\cdot B)' = (A + B)\cdot(A' + B') = A\cdot B' + B\cdot A' \qquad (2.3'')$$

which is implemented by the second circuit in Fig. 2.5. The circuit that forms the sum bit of two input bits (i.e. Eqs. (2.3) or (2.3″)) is given a special name the *exclusive-OR*. The symbol and truth table for the exclusive-OR are shown in Fig. 2.6 and that compact notation is often used when discussing higher level logic circuits.

A *full-adder* must accommodate, in addition to the two input bits, a carry input from the addition of the bits in the preceding lower order. The truth table is then as shown in Fig. 2.7 where the input carry is designated by $C$ and the output carry by ₵. A full-adder circuit using the exclusive-OR notation is shown in Fig. 2.7 and the complete circuit diagram for a full-adder using enhancement FETs is given in Fig. 2.8. In analysing the circuit of Fig. 2.8 note that the elementary gates are NANDs,

Fig. 2.6. Symbol and truth table for the exclusive-OR.

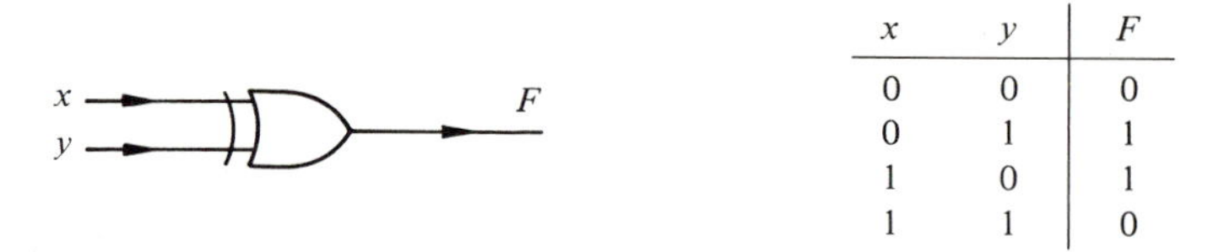


| x | y | F |
|---|---|---|
| 0 | 0 | 0 |
| 0 | 1 | 1 |
| 1 | 0 | 1 |
| 1 | 1 | 0 |

Fig. 2.7. Truth table and circuit for the full-adder.

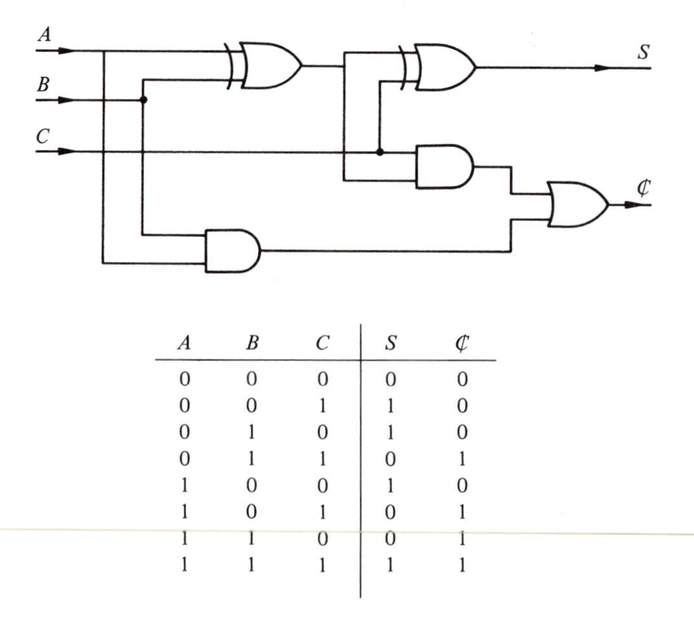


| A | B | C | S | ₵ |
|---|---|---|---|---|
| 0 | 0 | 0 | 0 | 0 |
| 0 | 0 | 1 | 1 | 0 |
| 0 | 1 | 0 | 1 | 0 |
| 0 | 1 | 1 | 0 | 1 |
| 1 | 0 | 0 | 1 | 0 |
| 1 | 0 | 1 | 0 | 1 |
| 1 | 1 | 0 | 0 | 1 |
| 1 | 1 | 1 | 1 | 1 |

and that the OR function is accomplished by tying the drains to a common line; also note the inclusion of inverters at the output to assure that the circuit obeys the truth table of Fig. 2.8. Finally, we see again that a particular logic function can be realized by more than one specific circuit.

*Subtraction* is achieved by forming the twos-complement of the subtrahend and adding it to the minuend; the overflow bit is to be ignored. Forming the twos-complement is equivalent to complementing all the bits and then adding 1. As an example we consider the subtraction of decimal 171 from decimal 372, where in binary representation we have

$$\left.\begin{array}{rrr} 372 & 101\ 110\ 100 & A \\ -\underline{171} & -\underline{010\ 101\ 011} & -\underline{B} \\ 201 & 011\ 001\ 001 & C \end{array}\right\} \tag{2.4}$$

According to the proposed algorithm we complement $B$ and add to $A$

$$\left.\begin{array}{rr} 101\ 110\ 100 & A \\ +\underline{101\ 010\ 100} & +\underline{B'} \\ 1\ 011\ 001\ 000 & D \end{array}\right\} \tag{2.4'}$$

Fig. 2.8. Full-adder circuit using NAND gates made from enhancement FETs. (Adapted from W. C. Holton, The large scale integration of microelectronic circuits, *Scientific American*, September 1977.)

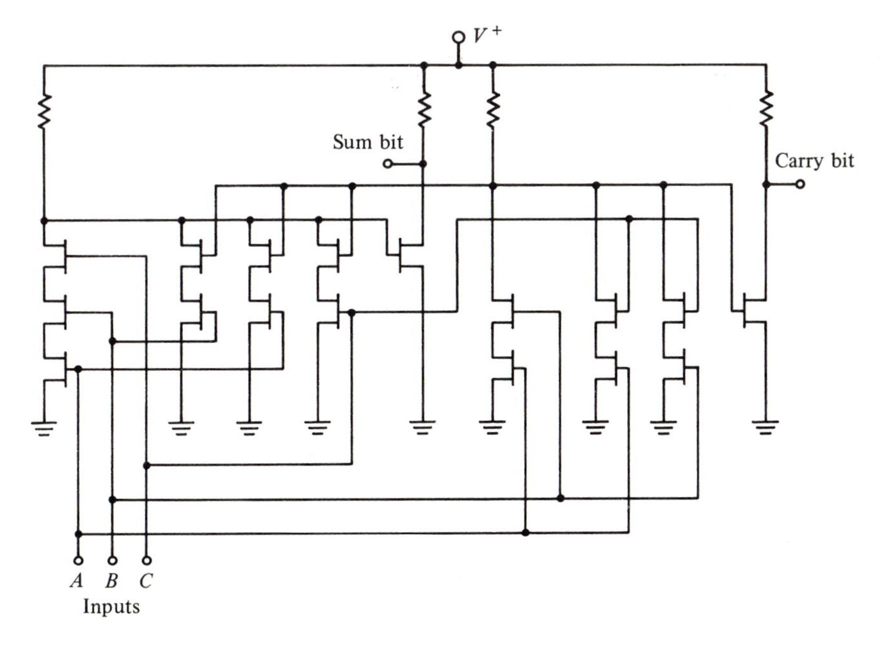

we also add 1 and ignore the overflow and obtain $C$

$$\left.\begin{array}{rr} 1\ 011\ 001\ 000 & D \\ \underline{000\ 000\ 001} & +\underline{1} \\ 011\ 001\ 001 & C \end{array}\right\} \tag{2.4''}$$

The proof of the algorithm is straightforward. Note that the complement of $B$ is given by

$$B' = (111\ 111\ 111 - B) \tag{2.5}$$

Therefore

$$\begin{aligned} A + B' + 1 &= A + (111\ 111\ 111 - B) + 1 \\ &= A - B + 1\ 000\ 000\ 000 \\ &= A - B + 000\ 000\ 000 = A - B \end{aligned} \tag{2.5'}$$

where in the last step the overflow bit was ignored. (See also Section 2.6($c$).)

*Multiplication* involves bit-shifting operations and addition. Shift registers are discussed in Section 2.4 and as their name indicates they are devices that shift the bit pattern to the left (or right) by any desired number of positions. For example, we consider the multiplication of the two decimal numbers 11 and 5, which are represented in binary form by 1011 and 101. When the bit of the multiplier is zero the partial product is zero whereas when the multiplier bit is one, the partial product consists of the multiplicand shifted left by as many times as indicated by the position of the multiplier bit; we carry out the operation as follows

| | | | |
|---|---|---|---|
| 1011 | multiplicand | 11 | |
| × 101 | multiplier | × 5 | = 55 |
| 1011 | no shift multiply by 1 | | |
| 0000 | shift left once, multiply by 0 | | |
| 1011 | shift left twice, multiply by 1 | | |
| 110111 | add, | | $= 55_{10}$ |

*Division* can be implemented by a similar algorithm where the divisor is shifted left and subtracted from the dividend.

As an example of a logic operation we discuss the comparison of two single-bit binary numbers $A$ and $B$. The *comparator* is a very important device because it allows a computer to branch to different locations in its program (the familiar 'IF' statement in programming) on comparison of two logical or arithmetic functions. We use the notation

| | | |
|---|---|---|
| $A$ .LT. $B$ | $A$ Less Than $B$ | $A < B$ |
| $A$ .EQ. $B$ | $A$ Equal to $B$ | $A = B$ |
| $A$ .GT. $B$ | $A$ Greater Than $B$ | $A > B$ |

The truth table for the 1-bit comparator is specified by Eqs. (2.6) and involves three simple Boolean functions

$$A \text{ .LT. } B = A' \cdot B \tag{2.6}$$

$$A \text{ .EQ. } B = A \cdot B + A' \cdot B' \tag{2.6'}$$

$$A \text{ .GT. } B = A \cdot B' \tag{2.6''}$$

The circuit for a comparator that obeys Eqs. (2.6) is shown in Fig. 2.9. It is straightforward to extend these operations to the comparison of binary numbers with any number of bits.

Note that the operation $A$ .EQ. $B$ represented by the circuit of Fig. 2.9 is

$$F = [(A \cdot B') + (A' \cdot B)]'$$

With the help of the De Morgan's theorems this is equivalent to

$$F = (A \cdot B')' \cdot (A' \cdot B)' = (A' + B) \cdot (A + B') = (A' \cdot B') + (A \cdot B)$$

in agreement with the definition of Eq. (2.6).

Fig. 2.9. Truth table and circuit for a 1-bit comparator.

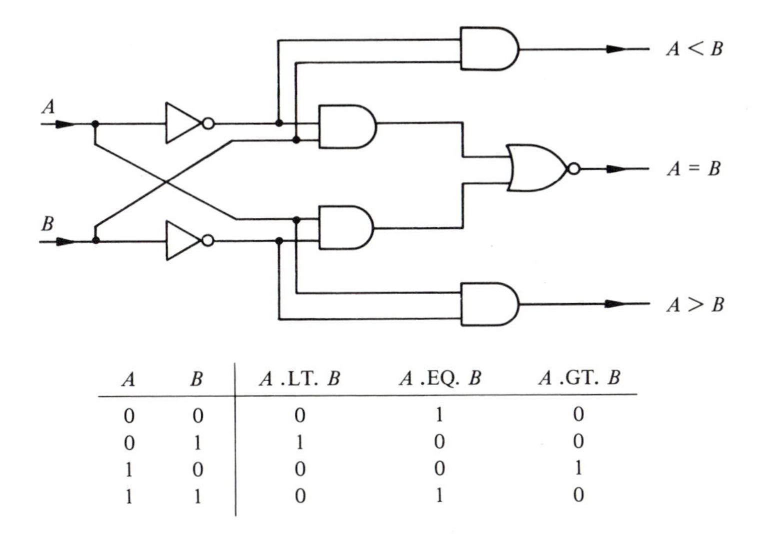


| $A$ | $B$ | $A$ .LT. $B$ | $A$ .EQ. $B$ | $A$ .GT. $B$ |
|---|---|---|---|---|
| 0 | 0 | 0 | 1 | 0 |
| 0 | 1 | 1 | 0 | 0 |
| 1 | 0 | 0 | 0 | 1 |
| 1 | 1 | 0 | 1 | 0 |

## 2.3 Decoders and multiplexers

The devices discussed in this section are built out of basic logic gates just as were the arithmetic units. Their purpose however is to interpret instructions, sent in binary form and to switch a signal along a particular path. Their operation and usefulness will become apparent as we discuss their performance and applications.

A *decoder* is a device that activates (sets high) a particular hardware line in response to a binary input. If the decoder has $n$ inputs it can have as many as $2^n$ output lines which is the number of combinations of $n$ objects. As an example we show in Fig. 2.10 the truth table and circuit for a 2-bit decoder, also referred to as a $2 \times 4$ decoder. The decoded outputs are labeled $D_0, D_1, D_2, D_3$ and are asserted respectively by the binary signals 00, 01, 10, 11. An *enable* lines has been included so that no output is generated unless the enable is true.

The presence of the enable line makes it possible to combine two $2 \times 4$ decoders into a $3 \times 8$ decoder. This is shown in Fig. 2.11 where the additional higher order bit acts on the enable line.

A *multiplexer* switches one of several inputs onto a single output line. The input to be selected is identified by the binary information on the control (or select) lines. A $4 \times 1$ multiplexer is shown in Fig. 2.12, where the inputs are labeled $I_0, I_1, I_2, I_3$ and the control lines by $S_0, S_1$. Depending on the gate selected by the $S_1S_0$ instruction, the input line attached to that gate is effectively connected to the output line whereas all the other input lines are isolated from the output.

The multiplexer is the key device in telephone communication. For instance when one lifts the receiver that particular telephone line (one of the inputs $I_0$ to $I_3$) is connected to a trunk line at the substation. Depending on the number dialed the trunk line is connected to a particular customer

Fig. 2.10. Truth table and circuit for a $2 \times 4$ decoder.

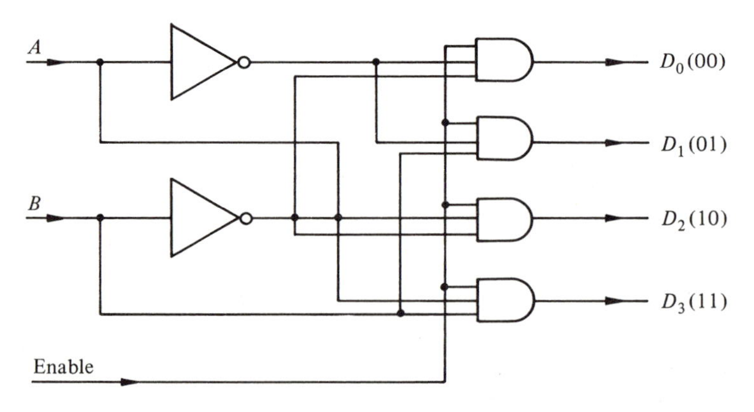


| $A$ | $B$ | $D_0$ | $D_1$ | $D_2$ | $D_3$ |
|---|---|---|---|---|---|
| 0 | 0 | 1 | 0 | 0 | 0 |
| 0 | 1 | 0 | 1 | 0 | 0 |
| 1 | 0 | 0 | 0 | 1 | 0 |
| 1 | 1 | 0 | 0 | 0 | 1 |

(one of the outputs $D_0$ to $D_3$ in Fig. 2.10), through a *demultiplexer*. In fact a decoder is also a demultiplexer, if the enable line is used as the signal input line.

Fig. 2.11. Combining two 2 × 4 decoders into a 3 × 8 decoder.

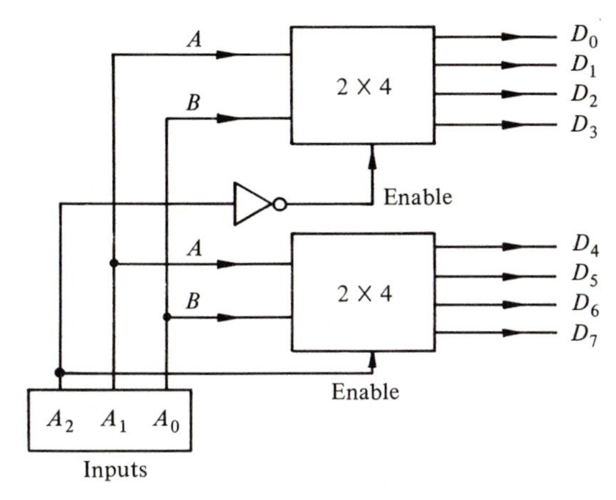


Fig. 2.12. Circuit for a 4 × 1 multiplexer.

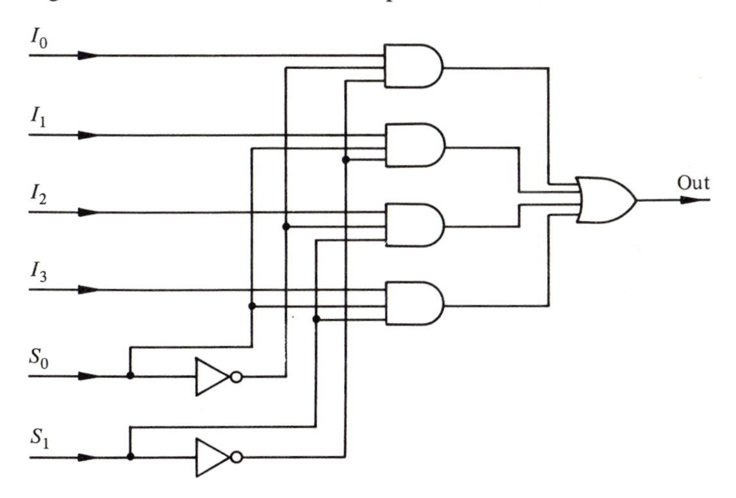


## 2.4 Flip-flops

The circuits that we discussed up to now have the property that their output depends on the instantaneous value of the input. In another class of circuits a momentary input fixes the output state; the output will remain in that state, even after the input is removed and until a rest command is received. We can think of such circuits as devices that maintain the 'memory' of the input signal. They are commonly related to as *latches*.

The simplest form of a latch is a *trigger* circuit: whenever the input exceeds a certain threshold the circuit outputs a square pulse of predetermined amplitude and width as shown by the typical waveform in Fig. 2.13. The simple trigger is a *monostable* device in the sense that it has only one stable state (in Fig. 2.13 this is the low voltage level) and under the influence of the signal it will give a high output for some predetermined time interval, but will then return to the low level. A bistable circuit has two stable states and can remain in either of these states indefinitely. Bistable circuits are referred to as *flip-flops* and always involve a certain amount of feedback.

We can make a flip-flop using two NAND or NOR gates which are interconnected. In Fig. 2.14 the output of the top NOR gate is fed back to form one of the inputs of the lower NOR gate, and vice versa. Such a circuit will be in one of two states: either $Q$ is high (and $Q'$ is low) or $Q'$ is high (and $Q$ low). Usually, both the $S$ and $R$ inputs are held low so that if $Q$ is high, the lower NOR gate has one input high and forces $Q'$ to be low. With $Q'$ low, the top NOR gate has both of its inputs low and therefore $Q$ remains high. Similarly, if $Q'$ is high and $Q$ is low, the circuit

Fig. 2.13. Input and output waveforms for a Schmidt trigger circuit.

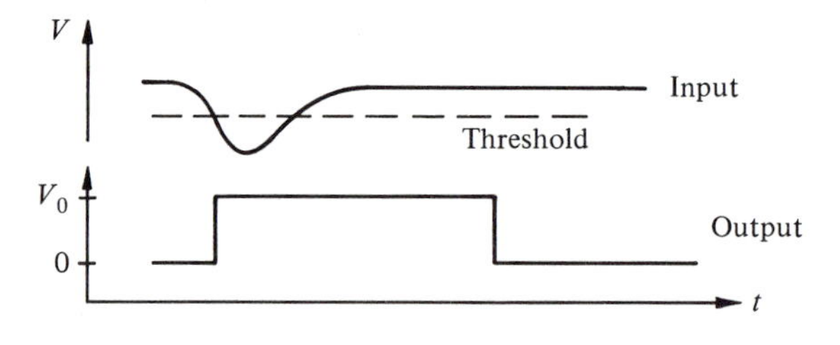


Fig. 2.14. Simple flip-flop using two NOR gates.

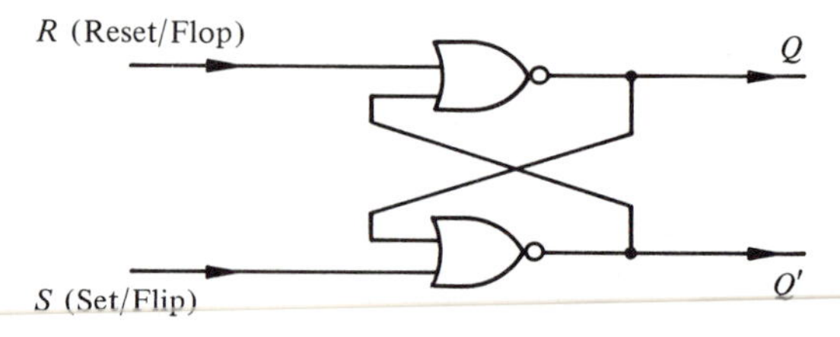

will remain in that state, unless an external signal is applied at the input. It is not possible for both $Q$ and $Q'$ to be high at the *same* time because the system becomes unstable and one or the other of the NOR gates will switch off allowing the circuit to settle in one of its stable states.

Suppose now that the circuit is in the state with $Q'$ high and that momentarily the $S$ input goes high. This will drive $Q'$ low; but as $Q'$ goes low it forces the top NOR gate off, namely $Q$ high. As $Q$ goes high the lower NOR gate is latched into its on position ($Q'$ low) even though $S$ may return to its low level. The circuit has been switched to the $Q$-high state. If instead the $R$ input was set high momentarily while the circuit was in the $Q$-low state no transition would occur. Thus the $S$ (set) and $R$ (reset) commands drive the flip-flop into its *set state* ($Q$-high) or its *reset state* ($Q$-low). Flip-flops (latches) are characterized by a table which gives the state of the flip-flop after a certain command is issued at its input, rather than by a truth table.

State of a flip-flop following the $R$–$S$ commands

| *Command* | $Q$ | $Q'$ |
|---|---|---|
| $S$ (momentarily high) | 1 | 0 |
| $R$ (momentarily high) | 0 | 1 |

Note that if both $R$ and $S$ go simultaneously high, both $Q$ and $Q'$ will want to go low, which is an undefined state for the latch. Furthermore it is not possible to predict in what state the latch will be when $R$ and $S$ are returned to low.

In complex logic circuits it is desirable that the changes of state and other operations occur through the entire system at well-defined time intervals. This is assured by using a clock which issues pulses at some fixed rate. A flip-flop which can undergo transitions only in synchronism with the clock pulse is shown in Fig. 2.15. Here we used NAND gates

Fig. 2.15. Clocked $R$–$S$ flip-flop.

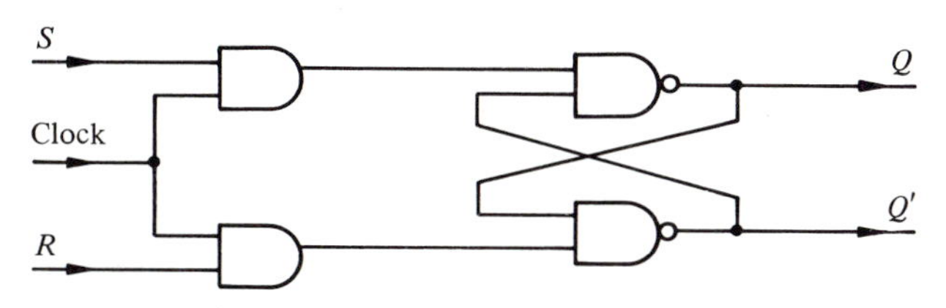

but the operation is the same as for the circuit of Fig. 2.14, except that the position of the $S$, $R$ inputs is reversed. The *characteristic table* gives the state of the $Q$ output at the time specified by the $i$, $(i+1)$, $(i+2)$, . . . etc. clock pulse. It is also convenient to have an *excitation table* which shows the status of the $Q$ output before and after a command; here we have marked by an $X$ the *don't care* conditions, namely when the output is independent of the state of that particular input line.

Characteristic table for a clocked *R–S* flip-flop

| $S$ | $R$ | $Q(t_{i+1})$ | *Function* |
|---|---|---|---|
| 0 | 0 | $Q(t_i)$ | No change |
| 0 | 1 | 0 | Clear |
| 1 | 0 | 1 | Set |
| 1 | 1 | | Not allowed |

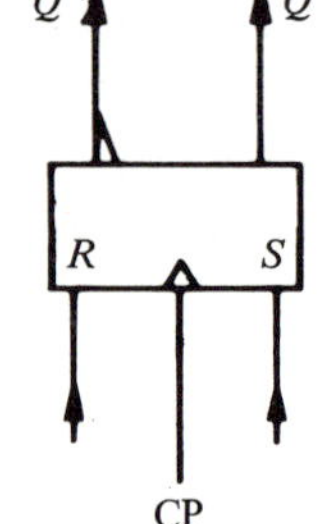


Excitation table for a clocked *R–S* flip-flop

| $Q(t_i)$ | $S$ | $R$ | $Q(t_{i+1})$ |
|---|---|---|---|
| 0 | 0 | $X$ | 0 |
| 0 | 1 | 0 | 1 |
| 1 | 0 | 1 | 0 |
| 1 | $X$ | 0 | 1 |

The asymmetries inherent in the *R–S* flip-flop are absent in a more elaborate type of flip-flop. This is referred to as a *J–K* flip-flop with $J$ the set and $K$ the clear inputs. In this case, if both the $J$ and $K$ inputs are simultaneously asserted the flip-flop undergoes a transition, i.e. it complements its output. The characteristic table, the symbol and the excitation table for the *J–K* flip-flop are shown below.

Characteristic table for a *J–K* flip-flop

| $J$ | $K$ | $Q(t_{i+1})$ | *Function* |
|---|---|---|---|
| 0 | 0 | $Q(t_i)$ | No change |
| 0 | 1 | 0 | Clear |
| 1 | 0 | 1 | Set |
| 1 | 1 | $Q'(t_i)$ | Complement |

Q' Q K J Clear CP Set

Excitation table for a *J–K* flip-flop

| $Q(t_i)$ | $J$ | $K$ | $Q(t_{i+1})$ |
|---|---|---|---|
| 0 | 0 | $X$ | 0 |
| 0 | 1 | $X$ | 1 |
| 1 | $X$ | 1 | 0 |
| 1 | $X$ | 0 | 1 |

## 2.5 Registers and counters

A set of latches can be used as a *register* which will hold any specific pattern of digits. Registers are used in all computers to hold data or instructions. The data can be transferred from one register to another or to and from arithmetic units. A simple 3-bit register built from $R$–$S$ flip-flops is shown in Fig. 2.16. The register can be loaded with the information on the three input lines $I_1, I_2, I_3$, when and only when the *load* command (line) is asserted; furthermore the loading is executed in synchronism with the clock pulse. The contents of the register appear as levels at the output lines $A_1, A_2, A_3$, which are connected to the $Q$ output of the flip-flop. The entire register can be simultaneously cleared by pulsing the *clear* line.

Information can be transferred between registers rapidly and efficiently, as shown in the following example which uses a two-phase clock. In this case clock pulses appear in two time sequences $\phi_1(t)$ and $\phi_2(t)$ which are displaced in time as shown in Fig. 2.17. The two signals never overlap, a condition that can be written in Boolean notation as $\phi_1(t)\cdot\phi_2(t)=0$. The transfer of the information from register $X$ to register $Y$ under clock control is realized as shown in Fig. 2.18($a$), which is based on MOS technology enhancement FETs. The transistors controlled by the clock pulses are referred to as *pass transistors* since they allow the signal to flow

Fig. 2.16. Three-bit register using $R$–$S$ flip-flops.

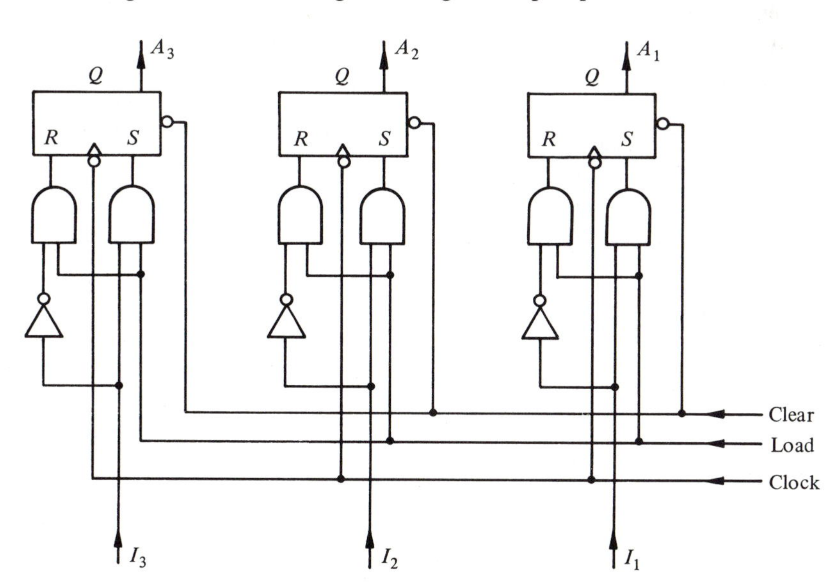

down the line. A simpler notation for the same circuit is shown in (*b*) of the figure, and an even simpler form, known as a *stick diagram* is shown in (*c*). Evidently, during the clock cycle $\phi_1$, the inverter *A* will be set; in the next cycle, i.e. $\phi_2$, the inverter *B* will be set and so on down the line. Thus the inverters can carry different information as a function of time.

A particularly useful form of register is the *shift register*, in which a bit pattern is shifted by one position to the right or left. The stick diagram for a 3-bit shift register is shown in Fig. 2.19 where in practice more bits are used. As compared to Fig. 2.18 the lines in the shift register contain two additional pass transistors (sticks) indicated by the heavier lines. The shift command is indicated by *S* and if *S* is not asserted, $\bar{S} \cdot \phi_2$ is true while

Fig. 2.17. Timing diagram for a two-phase non-overlapping clock.

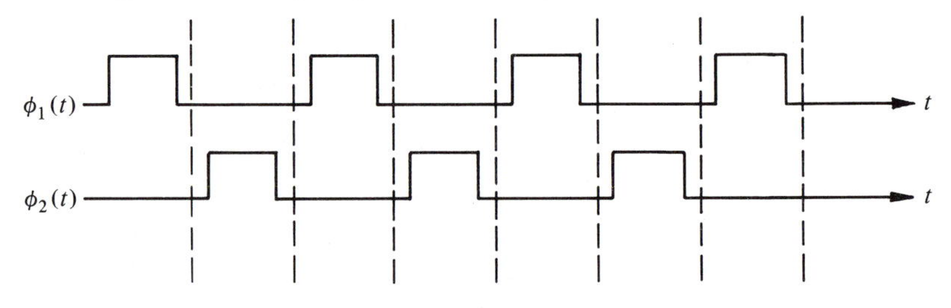


Fig. 2.18. Transfer of signals from *X* to *Y* under clock control: (*a*) circuit diagram, (*b*) mixed notation, (*c*) stick diagram.

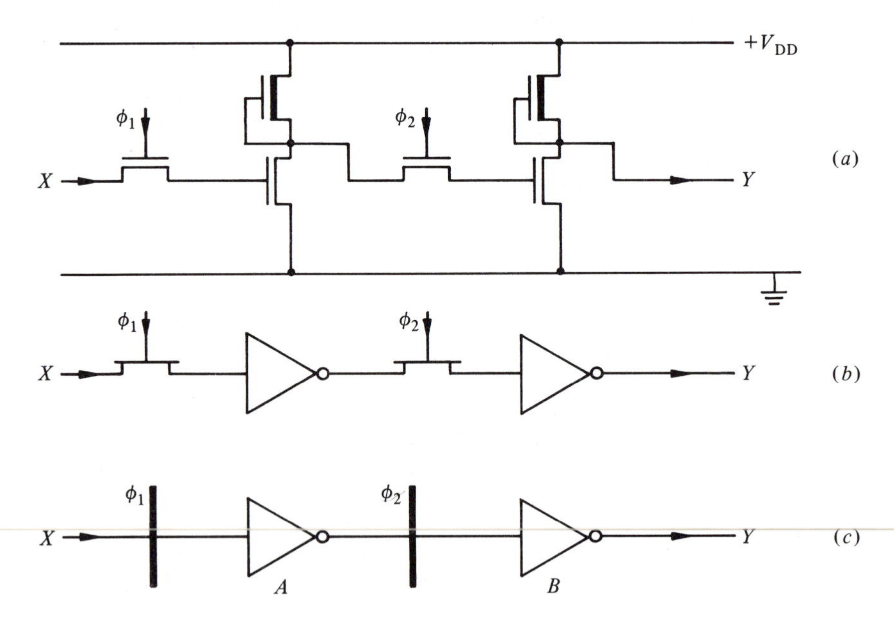

$S \cdot \phi_2$ is false, letting data flow along the horizontal path. When $S$ is true, the transistors attached to the $S \cdot \phi_2$ line conduct, whereas those connected to $\bar{S} \cdot \phi_2$ are turned off. As a result the signals flow along the diagonals and the bit pattern is shifted one position up $X_3 \rightarrow Y_2$, $X_4 \rightarrow Y_3$, $X_5 \rightarrow Y_4$, etc.

A related device is a *binary counter*. As the name indicates it counts the number of input pulses it has received since the last time it was cleared. A binary counter using $J$–$K$ flip-flops is shown in Fig. 2.20. Whenever an input signal arrives it asserts simultaneously both the $J$ and $K$ inputs of the first flip-flop. Thus the flip-flop complements itself. The second flip-flop does not see the input signal unless line $A_0$ is asserted; similarly the third flip-flop does not see the input signal unless $A_1$ *and* $A_0$ are asserted. As drawn, the counter changes states only in synchronism with

Fig. 2.19. Stick diagram of a shift register.

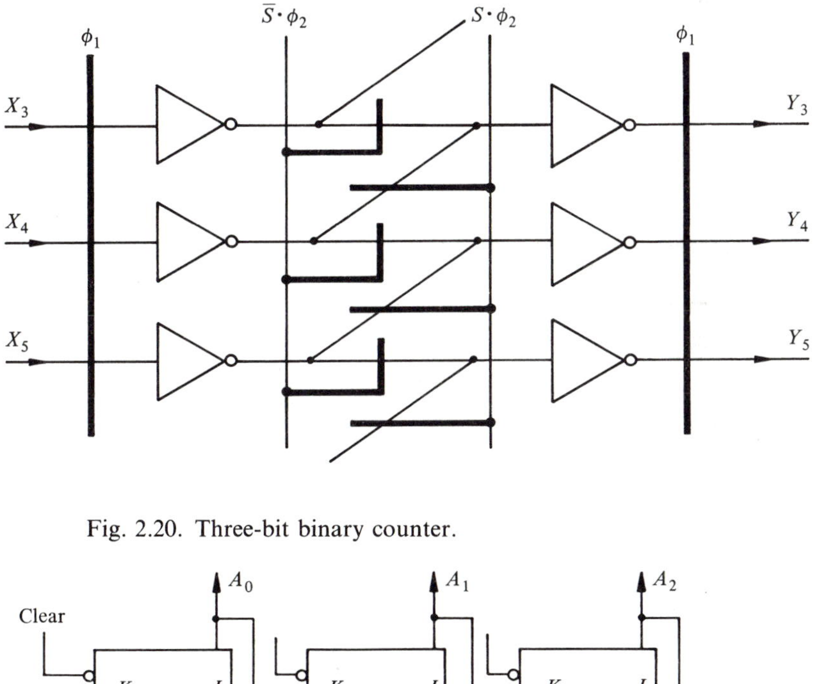

Fig. 2.20. Three-bit binary counter.

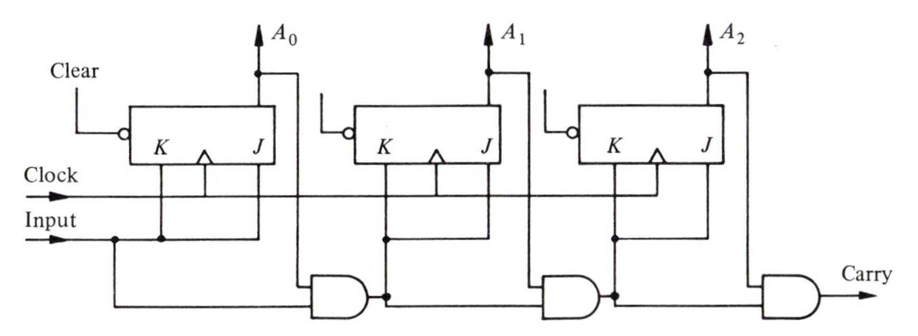

the clock pulse. If the counter was initially cleared the bit pattern changes as in the table below, namely as a binary count. By using appropriate feedback paths one can build decimal counters or counters in any basis.

Bit pattern for 3-bit binary counter

| | $A_2$ | $A_1$ | $A_0$ | |
|---|---|---|---|---|
| CLEAR | 0 | 0 | 0 | |
| INPUT | | | | |
| Yes | 0 | 0 | 1 | |
| Yes | 0 | 1 | 0 | |
| Yes | 0 | 1 | 1 | |
| Yes | 1 | 0 | 0 | |
| Yes | 1 | 0 | 1 | |
| Yes | 1 | 1 | 0 | |
| Yes | 1 | 1 | 1 | |
| Yes | 0 | 0 | 0 | Carry pulse out |

## 2.6 Data representation and coding

(a) *Number systems:* Numerical information can be expressed in any base system. In daily life we use the decimal (base ten) system whereas for digital logic the most natural system is binary (base two). The number of symbols required in any particular system equals the base $r$

| System | Base | Symbols |
|---|---|---|
| Binary | 2 | 0, 1 |
| Octal | 8 | 0, 1, 2, 3, 4, 5, 6, 7 |
| Decimal | 10 | 0, 1, 2, 3, 4, 5, 6, 7, 8, 9 |
| Hexadecimal | 16 | 0–9, A, B, C, D, E, F |

As an example consider conversion from hexadecimal to decimal

$$(\text{E4})_{16} = (\text{E} \times 16) + (4 \times 1) = 224 + 4 = 228_{10}$$

$$(\text{AE4})_{16} = [\text{A} \times (16)^2] + (\text{E} \times 16) + (4 \times 1) = 2788_{10}$$

Next we convert a decimal number $(47.8125)_{10}$ to binary. To convert the integer part we should find the largest power of 2 that fits in the number and this would give the position of the highest binary bit. The difference between the number and that highest power of 2 is then converted to binary and so on until the difference is either 0 or 1. An algorithm for the conversion process is indicated below on the left side: *Divide* the number by 2 save the remainder; in the next line divide the quotient by

2 and save the remainder; the quotient from the second line is divided by 2 in the third line, . . . and so on until the *quotient is zero*. The binary number is then made up by reading the remainders from the bottom up; the highest bit, is the last remainder. Thus

$$(47)_{10} = (101{,}111)_2$$

For the fractional part of the decimal number, the algorithm is shown on the right side: we *multiply by 2* and keep the integer part of the result; and continue multiplying until no more fraction is left. Note that this process *may not* terminate. The bit pattern is now read from the top down $(0.8125)_{10} = (0.1101)_2$.

| Integer part (read bottom up ↑) | Fractional part (read top down ↓) |
|---|---|
| $47 \div 2 = 23 + 1$ | $0.8125 \times 2 = 1.625 = 1 + 0.625$ |
| $23 \div 2 = 11 + 1$ | $0.625 \times 2 = 1.25 = 1 + 0.25$ |
| $11 \div 2 = 5 + 1$ | $0.25 \times 2 = 0.5 = 0 + 0.5$ |
| $5 \div 2 = 2 + 1$ | $0.5 \times 2 = 1.0 = 1 + 0$ |
| $2 \div 2 = 1 + 0$ | |
| $1 \div 2 = 0 + 1$ | |

Therefore

$$\begin{aligned}(47.8125)_{10} &= 101{,}111.110{,}1 \\ &= 2^5 + 2^3 + 2^2 + 2^1 + 2^0 + 2^{-1} + 2^{-2} + 2^{-4}\end{aligned}$$

as the reader can easily verify.

Octal numbers are often used instead of binary numbers because their bit image is the same as for the corresponding binary number, yet they can be expressed with three times fewer digits. Even more economic writing occurs with hexadecimal notation even though using letter symbols for numbers is less familiar. This correspondence can be seen by a simple example. Consider the 16-bit binary number

$$(X)_2 = 1{,}111{,}011{,}100{,}101{,}010$$

To obtain the octal representation we express every group of 3 binary bits (starting from the right) by its corresponding octal symbol. Thus by inspection

$$(X)_8 = (173{,}452)_8$$

For the hexadecimal representation we break the bit pattern into groups of 4 bits

$$(X)_2 = 1111{,}0111{,}0010{,}1010$$

so that by inspection

$$(X)_{16} = (\mathrm{F72A})_{16}$$

Finally in decimal representation the number we have been considering is

$$(X)_{10} = 2^{15} + 2^{14} + 2^{13} + 2^{12} + 2^{10} + 2^9 + 2^8 + 2^5 + 2^3 + 2^1$$
$$= (63{,}274)_{10}$$

The largest decimal number that can be encoded with 16 bits is

$$(M)_{10} = 2^{16} - 1 = (65{,}535)_{10}$$

Groups of 8 bits are referred to as forming 1 *byte*.

(b) *Codes:* Numerical data can also be represented by a *code* instead of by a number system. One common system is the so-called binary-coded-decimal (BCD). In this code 4 bits are used to represent the numbers from 0 to 9. Decimal numbers are then constructed by adjoining such groups of 4 bits (half-bytes). For instance the number $(189)_{10}$ is represented in BCD by

$$(189)_{10} = (1 \times 100) + (8 \times 10) + (9 \times 1)$$
$$= [0001{,}1000{,}1001]_{\text{BCD}}$$
$$= \quad [110{,}001{,}001]_{\text{BCD}}$$

Note that this is quite different from the binary bit pattern of the number

$$(189)_{10} = (10{,}111{,}101)_2$$

Codes are generally less economical in terms of bits than the base-$2^n$ number systems but they are simpler. Furthermore, to encode alphabetic letters we must use a code if we wish to store and manipulate them by digital techniques. Various codes are in existence, one example being the telegraph code introduced by Samuel Morse. Another code that was quite popular and used on 'IBM' punched cards was the Hollerith code.

Today most devices use the ASCII code (American Standard Code for Information Interchange). This is a 7-bit code and therefore can encode up to $2^7 = 128$ different symbols. These include the 26 upper case and 26 lower case letters, the 10 numbers, 11 special symbols, 23 format controllers such as carriage return, indent, skip etc., as well as data flow control signals. When a parity bit is included, every ASCII symbol occupies one byte. A partial list of the ASCII code is shown in Table 2.2.

Another example of a code is the *gray code* which is well suited for

Fig. 2.21. Encoding of shaft rotation in gray code.

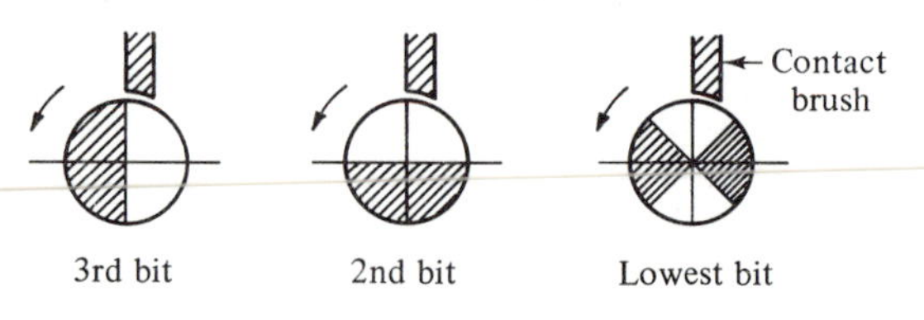

encoding the rotation of a shaft. Consider for instance a shaft onto which the three wheels shown in Fig. 2.21 are rigidly attached. The shaded areas of each wheel will give rise to a '1' when they are in contact with the corresponding brush which is located near the vertical. The position of the shaft can then be encoded modulo 45° as shown in the table. In the gray code only *one* bit changes at any transition and this helps resolve ambiguities when a brush first makes contact with the conducting part of the wheel.

| Angle | Encoded signal | Angle | Encoded signal |
|---|---|---|---|
| 0°–45° | 000 | 180°–225° | 110 |
| 45°–90° | 001 | 225°–270° | 111 |
| 90°–135° | 011 | 270°–315° | 101 |
| 135°–180° | 010 | 315°–360° | 100 |

Table 2.2. *American National Standard Code for Information Interchange (ASCII)*

| Character | Binary code | Character | Binary code |
|---|---|---|---|
| A | 100 0001 | 0 | 011 0000 |
| B | 100 0010 | 1 | 011 0001 |
| C | 100 0011 | 2 | 011 0010 |
| D | 100 0100 | 3 | 011 0011 |
| E | 100 0101 | 4 | 011 0100 |
| F | 100 0110 | 5 | 011 0101 |
| G | 100 0111 | 6 | 011 0110 |
| H | 100 1000 | 7 | 011 0111 |
| I | 100 1001 | 8 | 011 1000 |
| J | 100 1010 | 9 | 011 1001 |
| K | 100 1011 | | |
| L | 100 1100 | | |
| M | 100 1101 | blank | 010 0000 |
| N | 100 1110 | . | 010 1110 |
| O | 100 1111 | ( | 010 1000 |
| P | 101 0000 | + | 010 1011 |
| Q | 101 0001 | $ | 010 0100 |
| R | 101 0010 | * | 010 1010 |
| S | 101 0011 | ) | 010 1001 |
| T | 101 0100 | − | 010 1101 |
| U | 101 0101 | / | 010 1111 |
| V | 101 0110 | , | 010 1100 |
| W | 101 0111 | = | 011 1101 |
| X | 101 1000 | | |
| Y | 101 1001 | | |
| Z | 101 1010 | | |

(c) *Data representation:* We have indicated how integers can be expressed in different bases or encoded. Numbers, however, can be either positive or negative and furthermore it is often desirable to use exponential notation, also referred to as the *floating point* representation, where very large and/or very small numbers can be expressed with a limited number of digits.

To indicate negative numbers the left-most bit of the computer word is frequently used as the *sign-bit*. A '0' indicates positive numbers and '1' negative numbers. For instance in a 16-bit machine

$$0\ 100\ 000\ 001\ 001\ 001 = +(16457)_{10}$$

$$1\ 100\ 000\ 001\ 001\ 001 = -(16457)_{10}$$

A more efficient format for manipulating negative numbers is to represent them by their *complement*. For binary numbers their *ones-complement* consists in complementing every bit. The *twos-complement* consists in subtracting the number from zero: this is equivalent to forming the ones-complement and then adding 1 to the result. For example in a 16-bit machine the number $(16{,}457)_{10}$ used above, has

$1^s$-complement 1 011 111 110 110 110

$2^s$-complement 1 011 111 110 110 111

In *floating point* any integer number is represented by the *mantissa* and the *exponent*. Thus, we need two registers to store the number; as an example consider

$$A = (457.63)_{10}$$

We write $A$ as a pure decimal

$$A = 0.45763 \times 10^3$$

and therefore the two registers will contain

| sign bit | | |
|---|---|---|
| | 0. . . .45763 | mantissa |
| | 0. . . . . . . .3 | exponent |

The same procedure is applicable to binary numbers. For instance $(8.375)_{10}$ is expressed in binary as

$$(8.375)_{10} = (1000.011)_2 = (0.1000011)_2 \times 2^4$$

and therefore the two registers will contain

| sign bit | | |
|---|---|---|
| | 0. . . . .1 000 011 | mantissa |
| | 0. . . . . . . . . .100 | exponent |

The precision with which a number is represented depends, of course, on the number of bits available. For instance a 32-bit machine has a precision of 10 decimal digits. Often, when special precision is needed in a particular calculation, two computer words can be used to represent a single number; such calculations are said to be executed in *double precision*. On the other hand, the largest or smallest number that can be represented is dependent on the bits assigned to express the exponent.

(d) *Error checking:* We have seen that a binary bit pattern can be used to represent numbers or characters. It is possible to add an extra bit to this pattern so as to make the *total* number of *ones* in the word always odd (or always even). This is referred to as the *parity* of the word. As an example we consider the word 'WHAT' encoded in ASCII (see Table 2.2); we have added a parity bit to make all letters to be represented by bytes of odd parity.

| Message | ASCII | Parity bit | Number of 1s (odd) |
|---|---|---|---|
| W | 101 0111 | 0 | 5 |
| H | 100 1000 | 1 | 3 |
| A | 100 0001 | 1 | 3 |
| T | 101 0100 | 0 | 3 |

If this message is transmitted over a teletype line, or is stored on tape, upon retrieval, the parity of each word is checked. If the parity is not odd a mistake in transmission must have occurred. A simple odd parity generator, for 3-bit words is shown in Fig. 2.22(*a*). It uses exclusive-ORs and the reader can construct the truth table and verify that either 1 or 3 of the output lines $A$, $B$, $C$ and $P$ are always asserted. An odd parity checker circuit is shown in (*b*) of the figure. Parity generating and checking circuits are now incorporated in most VLSI packages.

The form of parity discussed here is known as transverse parity and is

Fig. 2.22. Odd parity generating and checking circuits for 3-bit words: (*a*) generator, (*b*) checker.

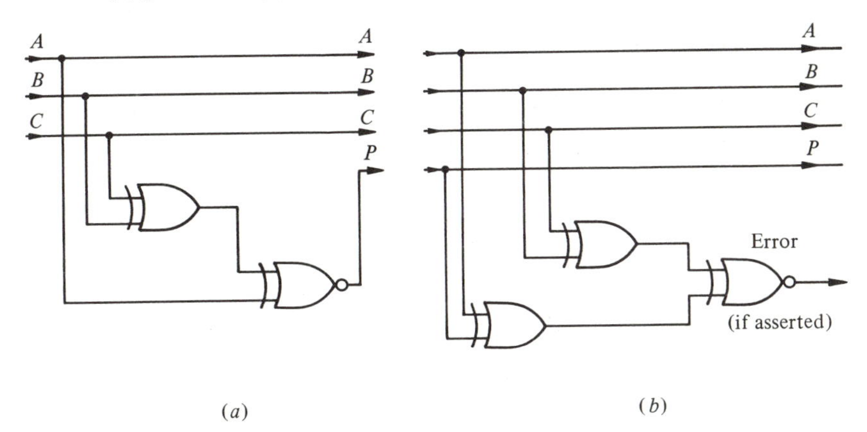

extensively used when data is written on magnetic tape. It is the simplest error checking method; for instance it would not detect an error if an even number of bits were 'dropped'. More elaborate checking methods are available by using complicated codes and can even locate the bit that was in error.

## 2.7 Computer memories

A computer is a high speed electronic calculating machine. Before the advent of electronics, mechanical calculators were in use and the first design of such a device is attributed to Blaise Pascal (1623–62). The concepts used in modern computers first appear in a proposal by Charles Babbage (1792–1871) who suggested the *storage* of intermediate results in the calculation. The memory of Babbage's machine would have consisted of wheels with 10 positions for each decimal digit. He proposed a memory for 1000 numbers each with 50 digit precision which would require 50 000 wheels. It took almost 100 years before these ideas could be used in practice in electronic computer memories. The memory must hold the program that the computer executes, the input and output data as well as intermediate results. It must be possible to store and retrieve data from specific memory locations, every location being assigned a specific address.

In principle, a computer memory could consist of a set of registers such as those used for transferring or holding data. However, in the early computers electronic storage was impractical and the first commercial memories used small cylindrical ferrite cores. The cores could be driven into saturation with the magnetization clockwise or counterclockwise indicating the two binary states. Even though ferrite core memories are not any more in use, magnetic recording on disk and tape is the principal technique for mass storage of data, as discussed in Section 2.8.

Flip-flops or latches are well suited as elements of an electronic memory. However it is simpler, and therefore less expensive, to use a single capacitor as a memory cell. If the capacitor is charged the cell is assumed to be in the '1' state whereas when the capacitor is discharged the cell is in the '0' state. Today a charge of $Q = 500$ fC can be easily measured so that for a system using 5 volts the capacitance must be

$$C = \frac{Q}{V} \sim \frac{500 \times 10^{-15}}{5} = 0.1 \times 10^{-12}\ \text{F} = 0.1\ \text{pF}$$

Such a capacitance covers only a small area and can be easily implemented in VLSI technology. For instance a $10^6$ bit memory can be placed on a $3 \times 3\ (\text{mm})^2$ chip.

The memory cells are arranged in an array and are addressed by two simultaneous signals as shown in Fig. 2.23. Here the memory capacitor is charged through a pass transistor. One of the address lines connects the source of the FET to the ground, whereas the other address line enables the gate of the FET. The memory capacitor, however, will not hold its charge forever and therefore the charge must be *regenerated* at fixed time intervals. Typically a read or write operation takes about 200 nsec and the charge must be regenerated at time intervals of order of 1–2 milliseconds, leading to dead time of the order of few percent.

In memories such as the one shown in Fig. 2.23 any cell can be accessed at random. They are referred to as RAMs for random access memory and a 1 Mbyte card can be purchased today for a few hundred dollars. To address a 1 Mbyte RAM, $1024 \times 1024$ address lines are required; these can be encoded using 20 bits. Therefore, it suffices to bring into the RAM $(10 + 10) = 20$ lines of binary information to address the $10^6$ memory locations. The block diagram of a memory unit with its supporting circuitry can be represented as in Fig. 2.24.

A variant of the RAM is the *read-only memory* (ROM). In this case the contents of the memory are introduced at the manufacture stage and cannot be altered. However the memory is addressable and the contents of each location can be read out. Such memories are typically used in small pocket calculators. In more advanced types of ROM any particular program can be 'burnt' into the ROM, i.e. the memory can be permanently loaded according to the user's specifications. Memories using flip-flops

Fig. 2.23. Schematic of an electronic memory array.

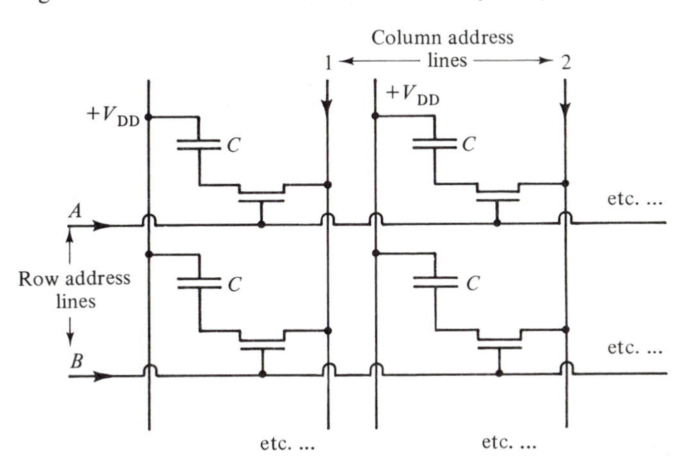

are more versatile than the simple capacitor memories that we discussed, but they are more expensive as well. A block diagram of a memory cell using an $R$–$S$ flip-flop is shown in Fig. 2.25.

Random access memories are indispensable for the operation of a computer but when there is need to store large amounts of data it becomes necessary to use cheaper and more compact devices such as magnetic tape, or magnetic disks, where the access is serial. A memory intermediate in access and cost between the RAMs and the mass storage devices is the CCD: the abbreviation stands for *Charged Coupled Device.* This is a two-dimensional register through which the data is continuously shifted at high rate, so that data can be accessed fairly quickly (even though not randomly); yet CCDs can be manufactured very compactly.

A schematic of a 64 × 64 bit CCD is shown in Fig. 2.26 and operates under clock control. The data comes in as a 64 bit word and is transferred to the first row of the CCD. At the next clock cycle the 64 bit word is pushed to the second row and new data can be stored in the first row. When a word reaches the last row, it exits the CCD and can be sent out

Fig. 2.24. Block diagram of computer memory.

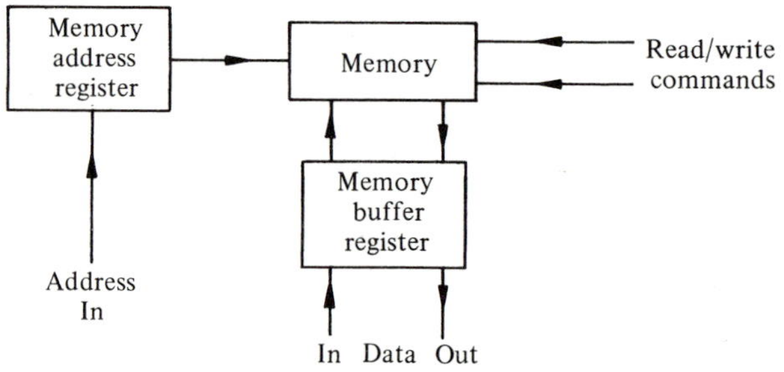

Memory address register
Memory
Read/write commands
Memory buffer register
Address In
In Data Out

Fig. 2.25. Read/write memory cell using an $R$–$S$ flip-flop.

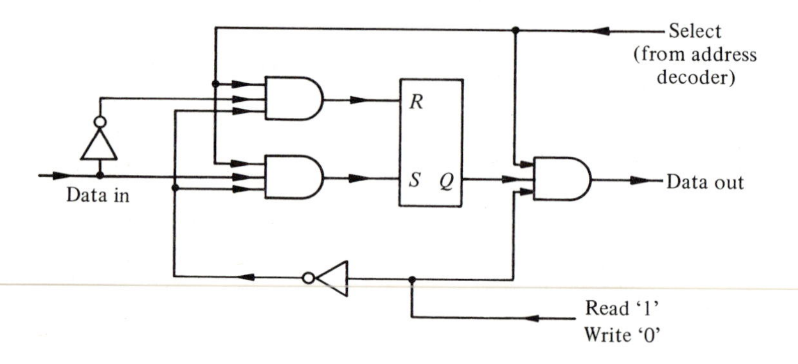

via the output line or otherwise it is recirculated to the first row of the CCD. To estimate the access time consider a CCD with a shift rate of 80 kHz. Then any one row takes at most

$$\Delta t_{max} = 64 \times \frac{1}{(8 \times 10^4)} = 8 \times 10^{-4}\ \text{s}$$

before appearing at the output. Since each row contains 8 bytes the rate of access is

$$\text{Rate} = 8 \times (1/\Delta t_{max}) = 10^4\ \text{bytes/s}$$

If blocks of data are transferred the rate is much higher, approaching $8 \times (8 \times 10^4) = 6.4 \times 10^5$ bytes/s.

Fig. 2.26. Block diagram of a 64 × 64 CCD.

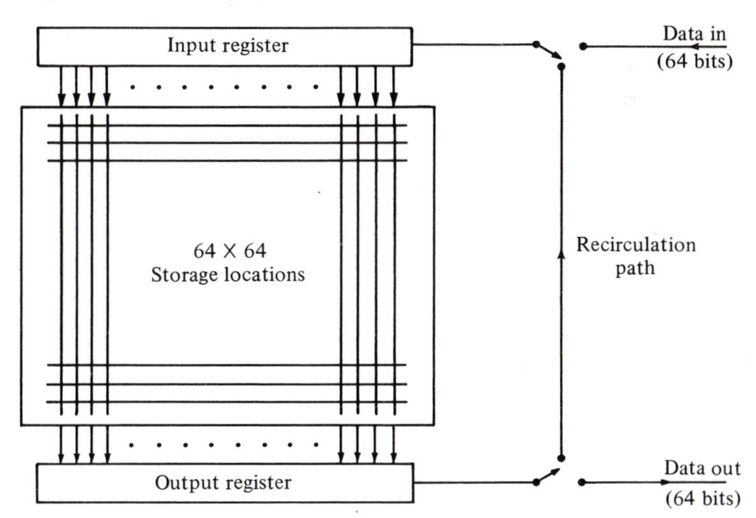


## 2.8 Magnetic storage

The invention of magnetic tape in the late 1950s has made possible the storage of large amounts of information in highly compact form. For instance, a reel of magnetic tape is typically 2400 feet long, $\frac{1}{2}$ inch wide and a total of 9 tracks are encoded. The 9 tracks correspond to 8 bits (i.e. one byte) of information and one transverse parity bit. The density that can be written/read by modern tape drives is 6250 bits per inch (bpi). Thus if we use one byte to encode an alphanumeric character the tape can hold $6.25 \times 10^3 \times (2.4 \times 10^3 \times 12) = 1.8 \times 10^8$ characters. The average word in the English language has 6 letters and a typical book (500 pages) contains 200 000 words so that one reel of tape (2400 feet) written at 6250 bpi can hold

$$\frac{1.8 \times 10^8}{6 \times (2 \times 10^5)} = 150 \text{ books!}$$

Storage of data in digital form not only has large capacity but also provides rapid access and ease of information retrieval as compared to written records. First of all the density of information storage is high: given the width of the tape the density of bits is

$$(6.25 \times 10^3) \times 8/0.5 = 10^5 \text{ bits/square inch}$$

or $10^4$ characters/square inch. Such density can be obtained on microfilm but in that case it cannot be retrieved automatically. Secondly, magnetic tape is an extremely cheap storage device; a reel of $\frac{1}{2}$-inch tape costs less than $10. As to the speed with which information can be retrieved, a modern tape drive will operate at 200 inches per second (ips). Television images (so-called video) are stored at even higher density because one can tolerate a larger fraction of errors and because the format is encoded in a highly repetitive fashion.

Magnetic recording is based on the properties of ferromagnetic materials. Certain materials can be magnetized, namely the atomic magnetic dipole moments can become aligned along a particular direction. This gives rise to a macroscopic *magnetization* **M** inside the material. The magnetization is defined as the net dipole magnetic moment per unit volume

$$\mathbf{M} = \frac{1}{V} \sum_i \boldsymbol{\mu}$$

and in the MKS system it is measured in amperes/meter. Often the alignment of the magnetic dipoles is due to the presence of an *external magnetizing field* **H**; **H** is also measured in amperes/meter and is generated by currents. The *magnetic field* **B** is the vector sum of **M** and **H**

$$\mathbf{B} = \mu_0(\mathbf{H} + \mathbf{M})$$

and in the MKS system it is measured in tesla (V-s/m$^2$). While **M** is confined inside the material, **B** can exist in free space, and

$$\mu_0 = 4\pi \times 10^{-7} \text{ V-s/A-m}$$

is the magnetic permeability of the vacuum. In practical applications, cgs–emu units are still frequently used, and therefore we give the relevant conversion factors

| | |
|---|---|
| Magnetic field **B** | 1 gauss $= 10^{-4}$ T |
| Magnetizing field **H** | 1 oersted $= 10^3/4\pi$ A/m |
| Magnetization **M** | 1 emu/cm$^3$ $= 10^3$ A/m |

For most materials the magnetization is linearly related to the external magnetizing field

$$\mathbf{M} = \chi \mathbf{H}$$

where $\chi$ is the magnetic susceptibility $|\chi| \ll 1$, and $\chi$ is positive for paramagnetic materials, negative for diamagnetic materials. In *ferromagnetic* materials $\chi \gg 1$, and the magnetization can persist even after the external field **H** is removed. The relation between **M** and **H**, or **B** and **H** is complicated and takes the form shown in Fig. 2.27(*a*).

The graphs of Fig. 2.27 are called the *hysteresis* loops. When the magnetizing field $H$ is first applied, the material follows the 'virgin curve' (*a*) until saturation is reached for some value $B_s$. As $H$ is reduced the upper curve (*b*) is followed and even for $H = 0$ the magnetic dipoles remain partially aligned and give rise to a *remanent magnetization* $B_r$. As $H$ is reversed the magnetic field will reach zero for some value $H_c$ the *coercive field* strength; when $H$ is further increased in the reverse direction saturation is reached again. If now $H$ is decreased the material follows the lower curve (*c*) to complete the loop.

The magnetization curve differs for various ferromagnetic materials and by special preparation can take the form of Fig. 2.27(*b*). In this case the ferromagnetic material has many of the properties of a *bistable* system, its two states being characterized by $B = +B_r$ or $-B_r$; furthermore, for all but for a small range of $H$ the ferrite is in one of those states. The area enclosed by the magnetization curve is a measure of the energy stored in the aligned magnetic dipoles; this is the energy that must be provided in order to flip the state of magnetization. Such materials could be used as memory cells for binary digits if we are able to switch them from one

Fig. 2.27. Magnetization curves for ferromagnetic materials: (*a*) soft iron, (*b*) special high remanence material.

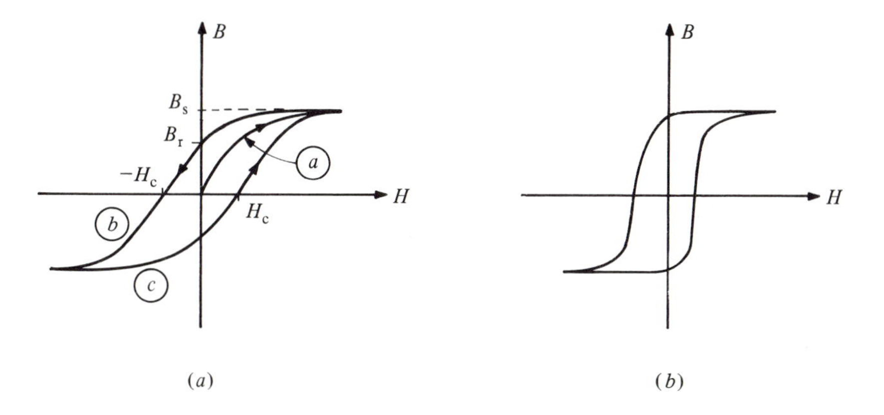

state to the other and can read their state of magnetization. The switching is accomplished by applying an external magnetic field, whereas reading is achieved by sensing the field outside the material with a small coil.

Magnetic tape is made by coating a thin layer of iron oxide ($\gamma$-$Fe_2O_3$) in the form of fine particles onto a plastic backing. The coating is approximately 10 $\mu$m thick and the size of the particles is about 0.5 $\mu$m. The material is ferromagnetic with a coercive field of order 300 oersted and a remanent magnetization of 1500 gauss. The tape is transported over the head which serves to write onto, and read from the tape. The head consists of a slotted electromagnet with sufficient leakage field so as to magnetize the tape. Similarly, changes in the magnetization of the tape, change the flux through the gap of the electromagnet and thus induce an emf in the read coil as indicated in Fig. 2.28(*b*). There is a variety of orientations that can be chosen to magnetize the tape. For digital tapes, longitudinal recording is used and the magnetization is saturated in one

Fig. 2.28. Sketch of magnetic tape head: (*a*) positioning of the tape, (*b*) field lines in the gap region and through the core.

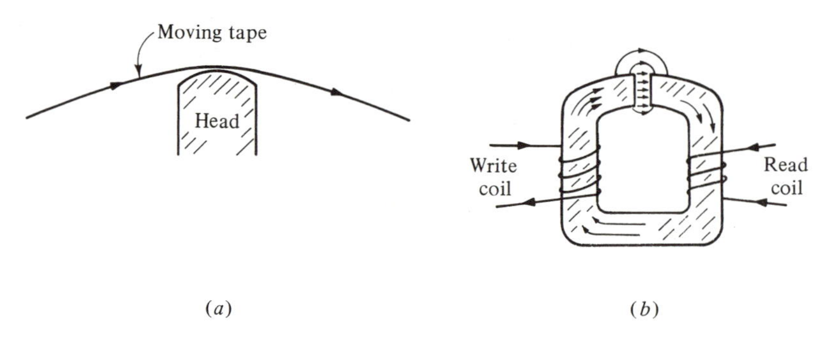


Fig. 2.29. Magnetic disk and movable head assembly.

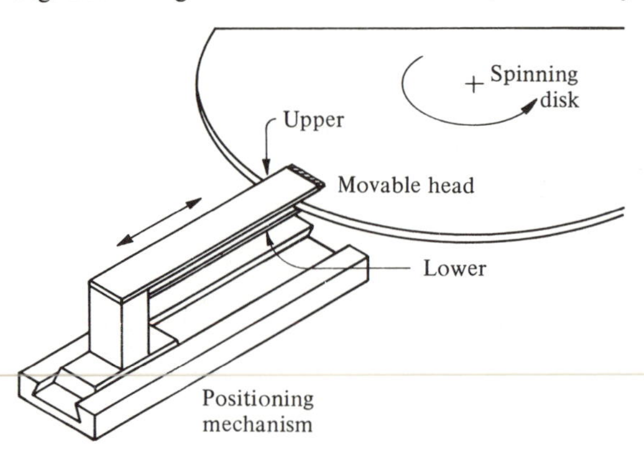

or the other direction. For analogue tapes as in cassette recorders the magnetization is in the linear region of the hysteresis curve.

A form of mass storage that provides much faster access than tape is the magnetic disk. Typically an aluminum disk, 14 inches in diameter, is coated with ferromagnetic material. The disk spins at 3000 rpm and the head can be moved radially to access different tracks as shown in Fig. 2.29. The data is written on as many as 1256 circular tracks and is subdivided into sectors, typically 50. For large disks the densities are 512 bytes per sector leading to approximately 25 Mbytes per disk surface. More often than not, both sides of the disk are used and a pack of 8 disks with 16 read/write heads can be stacked together. Such an assembly can

Fig. 2.30. Formats for digital encoding: (*a*) return to zero (RZ), (*b*) non-return to zero inverse (NRZI), (*c*) phase encoding.

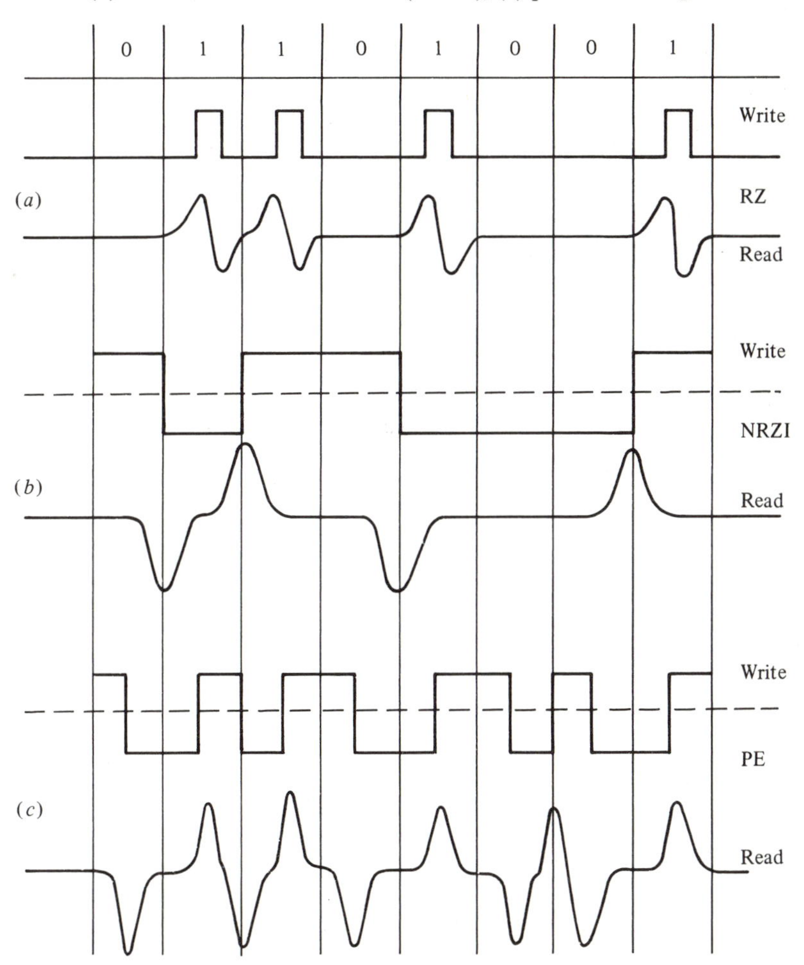

store some 500 Mbytes of information. The access time for a sector (which defines a block) is about 30 ms. Disks are used to transfer large blocks of data or program code in and out of the computer memory, greatly increasing the capability of the computer. Floppy disks, common on smaller computers use a flexible backing and therefore operate at slower speed (300 rpm) and can now hold a Mbyte of data or even more.

There are various methods for encoding the binary information onto the tape or disk. For instance one orientation of magnetization could signify a '1' whereas the other a '0'; this code is called RZ (return to zero) and is shown in the first two rows of Fig. 2.30 for the binary sequence 01101001. The first line represents the magnetization on the tape and the second line shows the signal that is picked up in the read head. The NRZI (non-return to zero inverse) method is indicated in the next two lines and here a transition indicates a 'one'; no transition in a cell is a 'zero'. In the PE (phase encoding) method the cell available for each bit is split into two parts: a 'zero' has a transition down in the middle of the cell whereas a 'one' has a transition up. In PE only the transitions in the middle of the cycle are significant; it is much easier to maintain the timing and synchronism between the tape motion and the read process in PE encoding which is the most widely used method.

Present magnetic storage devices operate with a high signal to noise ratio ($S/N$) and are therefore quite reliable. In principle the area reserved for encoding a single bit could be reduced, thus increasing the density of the stored data but also reducing the $S/N$ ratio. It appears however that it will be easier to achieve higher densities by using optical storage as exemplified by compact disk technology which we discuss next.

## 2.9 The compact disk

Compact disks are used to store audio information in digital form and are widely used since they were first introduced in 1982. It is estimated that two million disks are sold annually. In the phonograph and the cassette player audio was recorded in analogue form. Analogue to digital conversion is discussed in Chapter 3, from where it can be appreciated that digital recording offers significant advantages in the fidelity and quality of the reproduced sound.

Compact disks are equivalent to 'read-only' memories, since the information is written onto the disk at the time of manufacture and cannot be altered. On the other hand they have much higher storage density; a 12 cm disk could hold as much as 800 Mbytes of data, which is equivalent to some 2000 floppy disks. Such high density of information storage is achieved because the disk is read out optically, using a beam from a solid state laser. The beam can be focused to a spot of rms radius of 1 $\mu$m which sets the ultimate resolution limit.

The disk is constructed of plastic and the information is placed on the disk in the form of pits 0.5 $\mu$m wide and 2–3 $\mu$m long. The pits are laid out on a spiral track, consecutive tracks being separated by $\sim 2$ $\mu$m as shown in Fig. 2.31(*a*). After the pits are impressed in the plastic a reflective coating is placed on the disk, as shown in (*b*) of the figure. The optical read-out beam enters the disk from the opposite side and if it is directed onto the flat 'land' area between pits it is totally reflected. If the beam strikes a pit area, which now appears as a bump, the optical path of the reflected light is shorter by twice the pit depth; this has been chosen so that the light reflected from the bump arrives at the detector with a $\lambda/2$ phase advance and interferes destructively with the light reflected from the land area. Thus as the disk rotates past the optical beam, the pits appear as no reflected light and are characterized as '1' whereas the flat land appears as reflected light, '0', providing the binary information to the digital circuitry.

Fig. 2.31. Encoding of a compact disk: (*a*) plan view showing the tracks and a sequence of 'pits', (*b*) elevation profile and details of the illumination of the tracks.

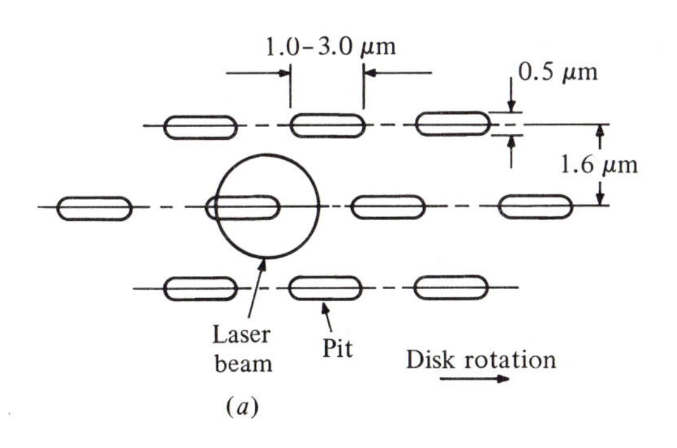


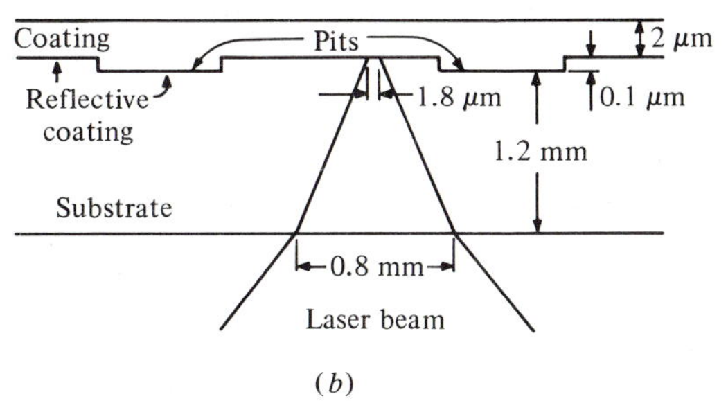

The optical read out technique used here is based on the coherent nature of the optical beam which is obtained from a solid state laser (see Chapter 4). These lasers are made by depositing layers of compounds of GaAs with different doping and are extremely compact. A schematic of the optics is shown in Fig. 2.32. The reading speed is maintained constant at 1.25 m/s which corresponds to between 4 and 8 revolutions per second. The disks hold 74 minutes of data, namely 5.5 km of track. The information is encoded on the disk by one of the codes discussed in the previous section. In view of the high information density, sound in the full audio range of 20 Hz to 20 kHz can be reproduced with a dynamic range of 90 db.

Research on erasable optical disks is being actively pursued. This is based on magneto-optic effects in thin films of GdFe and similar materials. Such disks could exceed the storage density of magnetic disks by a factor of 10–100 and would also offer faster access to the data.

Fig. 2.32. Schematic of the optics used for the read out of the optical disk. (From T. Rossing, The compact disc digital audio system, *The Physics Teacher* (1987) by permission.)

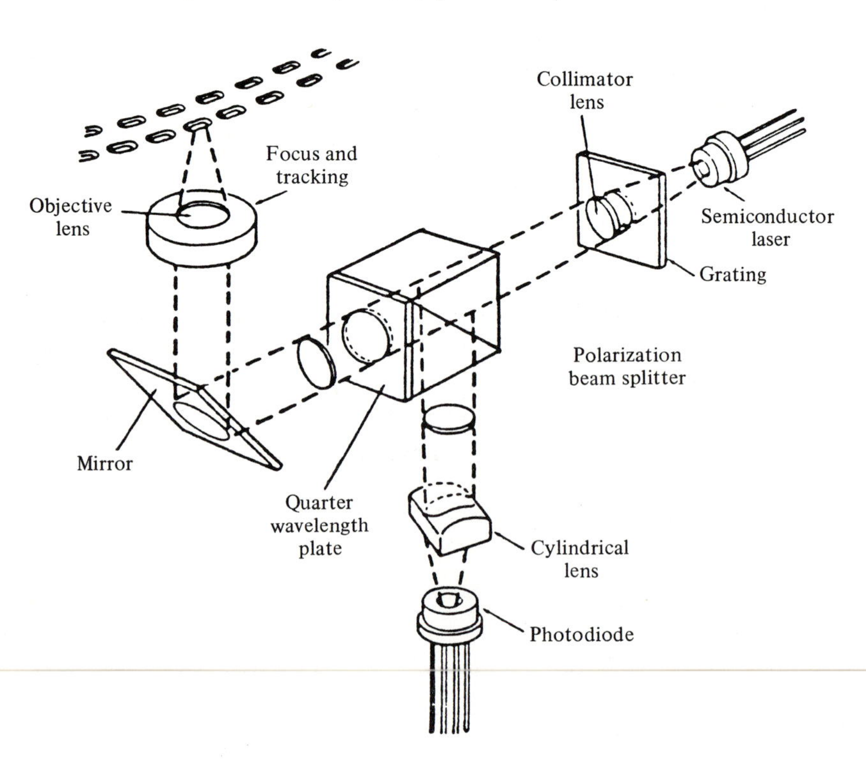

## 2.10 Computer architecture

The fundamental principle of a computer is to use a *finite* number of logic circuits and devices and change the *interconnections* between them so that they can perform any one of a variety of logical or arithmetic operations. The interconnection of the circuits and memory cells is done by the *program* which is provided to the computer by its user. A major advance in the development of computers was the suggestion by John von Neumann that binary format be used to specify the instructions that the computer should execute.

To illustrate how a computer works, we will use as an example a 3-bit machine. With three bits only 8 memory locations can be addressed directly and in our example the instruction set will contain only five operations. Nevertheless the principles and methods of operation are exactly the same as in larger machines which use from 8 up to 60 bits. This example has appeared in an article by W. C. Holton in the September 1977 issue of *Scientific American.*

A computer operates under clock control which defines the timing for each 'cycle'. Cycles are divided into two parts

FETCH 1st half-cycle
EXECUTE 2nd half-cycle

During the fetch part of the cycle the computer retrieves the instruction which is then carried out during the execute half-cycle. Even the half-cycles are further subdivided into two clock periods each, during which very specific operations are carried out.

In the fetch half-cycle we have

1st period: The computer finds the memory address from which to fetch the instruction; this is provided by the program counter.
2nd period: The instruction is taken from memory, placed on the bus and routed to its destination.

During the execute half-cycle we have

3rd period: Find and retrieve the data from memory or the appropriate register.
4th period: Perform the designated operation.

It is also possible that in the 3rd period a designated operation is carried out and then the 4th period is devoted to storing the data. Some complex operations may take more than one computer cycle.

The hardwired connections between the various units of our example computer are shown in Fig. 2.33. Note the presence of four registers which are all connected to the (3-line) bus via appropriate buffers. At any one time only the *two* registers, between which data transfer takes place, should be connected to the bus; this is accomplished by the buffers which are activated by the control lines – and are also under clock control. The role of each of the registers used in the machine is as follows:

(a) The instruction register contains the instruction that is to be executed. It is connected to a decoder which activates the appropriate control lines – one for each of the five instructions of the set.
(b) The program counter keeps track of the place at which the current operation is within the program. It is incremented every half-cycle unless it is inhibited.

Fig. 2.33. Elements and interconnections of a hypothetical 3-bit computer. (Adapted from W. C. Holton, The large scale integration of microelectronic circuits, *Scientific American*, September 1977.)

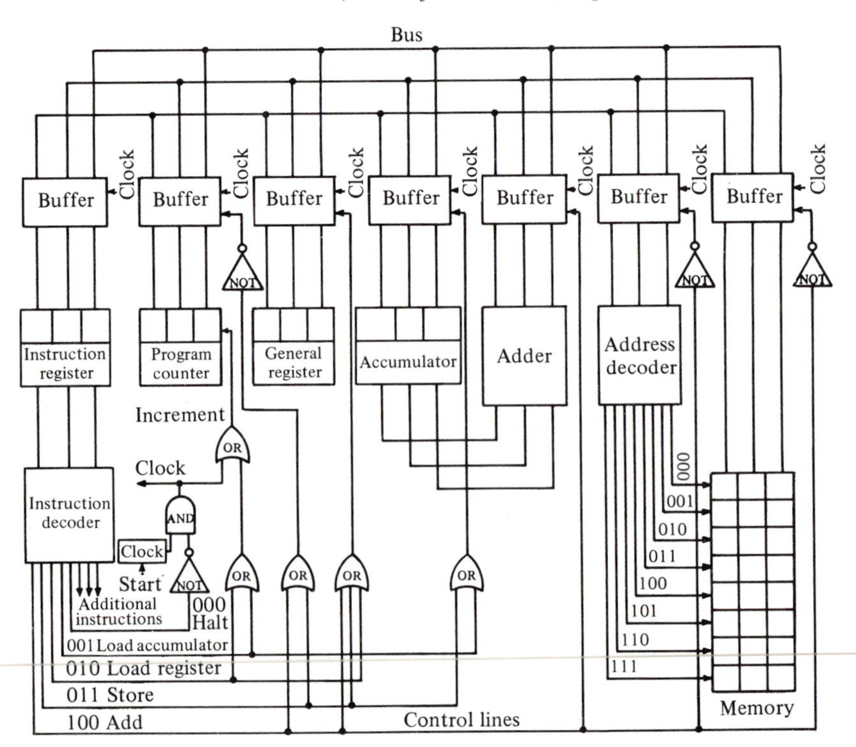

(c) The general register is used to temporarily hold information during arithmetic operations.
(d) The accumulator is the register which holds data that is passed to and from the adder.

In addition to the registers and the memory stack, the computer contains the arithmetic and logic unit (ALU) which in our example is a simple adder. We also note in Fig. 2.33 two decoders: the memory decoder is used to enable the particular memory cell that is being addressed. The instruction decoder enables the various buffers thus interpreting the instruction that is being carried out. The ALU is connected to the bus and can also be thought of as a register since information must be communicated to it from the bus. The contents of the memory are transferred to, or received from, the bus through the memory buffer which is enabled by the clock and the relevant control lines.

The instruction set, as also indicated in the diagram, includes the following five commands

000 Halt
001 Load accumulator (with the contents of the next memory location)
010 Load general register (with the contents of the next memory location)
011 Store the contents of the accumulator in memory (at location held in general register)
100 Add (the contents of the general register and accumulator, and retain the result in the accumulator)

The implementation of these instructions by the hardware can be seen in the diagram.

Let us then write a program to add two numbers, contained in memory, and store the result in memory. Such a program is indicated below and will have to be loaded in the corresponding memory locations.

| Memory location | Content | Instruction |
|---|---|---|
| 000 | 010 | Load general register with *A* |
| 001 | 011 | Data *A* ($011_2 = 3$) |
| 010 | 001 | Load accumulator with *B* |
| 011 | 001 | Data *B* ($001_2 = 1$) |
| 100 | 100 | Add and place in accumulator |
| 101 | 011 | Store in memory location 011 |
| 110 | 000 | Halt |

We can follow the execution of the program with the help of the 'timing diagram' shown in Table 2.3. Initially the program counter is set to 000 and the computer is allowed to run. During the first cycle the following events occur:

> 1st period: The program counter is connected to the bus and therefore the 000 memory location is addressed.
> 2nd period: The contents of memory location 000 are transferred to the instruction register which now contains 010.
> 3rd period: The next memory location (001) is enabled.
> 4th period: The contents of memory location 001 are transferred into the general register.

In the second cycle the same sequence is repeated to load the accumulator with the contents of memory location 011. In the third cycle the contents of the general register and of the accumulator are added and returned to the accumulator. In the fourth cycle the contents of the accumulator are transferred to memory location 011. The fifth cycle is used to interpret the 'halt' instruction and the computing halts. At this point the contents of the memory are the same as before the execution of the program except for location 011 which now contains 100 instead of 001. Note that in general, during the 1st clock period the contents of the program counter go on to the bus and are used to select the memory address. During the 2nd clock period the contents of that memory location are placed on the bus and must be interpreted as an instruction, i.e. placed into the instruction register.

The example shows how the program stored in memory can control the operations that are carried out by the computer. In addition we need input/output (I/O) devices if a human operator is to communicate with the computer. These functions are performed by terminals which have keyboards and alphanumerics or graphic displays and by hardcopy devices. A computer can also communicate directly with electromechanical devices from which it receives data and to which it issues control signals. Microprocessors, minicomputers, hand calculators, supercomputers, are all based on similar architecture. The principal differences are in the length of the word (i.e. the number of parallel bits used), the versatility of the instruction set, the cycle speed, and the size of the available direct access memory.

| Machine cycle | Clock period | Program counter | Information on the bus | | Meaning of instruction | Instruction register | Memory address |
|---|---|---|---|---|---|---|---|
| 1 FETCH | 1 | 000 | Contents of program counter | 000 | | — | 000 |
| | 2 | 000 | Instruction: load register | 010 | Load contents of next | 010 | 000 |
| EXECUTE | 3 | 001 | Contents of program counter | 001 | memory location into | 010 | 001 |
| | 4 | 001 | Data | 011 | general register | 010 | 001 |
| 2 FETCH | 1 | 010 | Contents of program counter | 010 | | 010 | 010 |
| | 2 | 010 | Instruction: load accumulator | 001 | Load contents of next | 001 | 010 |
| EXECUTE | 3 | 011 | Contents of program counter | 011 | memory location into | 001 | 011 |
| | 4 | 011 | Data | 011 | accumulator | 001 | 011 |
| 3 FETCH | 1 | 100 | Contents of program counter | 100 | | 001 | 100 |
| | 2 | 100 | Instruction: add | 100 | Add contents of general | 100 | 100 |
| EXECUTE | 3 | 100 | idle | — | register and accumulator | 100 | 100 |
| | 4 | 100 | Contents of general register | 011 | and retain result | 100 | 100 |
| 4 FETCH | 1 | 101 | Contents of program counter | 101 | | 100 | 101 |
| | 2 | 100 | Instruction: store | 011 | Store contents of | 011 | 101 |
| EXECUTE | 3 | 101 | Contents of general register | 011 | accumulator in memory at | 011 | 011 |
| | 4 | 101 | Contents of accumulator | 000 | location specified by | 011 | 011 |
| | | | | | contents of general register | | |
| 5 FETCH | 1 | 110 | Contents of program counter | 110 | | 011 | 110 |
| | 2 | 110 | Instruction: halt | 000 | Stop operations | 000 | 110 |
| EXECUTE | 3 | 110 | — | | | | |
| | 4 | 110 | — | | | | |

## Exercises

### *Exercise 2.1*

(a) Draw the diagram of gates necessary to perform the following two operations

$$X = A \cdot B + C \cdot D, \qquad Y = A' \cdot B + C' \cdot D$$

($A$, $B$, $C$, $D$ are inputs; $X$, $Y$ are outputs).

(b) Construct the truth table.

(c) For what values of the input is $X \cdot Y = 1$.

### *Exercise 2.2*

A 12-bit register holds a decimal floating point number represented in binary. The mantissa occupies 8 bits and is assumed to be a *normalized integer*. Negative numbers in the mantissa and exponent have their left-most bit in the 1-state.

(a) What are the largest and smallest positive quantities that can be represented (excluding zero).

(b) Give the binary image in that register of the decimal numbers

$$(12.2)_{10} \qquad \text{and} \qquad (-122)_{10}$$

### *Exercise 2.3*

(a) Design a flip-flop circuit using NAND gates. At what level are the 'set' and 'reset' lines kept normally?

(b) Construct the characteristic table.

(c) How does it differ from a flip-flop with NOR gates?

### *Exercise 2.4*

Construct the equivalent gate diagram for the full adder shown in Fig. 2.8. Construct the truth table and show that it is indeed that of a full adder.

### *Exercise 2.5*

(a) Convert decimal 225.225 to octal and hexadecimal.

(b) Represent your first name, middle initial and last name in binary using ASCII; include blanks between names and a period after the middle initial.

# PART B

# COMMUNICATIONS

Communication implies the transmission of messages and is the basis of human civilization. Speech, smoke signals, or written notes are all forms of communication. We will be concerned principally with communication over large distances, often refered to as telecommunications. Telecommunications are based on the transmission of electromagnetic (em) waves from a sending to a receiving station. The em wave can propagate either in a guided structure such as a pair of conductors, a waveguide or an optical fiber or it can propagate in free space. As technology progressed, higher frequency em waves became available and they offer important advantages as information carriers.

In Chapter 3 we introduce some general principles of information transmission. We examine the analysis of an arbitrary signal into a Fourier series, methods for modulating the carrier, and the sampling theorem for digital encoding of analog signals. The topic of noise in communication channels and of the expected level of random noise is treated next. Finally a brief overview of information theory is given. Information theory assigns a quantitative measure to the information contained in a message and is used to define the capacity of a communication channel.

Chapter 4 is devoted to the problems of the generation, propagation and detection of electromagnetic radiation at different frequencies. The physical laws governing these phenomena are Maxwell's equations and are universally valid. Different frequencies however present different problems in their transmission through the atmosphere and in their propagation along guided structures. We thus treat separately the reflection of radio waves from the layers of the ionosphere, the propagation and focusing of microwaves and the topic of optical fiber communications. The laser and the properties of laser radiation are discussed in the concluding sections.

# 3

# THE TRANSMISSION OF SIGNALS

## 3.1 The electromagnetic spectrum and the nature of the signals

Information is transmitted over long distances by electromagnetic waves which are modulated by the information signal. It is interesting that the first telecommunication system, the telegraph, invented by Samuel Morse, used a binary encoding system. In that case the carrier was a direct current and two types of symbols were transmitted: short and long dashes. Higher frequency carriers can be modulated at higher speed and thus can carry more information.

The relation between the frequency $f$ and the wavelength $\lambda$ of a wave is

$$f\lambda = c \qquad \text{or} \qquad f = c/\lambda \tag{3.1}$$

where $c$ is the velocity of propagation of the wave. For em waves in free space

$$c = 3 \times 10^8 \text{ m/s}$$

In a material of refractive index $n(\omega)$ the velocity of propagation of an em wave of angular frequency $\omega$ becomes

$$c'(\omega) = \frac{c}{n(\omega)} \tag{3.2}$$

and

$$\lambda' = \frac{c'}{f} = \frac{c}{nf} \tag{3.2'}$$

Since for visible light the refractive index for most materials is larger than one, the wavelength inside the material appears shorter than in free space. Frequency is expressed in hertz (Hz) which measure the number of oscillations (or cycles) completed in one second. Angular frequency $\omega$ is defined through

$$\omega = 2\pi f$$

and is expressed in radians/s.

The em spectrum and the designation of the radiation at different frequencies is indicated in Fig. 3.1. Such apparently disparate phenomena as radio-waves, visible light or X-rays are all em radiation, subject to the same physical laws and differ only in frequency. The corresponding wavelength is shown by the lower scale of the figure. Television occupies the VHF and part of the UHF bands. Microwave links are currently used extensively in civilian communications including satellite traffic. The infrared band is reserved for military communications and optical fibers transmit visible light. The UV and X-ray bands are not used much for communications because of the technical difficulties inherent in the modulation and propagation through the atmosphere of waves within these bands. Other methods of communication using neutrinos or even seismic waves on the earth's surface have been proposed for military purposes but have not as yet found any practical application.

The message that we wish to transmit may be stationary, such as a graph or typed page, or time dependent as for instance a spoken message. Even when the message is stationary it will have to be transmitted as a function of time. Thus we will treat all signals as being functions of time and we classify them as analog signals if they are continuous functions (Fig. 3.2($a$)) or as *digital* signals if they are discrete functions of time. The digital signals shown in Fig. 3.2($b$) have only two levels (they are binary) but in general could be more complex.

In order to transmit the signal we must impose it onto a carrier. This is because the carrier, usually a high frequency em wave, has much better

Fig. 3.1. The electromagnetic spectrum.

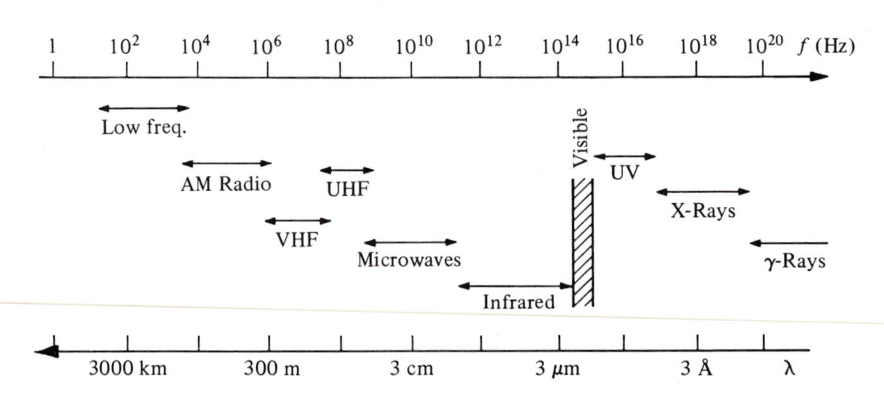

transmission characteristics than the signal which is at low frequency. Furthermore, a high frequency carrier can support simultaneously many low frequency signals. The signal is imposed onto the carrier by modulating either the amplitude or frequency of the carrier. Since the carrier is a sinusoidal wave of well-defined frequency it is convenient for calculational purposes to decompose the signal into a sum of sinusoidal (or harmonic) waves of definite frequencies. When the signal is periodic, such a sum is called a Fourier series, whereas an aperiodic signal is represented by a Fourier integral. Fourier decomposition is an indispensable tool in discussing communications because it provides the connection between the time and frequency domains.

## 3.2 Fourier decomposition

The Fourier theorem, named after its discoverer the French mathematician J. Fourier (1768–1830), states that any arbitrary function of time can be represented by a linear superposition of sine and cosine functions with angular frequencies $\omega$ varying from zero to infinity. If the function is periodic in time at frequency $\omega_0$ then the frequencies $\omega$ contributing to the linear superposition are only the *harmonics* of $\omega_0$, namely

$$\omega = \omega_0, 2\omega_0, 3\omega_0, \ldots, n\omega_0, \ldots \qquad n \text{ an integer}$$

As a first example, we consider the series of square pulses shown in Fig. 3.3 where it is assumed that the pulse train extends from $t = -\infty$ to $t = +\infty$, that is, the function $V(t)$ is truly periodic in time with a period $T$. Then the fundamental frequency is

$$f_0 = \frac{1}{T} \qquad \omega_0 = 2\pi f_0 = \frac{2\pi}{T} \tag{3.3}$$

Fig. 3.2. Time-dependent voltage conveys a signal: (*a*) in analog form, (*b*) in digital form.

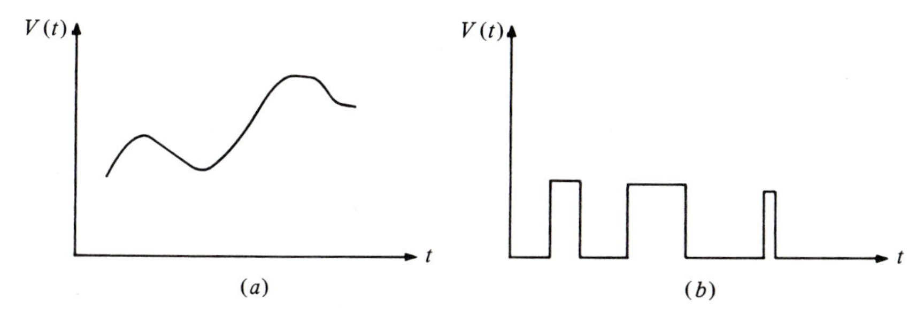

We designate the width of the pulse by $\Delta\tau < T$ and its amplitude by $V_0$. According to the Fourier theorem we can write

$$V(t) = A_0 + \sum_{n=1}^{\infty} A_n \cos(n\omega_0 t) + \sum_{n=1}^{\infty} B_n \sin(n\omega_0 t) \tag{3.4}$$

The $t = 0$ point on the time axis is, in general, arbitrary and we can assume that it is in the middle of one of the square pulses. Then the function $V(t)$ is symmetric with respect to the variable, namely $V(-t) = V(t)$ and therefore the terms in Eq. (3.4) containing sines must vanish; this is so, because $\sin(n\omega_0 t) = -\sin(-n\omega_0 t)$. Thus we set all $B_n = 0$ and the Fourier series reduces to

$$V(t) = A_0 + \sum_{n=0}^{\infty} A_n \cos(n\omega_0 t) \tag{3.4'}$$

The coefficients $A_n$ can be determined if $V(t)$ is known. As shown in Appendix 1 one finds that

$$A_0 = \frac{1}{T}\int_{-T/2}^{+T/2} V(t)\,\mathrm{d}t \tag{3.5}$$

$$A_n = \frac{2}{T}\int_{-T/2}^{+T/2} V(t)\cos(n\omega_0 t)\,\mathrm{d}t \tag{3.5'}$$

If we introduce the function $V(t)$ shown in Fig. 3.3 into Eqs. (3.5) we can find its Fourier amplitudes. The result for the first 12 coefficients $A_n$ (for the particular choice $\Delta\tau/T = 0.2$) is shown in Fig. 3.4. It is important to realize that the representation of the signal by specifying $V(t)$ (as in the graph of Fig. 3.3) or by specifying all its Fourier amplitudes $A_n$ (as in Fig. 3.4) are completely equivalent. We can view a signal either in the *time domain* or in the *frequency domain.* Note that only few amplitudes are dominant, in fact the dominant amplitudes are contained in the frequency interval $\Delta\omega$, which in this case appears to satisfy the condition

$$\Delta\omega \simeq 5\omega_0 = \frac{T}{\Delta\tau}\,\omega_0 = \frac{2\pi}{\Delta\tau}$$

Fig. 3.3. Periodic signal consisting of a sequence of square pulses of width $\Delta\tau$ and of period $T$.

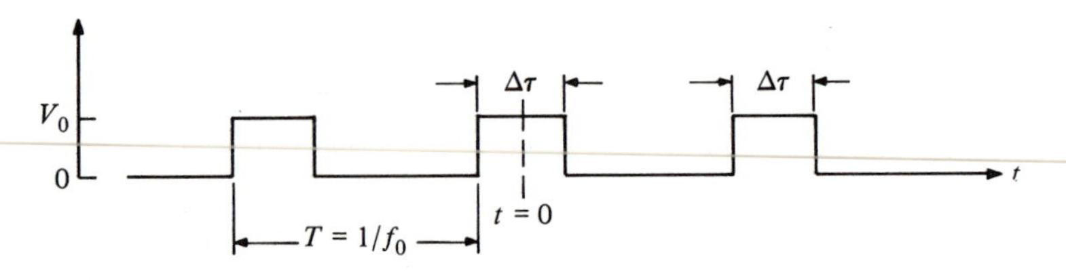

where we used $T/\Delta\tau \sim 5$. We conclude that

$$\Delta\omega\Delta\tau \sim 2\pi \tag{3.6}$$

a result which is generally valid, provided $\Delta\omega$ and $\Delta\tau$ are appropriately defined.

The result of Eq. (3.6) is of fundamental importance for all wave phenomena. It shows that the bandwidth $\Delta f = \Delta\omega/2\pi$ is the inverse of the pulse duration $\Delta\tau$. If we wish to transmit many pulses per unit time we must make the pulses short; this implies that $\Delta f$ will be large, namely that high frequencies will be involved in the transmission of short pulses. In a different context, in the quantum mechanical description of physical phenomena, Eq. (3.6) leads to the uncertainty relation between complementary variables such as momentum and position, or time and energy.

When the signal is not periodic, there is no fundamental frequency $\omega_0$ to provide the basis for discrete harmonics. Instead we can think of the signal as having a very long period, $T \to \infty$; thus $\omega_0 \to 0$ and the harmonics contain all frequencies, they form a continuous spectrum. In this case the signal is represented by a Fourier integral. This is discussed in Appendix 1 but it is convenient to use exponential notation

$$V(t) = \frac{1}{\sqrt{(2\pi)}} \int_{-\infty}^{+\infty} A(\omega) \mathrm{e}^{-\mathrm{i}\omega t} \, \mathrm{d}\omega \tag{3.7}$$

where the function $A(\omega)$ is given by the inverse expression

$$A(\omega) = \frac{1}{\sqrt{(2\pi)}} \int_{-\infty}^{+\infty} V(t) \mathrm{e}^{\mathrm{i}\omega t} \, \mathrm{d}t \tag{3.8}$$

Note that $A(\omega)$ is in general a complex function and has dimensions $[V] \times [\text{time}]$.

Fig. 3.4. The first 12 Fourier coefficients for the signal shown in previous figure when $\Delta\tau/T = 0.2$.

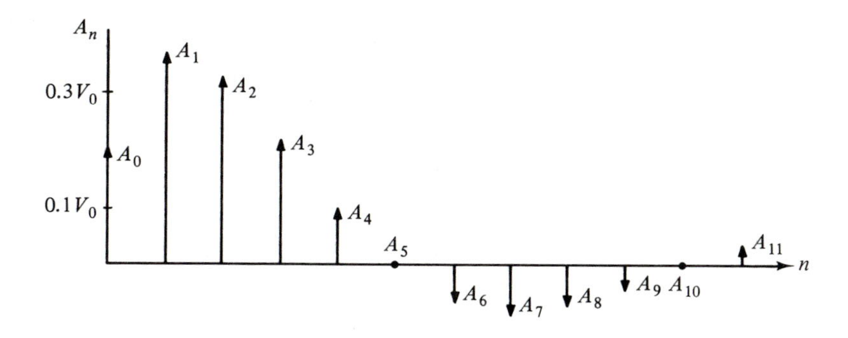

When $V(t)$ is symmetric, $V(t)=V(-t)$, Eqs. (3.7, 8) simplify to

$$V(t)=\frac{1}{\sqrt{(2\pi)}}\int_{-\infty}^{+\infty} A(\omega)\cos\omega t\,\mathrm{d}\omega \tag{3.7'}$$

$$A(\omega)=\frac{1}{\sqrt{(2\pi)}}\int_{-\infty}^{+\infty} V(t)\cos\omega t\,\mathrm{d}t \tag{3.8'}$$

and if in addition $V(t)$ is real, $A(\omega)$ is real and symmetric in $\omega$. This is indicated for an isolated square pulse in Fig. 3.5, where the representation in the time domain and in the frequency domain are shown. The two representations are completely equivalent. We again note that the product $\Delta\omega\Delta\tau \sim 2\pi$ as given in Eq. (3.6). The appearance of negative frequencies in $A(\omega)$ is a mathematical convenience and physically it implies a wave of frequency $\omega$ but with reversed phase.

We have seen that the representations of a communication signal in the time or in the frequency domain are equivalent. Since the frequency of a signal is often the determining factor in its propagation characteristics it is important to know the structure of the signal in the frequency domain. For instance a pure sinusoidal signal has only one frequency component, at its own frequency $\omega_0$. A very narrow pulse approaching a $\delta$-function in time (see Appendix 1), contains all frequencies with equal amplitudes. In reality no pulse is perfectly sharp, and we can always assign to it a width $\Delta\tau$; then only frequencies up to $\omega \lessapprox (2\pi/\Delta\tau)$ are contained in its Fourier representation.

Fig. 3.5. Square pulse and its Fourier transform: (*a*) voltage as function of time, (*b*) Fourier amplitude as a function of frequency.

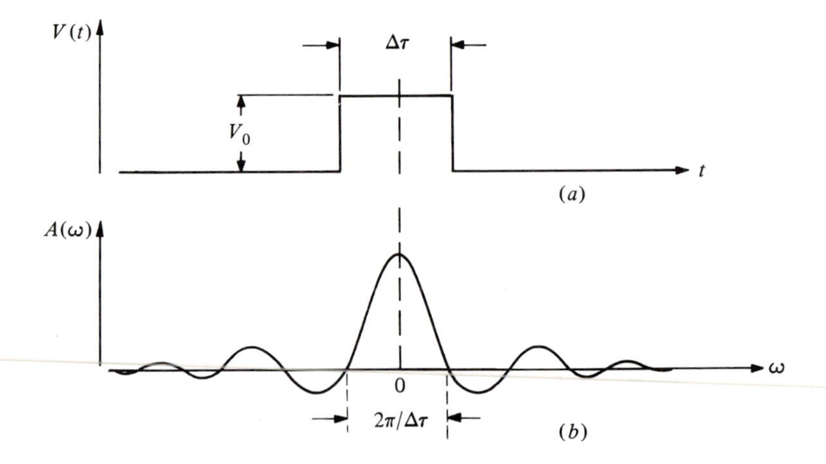

## 3.3 Carrier modulation

We now examine how information can be imposed on a carrier wave. For instance, voice communication involves frequencies, $f$, in the range $1 < f < 15$ kHz; such low frequencies cannot be transmitted efficiently over long distances and this is why typical radio carrier waves have frequencies $f_c \sim 600$–$1600$ kHz. We write the carrier wave as

$$y_c(t) = A_c \cos(\omega_c t + \phi_c) \tag{3.9}$$

and express the analog signal that we wish to transmit by $g(t)$. We can use $g(t)$ to modulate the *amplitude* of the carrier so that the high frequency wave takes the form

$$y(t) = [A_c + Kg(t)] \cos(\omega_c t + \phi_c) \tag{3.10}$$

as shown schematically in Fig. 3.6.

If the signal $g(t)$ has a pure sinusoidal form at frequency $\omega_m$

$$g(t) = \cos(\omega_m t) \tag{3.11}$$

then by elementary trigonometry the form of the amplitude modulated carrier can be expressed as a sum of three terms; we have set $\phi_c = 0$ in Eq. (3.10).

$$\begin{aligned} y(t) &= A_c \cos(\omega_c t) + K \cos(\omega_m t) \cos(\omega_c t) \\ &= A_c \cos(\omega_c t) + \frac{K}{2} \cos[(\omega_c + \omega_m)t] + \frac{K}{2} \cos[(\omega_c - \omega_m)t] \end{aligned} \tag{3.12}$$

Eq. (3.12) is equivalent to a Fourier expansion and shows that in addition to the carrier frequency $\omega_c$ the signal contains two pure cosine waves at frequencies $(\omega_c + \omega_m)$ and $(\omega_c - \omega_m)$. These frequencies are called *sidebands*, because in practice $\omega_m \ll \omega_c$.

The magnitude of $K$ as compared to $A_c$ determines the depth of modulation. When $K = A_c$, or $K/A_c = 1$ we have 100% modulation; this situation is shown in the time and frequency domains in Fig. 3.7. When $K \gg A_c$ we have 200% modulation and there is phase reversal of the carrier

Fig. 3.6. Amplitude modulated carrier signal.

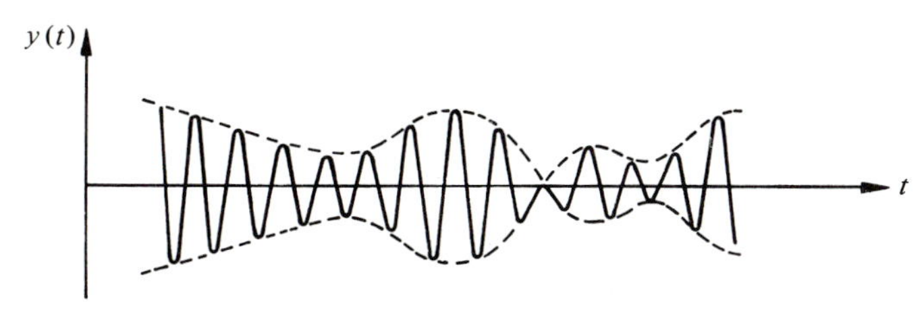

wave when $\cos(\omega_m t) < 0$; in this case the carrier is suppressed and only sidebands appear.

To retrieve the information at the receiving end we must 'demodulate' the carrier. This can be accomplished either by mixing the received signal with a local oscillator at the carrier frequency or by a square law detector. In the latter case the output of the detector is proportional to the square of the incoming amplitude; this contains the modulation signal $|g(t)|^2$ and can be separated (filtered out) from the other high frequencies. In square law detection the phase information of $g(t)$ is lost but this may not be important in many applications.

While we have discussed only the simplest form for $g(t)$, we know from the Fourier theorem that any arbitrary $g(t)$ can be decomposed into an integral over Fourier amplitudes. If the highest frequency contained in $g(t)$ is $\omega_{max}$, then the sidebands will be contained in the interval

$$\omega_c - \omega_{max} < \omega < \omega_c + \omega_{max}$$

Thus the transmission will take place over a frequency interval $2\omega_{max}$ centered at $\omega_c$. In practice, the carrier and even one of the sidebands are often suppressed in order to restrict the frequency band used for transmission. This explains why high frequency carriers can support many more information channels than lower frequencies do. For instance, for television transmission, a bandwidth of 6 MHz is required. At a carrier

Fig. 3.7. Carrier signal of angular frequency $\omega_c$ modulated sinusoidally at angular frequency $\omega_m$: (*a*) signal envelope as function of time, (*b*) the frequency spectrum contains the carrier frequency and two sidebands.

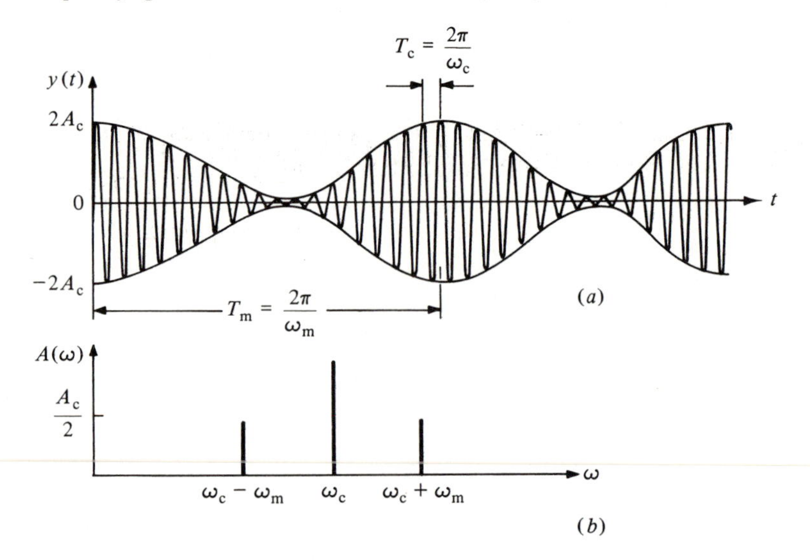

frequency of 200 MHz the channel width is comfortably smaller than the carrier frequency and spans 3% of the carrier. At a UHF frequency of 1 GHz it is possible to accommodate 5 television channels in a band which represents the same fraction of the carrier's frequency.

An alternate form of modulation is to vary the frequency of the carrier while the amplitude remains fixed. The unmodulated carrier is given as in Eq. (3.9)

$$y_c(t) = A_c \cos(\omega_c t + \phi_c) \tag{3.9}$$

When the frequency is modulated its value changes in time and

$$\omega(t) = \omega_c + g(t) \tag{3.13}$$

The transmitted wave has the *instantaneous* frequency $\omega(t)$ and the instantaneous phase angle $\phi(t)$ where $\mathrm{d}\phi/\mathrm{d}t = \omega(t)$. Thus

$$\phi(t) = \int_0^t \omega(t')\,\mathrm{d}t' + \phi_c = \omega_c t + \int_0^t g(t')\,\mathrm{d}t' + \phi_c \tag{3.14}$$

and therefore

$$y(t) = A_c \cos[\phi(t)] = A_c \cos\left[\omega_c t + \int_0^t g(t')\,\mathrm{d}t'\right] \tag{3.15}$$

where we have set $\phi_c = 0$. Eq. (3.15) shows that frequency modulation is equivalent to modulation of the initial phase of the carrier. Frequency modulation has several advantages over amplitude modulation, one of which is that $A_c$ remains constant. It does however require more complex equipment to modulate and demodulate the carrier.

Let us assume, as before, that $g(t)$ is a pure cosine function $g(t) = K_f \cos(\omega_m t)$. Inserting this form into Eq. (3.15) the transmitted wave has the form

$$y(t) = A_c \cos\left[\omega_c t + \frac{K_f}{\omega_m} \sin(\omega_m t)\right] \tag{3.16}$$

$K_f$ is dimensionless and the instantaneous angular frequency varies from $\omega_c - K_f$ to $\omega_c + K_f$. The modulating signal $g(t)$ and the transmitted signal are shown in the time domain in Fig. 3.8.

To discuss a frequency modulated (FM) signal in the frequency domain we define the *frequency deviation*, $f_a$, through

$$f_a = \frac{K_f}{2\pi} \tag{3.17}$$

and the *modulation index*, $m_p$, which is the ratio of the frequency deviation to the modulation frequency

$$m_p = \frac{f_a}{f_m} = \frac{K_f}{\omega_m} \tag{3.18}$$

In terms of the modulation index, Eq. (3.16) can be written as

$$y(t) = A_c \cos[\omega_c t + m_p \sin(\omega_m t)] \qquad (3.16')$$

By a theorem on special functions, the above expression can be expanded in a Fourier series

$$y(t) = A_c \sum_{n=-\infty}^{+\infty} J_n(m_p) \cos[(\omega_c + n\omega_m)t] \qquad (3.19)$$

Thus an infinite number of sidebands appear, symmetrically located above and below the carrier frequency $\omega_c$

$$\omega_n = \omega_c \pm n\omega_m \qquad n = 0, 1, 2, \ldots$$

The amplitude of the sidebands depends on the value of the Bessel functions $J_n$ evaluated at $m_p$. When $m_p < 1$ the amplitudes fall off quickly, an example being shown in Fig. 3.9 where only the $n = 0$, 1 and 2 sidebands are

Fig. 3.8. Frequency modulation: (*a*) modulating signal as function of time, (*b*) the modulated carrier as function of time.

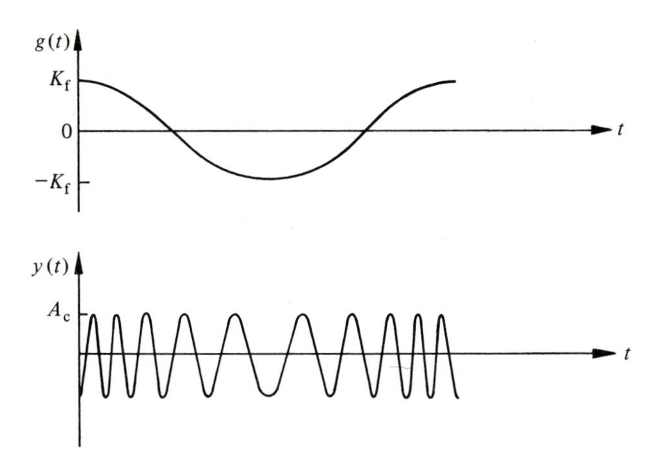


Fig. 3.9. Frequency spectrum of a sinusoidally frequency modulated signal; the relative amplitude of the sidebands depends on the modulation index $m_p$.

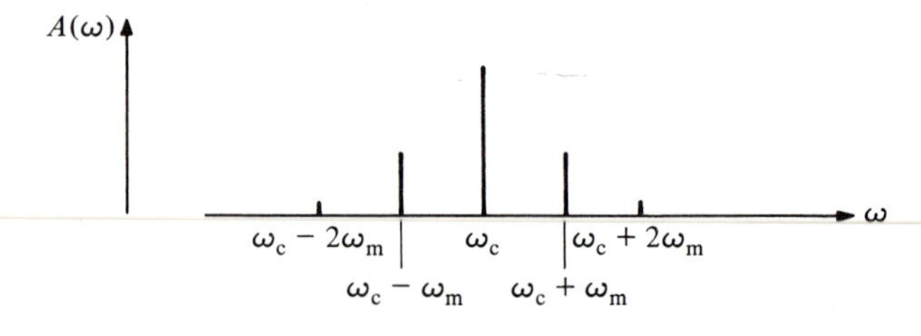

significant. For reference, the Bessel functions for the first few values of $n$ are shown in Fig. 3.10 as a function of their dimensionless argument, $x$.

The presence of several sidebands is an advantage in FM systems because they can be used to improve the quality of reception and in particular eliminate atmospheric noise. However FM requires a much wider passband than AM, and this is why FM is found only at the higher frequencies. As an example we consider a VHF radio channel that operates at a carrier frequency $f_c = 100$ MHz; it transmits audio so that $f_m = 2$–$15$ kHz, and uses a frequency deviation $f_a = 75$ kHz. Thus the modulation index is $m_p = 75/15 = 5$. For such a modulation index, approximately 16 sidebands are significant; thus the required bandwidth is $\pm 240$ kHz, as compared to $\pm 15$ kHz if it was an AM transmission.

In closing we mention one method of frequency modulating the carrier. For instance in the circuit of Fig. 3.11 if the capacitance is varied according to the signal $g(t)$ the resonant frequency of the circuit will change accordingly. There exist devices whose capacity is changed by the application of an electric signal; they are called varactors.

Fig. 3.10. Bessel functions $J_n(x)$ for $n = 0, 1, 2, \ldots$ plotted as a function of $x$.

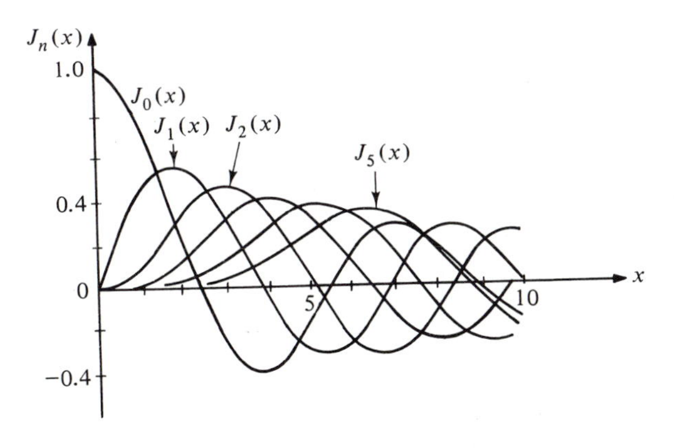


Fig. 3.11. Frequency modulating network where the capacity of the 'varactor' element depends on an externally applied voltage, i.e. $C = C_0 \cos \omega t$.

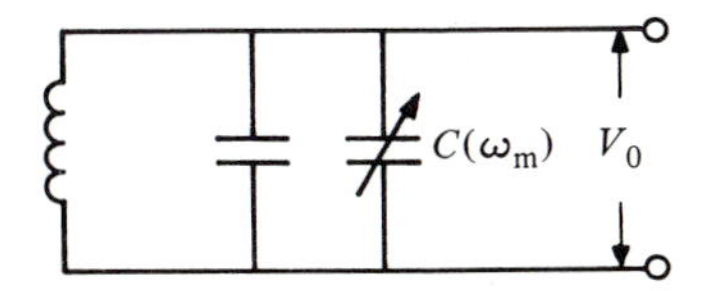

## 3.4 Digital communications

In the previous section we discussed the transmission of analog signals using the modulation of a carrier wave. It is possible to convey the same signal by representing it in digital form and transmitting the digital information over the communication channel. In fact most present day communication signals are transmitted in digital form because it is easier to recover errors and because high frequency technology has made possible digital encoding at extremely high rate.

Consider a continuous signal $v(t)$, of duration $\Delta t = T$ as shown in Fig. 3.12(*a*); we assume that the structure (i.e. complexity) of the signal is such that its frequency spectrum extends up to frequencies $f_{\max} = W$. This signal is sampled with a series of square pulses represented by the waveform $s(t)$ shown in Fig. 3.12(*b*); the sampling frequency is $f_s$. The sampled signal will then be given by the product (convolution) of $v(t)$ and the sampling signal $s(t)$

$$V(t) = v(t)s(t)$$

as shown in Fig. 3.12(*c*). The question is whether $V(t)$ is a faithful representation of $v(t)$, or more precisely what is the required sampling frequency $f_s$ such that $v(t)$ can be completely recovered from a knowledge

Fig. 3.12. Analog to digital conversion: (*a*) analog signal as a function of time, (*b*) sampling signal, (*c*) digitized output signal; this can be realized by multiplying the waveforms in (*a*) and (*b*).

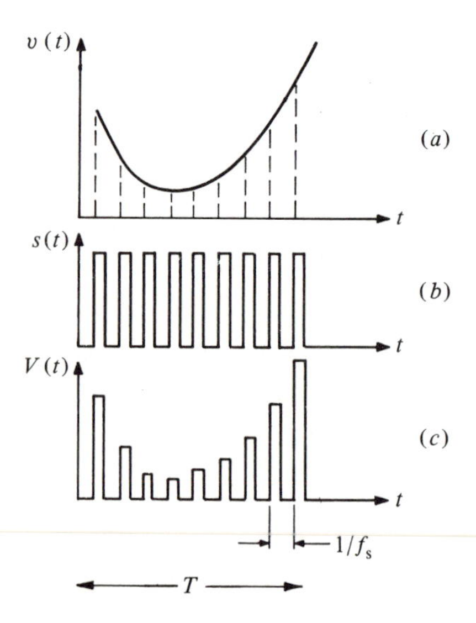

of $V(t)$. The answer is that when

$$f_s = 2W \tag{3.20}$$

the representation is faithful. This very important result is due to Nyquist and is known as the *sampling theorem*; we will not prove the theorem here but we can give a non-rigorous argument to indicate its plausibility. Since the function $v(t)$ has a duration of $T$ seconds we can think of it as a segment of a *periodic* function with period $T$; we ignore the region $t < -T/2, t > T/2$. Then there is a fundamental angular frequency $\omega_0$ that can be associated with $v(t)$, where

$$\omega_0 = 2\pi/T$$

Since $v(t)$ is treated as periodic it can be expanded into a Fourier series and by definition the highest significant amplitude is at angular frequency $\omega_{\max} = 2\pi W$. The number of the highest harmonic is $n = \omega_{\max}/\omega_0$, and the total number of amplitudes required to represent $v(t)$ is

$$(2n+1) = 2\left(\frac{\omega_{\max}}{\omega_0}\right) + 1 = 2\left(\frac{2\pi W}{2\pi/T}\right) + 1 = 2WT + 1 \tag{3.21}$$

(There are $n$ coefficients $A_n$ for the cosine terms, $n$ coefficients $B_n$ for the sine terms and the constant $A_0$; see Eq. (3.4).) Eq. (3.21) shows that to obtain a faithful representation of $v(t)$ we must make $(2WT+1)$ measurements in the time interval $T$. Thus the sampling rate must be

$$\frac{1}{T}(2WT+1) = 2W + \frac{1}{T} \simeq 2W$$

as given in Eq. (3.20).

The previous analysis assumes that every sampled point is measured with infinite precision. In practice, using a 6-bit word to encode an amplitude gives a dynamic range of 64 (18 db in amplitude) which is

Fig. 3.13. The effect of noise and distortion in the transmission of a digital signal; the original waveform can, however, be recovered.

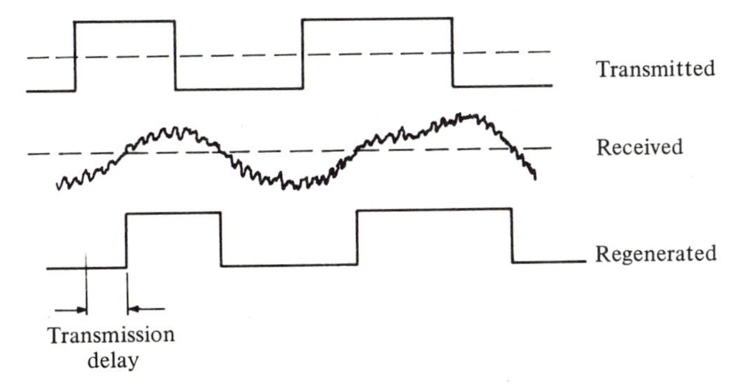

adequate for most applications; even a 3-bit word offers a dynamic range of 8. Conversion of an analog level to a digitized signal is accomplished electronically with a so-called A to D circuit. The simplest example of A–D is a digital voltmeter. Modern A–D chips can complete a cycle in 1 μs and thus could be used to sample at a rate of $f_s = 1$ MHz. For comparison, compact disk recorders sample at a rate of 44 kHz. This is twice the frequency range of audio as expected from the sampling theorem (Eq. (3.20)). Telephone communications are sampled at even slower rates.

The digital signal is transmitted in binary form and this is advantageous in terms of power requirements and error recovery. This latter point is illustrated in Fig. 3.13. Even when the signal is considerably distorted during transmission it is possible to reconstruct the information that was initially sent, by using a simple threshold circuit.

## 3.5 Noise in communication channels

So far we have considered idealized communication channels; in reality any channel contains some amount of noise. For instance if we try to talk at a crowded gathering we must raise our level of speech in order to be heard. This analogy can be carried even further because we can often carry on a meaningful conversation even if we do not hear exactly every word but can still successfully guess at it from part of its structure. Thus, information can be transmitted even in the presence of noise. For em signals noise can enter the system due to distortion in transmission, spurious noise at the receiver or even outright interference from unrelated sources.

One form of noise that is always present in any physical system, and therefore also in communication channels, is that due to the thermal motion of the molecules; the Brownian motion of suspended particles is due to the same cause. If we examine a constant voltage across a resistor, the deflection angle of a galvanometer, or many other physical quantities that are in equilibrium, we will observe that they fluctuate about their equilibrium position as shown in Fig. 3.14($a$). If we look in more detail by expanding the scale of $X(t)$ the fluctuations will be magnified as in Fig. 3.14($b$). We assume that the average value of $X(t)$ is well defined and designate it by $\langle X(t)\rangle$. Thus we can introduce the new variable

$$y(t) = X(t) - \langle X(t)\rangle \tag{3.22}$$

which has by definition zero mean, $\langle y(t)\rangle = 0$. Clearly the mean value of the signal does not carry information about the noise level. Instead we should introduce a measure which is positive definite and evaluate its mean value. The *mean square* value of $y(t)$ is defined through

$$\langle y^2(t)\rangle = \lim_{t\to\infty} \frac{1}{T}\int_{-T/2}^{T/2} y^2(t)\,dt \tag{3.23}$$

Note that $\langle y^2(t)\rangle$ is always positive and it is independent of time. Its magnitude is a measure of the fluctuations (or noise) on the signal $X(t)$. We note that $\langle y^2(t)\rangle$ is related to the signal $X(t)$ through

$$\begin{aligned}\langle y^2(t)\rangle &= \langle [X-\langle X\rangle]^2\rangle = \langle X^2\rangle - 2\langle X\rangle\langle X\rangle + \langle X\rangle^2\\ &= \langle X^2\rangle - \langle X\rangle^2\end{aligned} \tag{3.23$'$}$$

When we measure the variable $X$ at any particular time $t$, the result $X(t)$ will differ from the mean value $\langle X\rangle$ by the amount $y(t)$. Thus knowledge of $y(t)$ would enable us to determine the desired mean value $\langle X\rangle$ from a single measurement. However $y(t)$ represents *random* fluctuations and thus we cannot know *a priori* the function $y(t)$; but we can determine the *probability distribution* of $y(t)$. That is, there exists a function $f(y)$ such that $f(y_0)\,dy$ gives the *probability* that at any time $t$, the measured value of $y$, will be between $y_0$ and $y_0 + dy$. Thus we can write

$$P[y_0 < y < y_0 + dy] = f(y_0)\,dy \tag{3.24}$$

the function $f(y)$ is called a probability density function. To determine $f(y_0)$ experimentally we can divide the record of Fig. 3.14 into $N$ regions and count the number of times, $n_y$, that $y_0 < y < y_0 + dy$. Then $f(y_0)\,dy = n_y/N$ in the limit $N\to\infty$ and $dy\to 0$. Fluctuations are a stochastic, or random, phenomenon and are described by the function $f(y)$.

It is a remarkable fact of nature that the fluctuations in many phenomena, especially when several random causes contribute to the noise, follow a *Gaussian* distribution. Furthermore the Gaussian distribution is

Fig. 3.14. Random fluctuations about: (*a*) the mean value $\langle X\rangle$, (*b*) about zero, are shown as a function of time.

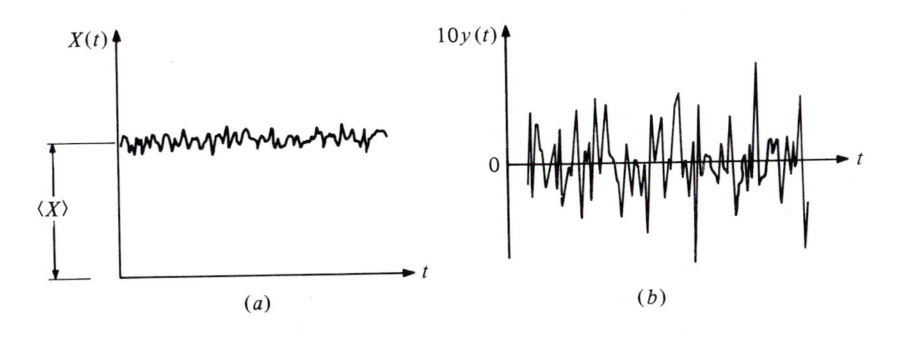

characterized by a *single* parameter, its *standard deviation* $\sigma$. The normalized Gaussian distribution function is

$$F(y) = \frac{1}{(2\pi)^{1/2}\sigma} e^{-y^2/2\sigma^2} \tag{3.25}$$

The square of the standard deviation, $\sigma^2$, is equal to the mean square value of the fluctuations

$$\sigma^2 = \langle y^2 \rangle = \langle X^2 \rangle - \langle X \rangle^2 \tag{3.26}$$

Note that $F(y)$ has dimensions of $1/y$ as required for a probability density. In terms of the variable $X$, the distribution function is

$$F(X) = \frac{1}{(2\pi)^{1/2}\sigma} e^{-(X-\langle X \rangle)^2/2\sigma^2} \tag{3.25'}$$

The normalized Gaussian distribution is plotted in Fig. 3.15 in terms of its dimensionless normalized argument $y/\sigma$. Given a normalized distribution function $f(y)$ the mean value of a function $\phi(y)$ is in general defined through

$$\langle \phi(y) \rangle = \int_{-\infty}^{\infty} f(y)\phi(y)\,\mathrm{d}y$$

The first few 'moments' of the Gaussian distribution are as follows

$$\int_{-\infty}^{\infty} F(y)\,\mathrm{d}y = \frac{1}{(2\pi)^{1/2}\sigma} \int_{-\infty}^{\infty} e^{-y^2/2\sigma^2}\,\mathrm{d}y = 1 \tag{3.27}$$

$$\langle y \rangle = \int_{-\infty}^{\infty} F(y)y\,\mathrm{d}y = 0 \tag{3.27'}$$

$$\langle y^2 \rangle = \int_{-\infty}^{\infty} F(y)y^2\,\mathrm{d}y = \sigma^2 \tag{3.27''}$$

Eq. (3.27′) follows immediately from symmetry whereas Eq. (3.27″) can be obtained by integration by parts followed by use of the normalization condition of Eq. (3.27).

Fig. 3.15. Normalized Gaussian distribution function plotted in units of its standard deviation $\sigma$.

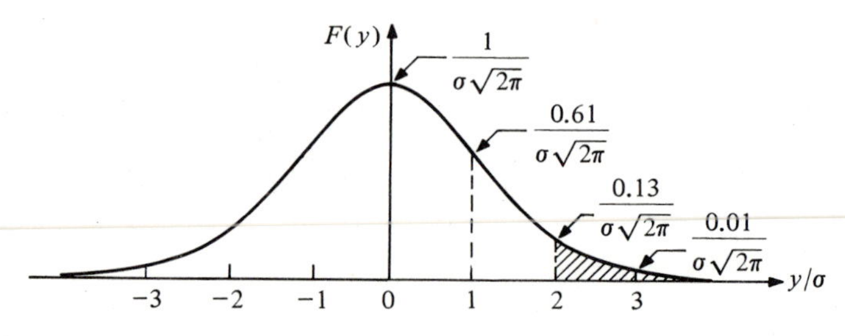

Knowing the distribution function makes it possible to calculate any pertinent probability. For instance the probability that the fluctuations will reach a value $y > y_0$ is given by the integral $P(y > y_0) = \int_{y_0}^{\infty} F(y')\,\mathrm{d}y'$. It is convenient to tabulate the values of this integral as a function of the normalized deviation $y_0/\sigma$; we give below the probabilities that $|y|/\sigma$ is larger than 0.5, 1, 2 and 3.

$$P(|y|/\sigma > 0.5) = 0.6175 \qquad P(|y|/\sigma > 1) = 0.3174$$

$$P(|y|/\sigma > 2) \;\; = 0.0456 \qquad P(|y|/\sigma > 3) = 0.0026$$

The probability that $y$ exceeds $y_0$, $(y_0 > 0)$, is half that of $|y|$ exceeding $y_0$; for instance the shaded area in Fig. 3.15 represents $P(y/\sigma > 2)$ and must equal 0.0228. The probability that $y/\sigma$ exceeds some very large number $m$ is finite but decreases very rapidly (exponentially) becoming zero for all practical purposes. The Gaussian function has universal application to many statistical phenomena in particular as they manifest themselves in most sciences.

If a function $z$ is the sum of two functions, $x_1$ and $x_2$ which are Gaussian distributed, then $z$ also is Gaussian distributed. Let

$$z = x_1 + x_2$$

and $\sigma_1, \sigma_2$ be the standard deviations of $x_1$ and $x_2$. Then, if $x_1$ and $x_2$ are uncorrelated, $\sigma_z$, the standard deviation, of the Gaussian distribution of $z$ is

$$\sigma_z = (\sigma_1^2 + \sigma_2^2)^{1/2} \tag{3.28}$$

We say that the distributions are combined in *quadrature*. Note that the standard deviation of $z$ is always larger than that of either $x_1$ or $x_2$.

Now we would like to examine the effects of the noise on a signal as they appear in the frequency domain. In general, in communications systems we are primarily interested in the transmitted power rather than in the amplitude of the signal. For instance, if an antenna receives a voltage $V$, the received power is

$$P = \frac{V^2}{R} = i^2 R \tag{3.29}$$

where $R$ is the characteristic impedance (resistance) of the antenna, and $i$ is the current flowing through it. Thus we are interested in a *quadratic* measure of the signal and of the noise. Such a measure can be expressed in a convenient form which has been adopted in the description of all fluctuation phenomena.

If $y(t)$ is the fluctuation signal we can evaluate its Fourier transform (see Eq. (3.8)) over the interval $T$

$$g_{\mathrm{T}}(\omega) = \frac{1}{(2\pi)^{1/2}} \int_{-T/2}^{T/2} y(t) e^{i\omega t} \, dt \tag{3.30}$$

where we use the subscript T to indicate the interval of integration. We now define the *power spectral density* or 'power spectrum' of the fluctuations (noise) by the relation*

$$G(\omega) = \lim_{T \to \infty} 2\pi \frac{\langle |g_{\mathrm{T}}(\omega)|^2 \rangle}{T} \tag{3.31}$$

The 'power' due to the fluctuations in a frequency interval $\mathrm{d}f$ is given by

$$G(\omega) \, \mathrm{d}f = \frac{1}{2\pi} G(\omega) \, \mathrm{d}\omega \tag{3.32}$$

We have used quotation marks in referring to power because the term is not exact. For instance if $y(t)$ represents a voltage then

$$P = \frac{V^2}{R} \qquad \text{and} \qquad P_N = \frac{\langle y(t)^2 \rangle}{R}$$

Thus in the frequency domain the noise power per unit frequency, $\mathrm{d}P_N/\mathrm{d}f$, is obtained from

$$\mathrm{d}P_N = \frac{1}{R} G(\omega) \, \mathrm{d}f$$

If $y(t)$ represents a current, then we multiply by $R$ etc. We will follow the adopted convention and treat $G(\omega)$ as 'power' *per unit frequency*. Note that $g_{\mathrm{T}}(\omega)$ has dimensions of ($y(t) \times$ time); thus $G(\omega)$ has dimensions of ($y(t)^2 \times$ time) or ($y(t)^2$/frequency).

Various spectra of $G(\omega)$ are shown in Fig. 3.16. In (*a*) we show a

Fig. 3.16. Examples of noise power spectra: (*a*) white noise, (*b*) resonant system, (*c*) $1/f$ noise.

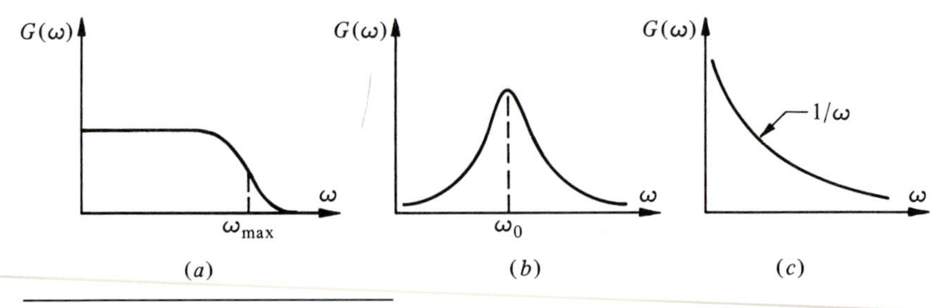


* The brackets indicate the ensemble average; note also that $G(\omega)$ is defined over positive and negative frequencies.

constant $G(\omega)$ which is referred to as random or *white noise*; however at some frequency $\omega_{\max}$ the spectrum must fall off because otherwise the total noise power would become infinite. In (*b*) is given the noise spectrum of a system which has strong resonant behavior. Yet another spectrum, that of the $1/f$ noise, is indicated in (*c*) of the figure.

The fluctuation spectra in the frequency and time domain are connected through the Wiener–Khintchine relationship. We first define the *autocorrelation* function $R(\tau)$, through

$$R(\tau) = \lim_{T\to\infty} \frac{1}{T}\int_{-T/2}^{T/2} y(t)y(t-\tau)\,\mathrm{d}t \tag{3.33}$$

The power spectrum can then be expressed as the Fourier transform of the autocorrelation function and vice versa†

$$G(\omega) = \int_{-\infty}^{\infty} R(\tau)\mathrm{e}^{\mathrm{i}\omega\tau}\,\mathrm{d}\tau \tag{3.34}$$

$$R(\tau) = \frac{1}{2\pi}\int_{-\infty}^{\infty} G(\omega)\mathrm{e}^{-\mathrm{i}\omega\tau}\,\mathrm{d}\omega \tag{3.35}$$

A discussion and proofs of these relations are given in Appendix 2. Here we give only an important result that can be derived from Eqs. (3.33, 3.35); as follows: We set $\tau = 0$ in Eq. (3.33) to obtain

$$R(0) = \lim_{T\to\infty} \frac{1}{T}\int_{-T/2}^{T/2} y(t)^2\,\mathrm{d}t = \langle y(t)^2\rangle \tag{3.36}$$

If we also set $\tau = 0$ in Eq. (3.35) we find

$$R(0) = \int_{-\infty}^{\infty} G(\omega)\frac{\mathrm{d}\omega}{2\pi} = \langle y(t)^2\rangle \tag{3.37}$$

Namely the total 'power' in the fluctuations can be obtained either by integrating the power spectral density over all frequencies, or by evaluating the mean square value of the fluctuations in the time domain. The two calculations yield the same result.

## 3.6 Sources of noise

*White noise*. In this case the noise is completely *random in time*. Thus the autocorrelation function must be zero unless $\tau = 0$. This is so because the noise never reproduces itself and the overlap of two different time intervals will always average to zero. When $\tau = 0$ we measure the mean square value of the noise, more precisely we have $R(\tau) \simeq A$ over a

† The normalization differs from that of Eqs. (3.7, 8) in order to conform with the usual definition of $R(\tau)$.

small interval $-\tau_c < \tau < \tau_c$. $\tau_c$ measures the time scale over which the noise changes and is called the *correlation time*. Furthermore $2\tau_c = \pi/\omega_{max}$, where $\omega_{max}$ is the highest frequency component contained in the noise. We write

$$R(\tau) = 2A\tau_c\delta(\tau) \tag{3.38}$$

Using this expression in Eq. (3.34) we immediately find

$$G(\omega) = 2A\tau_c = A\frac{\pi}{\omega_{max}} \tag{3.39}$$

Namely that the power spectrum is flat.

To obtain a convergent result we must assume that the spectrum rolls off at $\omega_{max}$, as shown in Fig. 3.16(*a*). Then the integral $\int_{-\infty}^{\infty} G(\omega)\,\mathrm{d}f = A$ as expected. A more realistic form for the autocorrelation of white noise is

$$R(\tau) = A\mathrm{e}^{-|\tau|/\tau_c} \tag{3.40}$$

and the resulting power spectrum

$$G(\omega) = A\frac{2\tau_c}{1 + (\omega\tau_c)^2} \tag{3.41}$$

When $\omega = 1/\tau_c$ the spectrum has fallen to half its flat value, and we need not introduce additional convergence conditions. The evaluation of the transform leading to Eq. (3.41) is elementary and is left to the reader.

*Shot noise*. In a diode, or in any electronic device, the current fluctuates due to the statistical distribution in the flow of the electrons. This gives rise to *shot noise* and we will analyse it in terms of the diode shown in Fig. 3.17(*a*). With sufficient time resolution we are able to detect the passage of individual electrons through the meter and we can plot their time of arrival as shown in Fig. 3.17(*b*). Since the electrons arrive at random times $t_n$ the instantaneous current can be written as

$$i(t) = q\sum_n \delta(t - t_n) \tag{3.42}$$

Fig. 3.17. Shot noise in a diode: (*a*) the current flowing through the detector fluctuates, (*b*) plot of the time of arrival of individual electrons.

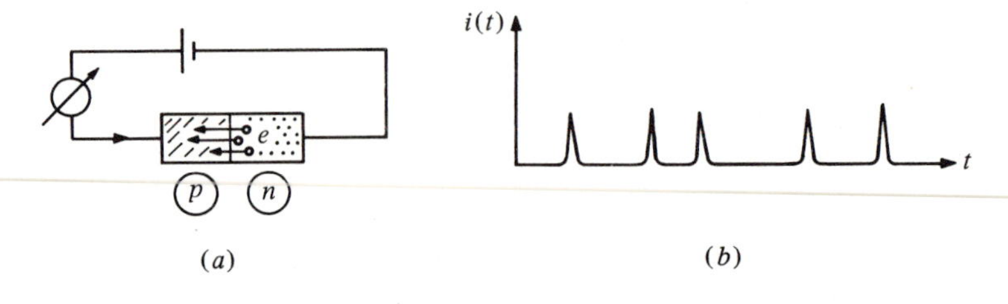

Here

| | |
|---|---|
| $q$ | is the electron charge |
| $t_n$ | is the time of arrival of the $n$th electrons |
| $R$ | is the number of electrons arriving per unit time |
| $i_0 = Rq$ | is the average current |

To obtain the power spectrum we evaluate the Fourier transform of $i(t)$

$$g_T(\omega) = \int_{-T/2}^{T/2} e^{-i\omega t} i(t)\,dt = q \sum_n e^{-i\omega t_n}$$

Here $n$ runs from 1 to $n_{\max} = RT$, the number of electrons that have arrived in the (long) time interval $T$. When we form $|g_T(\omega)|^2$ the cross-terms average out to zero so that

$$|g_T(\omega)|^2 = q^2 \sum_{n=1}^{n_{\max}} 1 = q^2 RT$$

and the power spectrum is given by

$$G(\omega) = \frac{1}{T} |g_T(\omega)|^2 = q^2 R = q i_0 \tag{3.43}$$

This spectrum is flat in frequency as expected for a random noise source; it has a physical cut-off at some high frequency. Since $2G(\omega)\Delta f$ gives the power in the frequency interval $\Delta f$ (the factor of 2 is introduced because we want to deal only with positive frequencies) the current fluctuations in a frequency interval $\Delta f$ are

$$\langle i^2 \rangle = 2 q i_0 \Delta f \tag{3.44}$$

This result was first derived by W. Schottky in 1915.

As an application of Eq. (3.44) consider the flow of direct current at an average value of 0.1 $\mu$A. If this current is viewed on an oscilloscope with a 1 MHz bandwidth ($\Delta f = 10^6$ Hz) the rms fluctuations will be of order

$$\sqrt{\langle i \rangle^2} = [2 \times 1.6 \times 10^{-19} \times 10^{-7} \times 10^6]^{1/2} = 1.8 \times 10^{-10}\ \text{A}$$

Namely

$$i_{\text{rms}}/i_0 \simeq 2 \times 10^{-3}$$

*Thermal noise.* This is the most important source because it is present in any physical system and is the limiting factor in measuring small signals. In electronic devices, for instance in a resistor, the electrons have random motion, but there is no net transport of charge; thus there is no dc current but an ac component is present. The thermal noise in a resistor is referred to as *Johnson noise*.

We will use a heuristic approach to derive the power spectrum as follows: when a current $i$ flows through a resistor $R$ the power dissipation is $P = i^2R$. The thermal energy per degree of freedom is $\frac{1}{2}kT$ and therefore the thermal power in a frequency band $\Delta f$ is $P = kT\Delta f$. If we equate the thermal power to the current dissipation we obtain for the current fluctuation

$$\langle i^2 \rangle = \frac{1}{R} 4kT\Delta f \tag{3.45}$$

The factor of 4 follows from a rigorous derivation of Eq. (3.45) which was first obtained by Nyquist. The thermal noise power that can be coupled out of a resistor $R$ into a matched load is given by

$$P_{\text{noise}} = kT\Delta f \tag{3.46}$$

Thus the power spectrum of the Johnson noise is flat and equal to

$$G(\omega) = \tfrac{1}{2}kT \tag{3.47}$$

In the above expressions $k$ is Boltzmann's constant: $k = 1.38 \times 10^{-23}$ J/K, and $T$ the *absolute* temperature in Kelvin.

As an application of Eq. (3.45) we take $R = 50\,\Omega$, $T = 300$ K and $\Delta f = 1$ MHz. This yields $\sqrt{\langle i \rangle^2} = 1.8 \times 10^{-8}$ A which exceeds by two orders of magnitude the rms current due to shot noise as calculated in the previous example. Measurements of Johnson noise can be used to accurately determine the value of Boltzmann's constant.

*Quantum noise.* At very high frequencies the quantization of electromagnetic energy leads to statistical fluctuations in the detection of individual photons; we refer to these fluctuations as *quantum noise*. Every photon carries energy $\hbar\omega$, so that if a signal of duration $T$ contains $N$ photons (the rate is $R = N/T$), the energy and power in the signal are

$$E = N\hbar\omega \tag{3.48}$$

and

$$P = \frac{N\hbar\omega}{T} = R\hbar\omega \tag{3.49}$$

We note that $1/T = \Delta f$ is the bandwidth of the measurement. If *no* photons are received in the interval $T$, the fluctuation is still *one* photon. This corresponds to a power level $\hbar\omega/T$, and therefore the noise power, $P_N$ is

$$P_N = \hbar\omega\Delta f \tag{3.50}$$

The power spectrum is $G(\omega) = (\hbar\omega)/2$ and is proportional to the frequency. At very high frequencies quantum noise can become dominant. If the number of photons received in the measuring time is larger than one, then

$$P_N = \frac{\sqrt{N}\,\hbar\omega}{T} = \frac{R\hbar\omega}{\sqrt{N}} \tag{3.50$'$}$$

*Noise in amplifiers.* As we have seen, noise will be present in any communication channel and will coexist with the signal. The ratio of noise power to signal power is called the *signal to noise ratio* and is written as $S/N$. An amplifier is characterized by a gain $G$; thus if the input signal power is $S_1$ the output power will be $S_2 = GS_1$. Similarly, we designate the input noise power by $N_1$ and the output noise power by $N_2$. However $N_2 \geqslant GN_1$ because, in general, an amplifier will contribute some additional noise. These definitions are indicated in Fig. 3.18.

The noise introduced by the amplifier can be characterized as an additive equivalent *input* noise power $N_e$, such that

$$N_2 = G(N_1 + N_e) \qquad N_e \geqslant 0; \qquad \text{or} \qquad N_e = \frac{N_2 - GN_1}{G} \tag{3.51}$$

We express $N_e$ by an equivalent amplifier *noise temperature* by analogy to the thermal noise power given by Eq. (3.46), namely

$$N_e = kT_e \Delta f \tag{3.52}$$

where $\Delta f$ is the bandwidth of the amplifier. The noise temperature depends on many parameters, including the frequency at which the amplifier operates. In a good amplifier $T_e$ can be *less* than the ambient input temperature.

An alternate convention for expressing the noise contribution of an amplifier is to give its *noise figure* ($NF$). The noise figure is given in db (decibel)

$$NF = 10 \log_{10}(F) \tag{3.53}$$

where $F$ is the noise factor. The noise factor, in turn, is defined as the ratio of the $(S/N)$ at the input of the amplifier to that at its output; by definition the noise factor is larger than one,

$$F = \frac{(S/N)\text{ input}}{(S/N)\text{ output}} = \frac{S_1/N_1}{S_2/N_2} = \frac{N_2}{GN_1} = \frac{G(N_1 + N_e)}{GN_1}$$

$$= 1 + \frac{N_e}{N_1} = 1 + \frac{T_e}{T_1} \tag{3.54}$$

In the last step we assumed that the input noise is thermal, and at temperature $T_1$. The difficulty that arises with the noise figure is that its value depends on some standard value for $T_1$. It is agreed that $T_1 = 290$ K.

Fig. 3.18. Amplifier of gain $G$ with signal and noise power $S_1$, $N_1$ at its input; the signal and noise power at the output are labeled $S_2$, $N_2$.

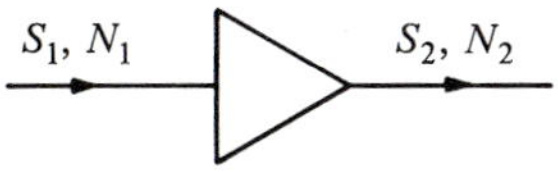


$$(S/N)\text{ input} \equiv \frac{S_1}{N_1}$$

$$(S/N)\text{ output} \equiv \frac{S_2}{N_2}$$

Thus for instance an amplifier with a noise temperature $T_e = 58$ K has a noise factor: $F = 1 + (58/290) = 1.2$, and a noise figure: $NF = 10 \log_{10}(1.2) = 0.8$ db.

When two or more amplifiers are cascaded the noise factor of the first stage usually dominates the overall noise in the system. This can be easily seen by constructing the noise factor $F$ for the combined system of the two amplifiers sketched in Fig. 3.19. It is easy to show that

$$T_N = T_{N1} + \frac{T_{N2}}{G_1} \tag{3.55}$$

Thus as long as $T_{N2}/G_1 \ll T_{N1}$ the noise contribution of the second amplifier becomes insignificant.

It is possible to detect signals even when the $(S/N) \ll 1$ provided the signal is present for an adequately long time interval. We can then make repeated measurements and average them. If $n$ averages are taken, the fluctuations in the noise level are reduced by a factor of $1/\sqrt{n}$ whereas the signal is unchanged. Thus it becomes possible to distinguish even small signal power over the 'smooth' noise level.

Fig. 3.19. Cascaded amplifiers.

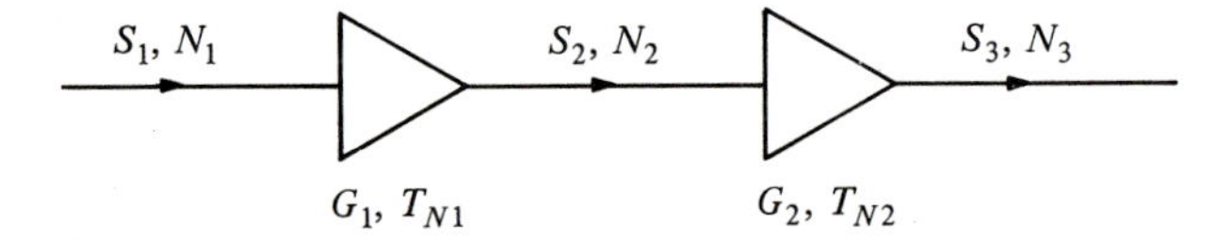


## 3.7 Elements of communication theory

In order to optimize a communication channel we must have a quantitative measure of the information that we wish to transmit. The efforts in this direction were pioneered at Bell Labs by J. W. Tuckey, R. L. V. Hartley, H. Nyquist and others, and placed in a formal context by C. E. Shannon.* Language messages have statistical properties which have been studied and are well known to cryptographers. Furthermore messages carry a varying degree of information and a communications channel has a given capacity for transferring information. Messages are not restricted to language but can be coded in a variety of forms; for instance a television picture. We will begin, nevertheless, by reviewing the

* C. E. Shannon, *The Mathematical Theory of Communication*, The University of Illinois Press, 1962.

statistical properties of the English language and use this analysis as an example of the more general concepts.

The letters of the alphabet form a *finite set* and in every language we can define the frequency $P(\mathrm{i})$ with which the letter 'i' appears. We can similarly define the frequency for digrams $P(\mathrm{ij})$, namely the probability that the combination 'ij' appears in the language; the conditional probability $P(\mathrm{i}\,|\,\mathrm{j})$ gives the frequency with which, once 'i' is found, it is followed by the letter 'j'. In summary

| | |
|---|---|
| $P(\mathrm{i})$ | frequency of letter i |
| $P(\mathrm{ij})$ | frequency of digram ij |
| $P(\mathrm{i}\,|\,\mathrm{j})$ | conditional probability that j follows i |
| $P(\mathrm{i})P(\mathrm{i}\,|\,\mathrm{j}) = P(\mathrm{ij})$ | |

These probabilities are normalized, in obvious manner, by

$$\sum_i P(\mathrm{i}) = 1 \qquad i, j = 1, \ldots, N\{\text{all letters of the alphabet}\}$$

$$\sum_j P(\mathrm{ij}) = P(\mathrm{i})$$

$$\sum_j P(\mathrm{i}\,|\,\mathrm{j}) = 1$$

Given a set of symbols we can select a sample from it according to the probability of each symbol. Such a sample is called stochastic. A stochastic sample of the letters of the alphabet will, in general, not be an intelligent message. This is so because intelligent messages contain information whereas a stochastic sample does not. We can improve on this selection process by using not only the probability of the letters, but also the higher order probabilities such as digrams, trigrams etc. In this case the sample may resemble a message in English but still it cannot contain information because it was constructed randomly.

As examples we give below messages which have been obtained by sampling the alphabet (including the blank space) through different approximations.

*Zero order* – All letters equiprobable

BTLCCVNFYRRKDXQR . . .

*First order* – Letters selected according to their frequency in the English language

TH EEI ALHENHTTPA NAH . . .

*Second order* – Letters selected according to their digram frequency. A practical way of doing this is to select the first letter, say E, and then open a book at random and find the letter E; enter in the sample the letter following E, say K. Now turn the page and begin at the top until you find K; enter in the sample the letter that follows K, say R, and so on

TEDELICEPY CHE SUS ORAINOS . . .

One can proceed to higher orders in forming such stochastic messages.

The same procedure can be used with words rather than letters. Here we have a much larger finite set of symbols and correspondingly the process is more laborious. For instance messages where the words have been selected from a first order and a second order approximation are given below.

*First order* word message

REPRESENTING AND SPEEDILY IS AN GOOD APT OR COME . . .

*Second order* word message

WHEN NOISE RATIO AT FIRST STEP, DEPENDS UPON . . .

Note that the second order message was constructed using a text on telecommunications and therefore is not completely random. It contains information as to the set of words from where it was drawn.

More details on the structure of the English language can be found in the references. Suffice it to say that in English there are

| | |
|---|---|
| 26 | letters |
| 16 357 | words (this is an approximation to the commonly used words) |
| 4.5 | letters/word |

The most common letter is E having a frequency of 0.13105 and the least common is Z with frequency of 0.00077. The frequency of occurrence of latters in English is reproduced in Table 3.1 below (taken from *Secret and Urgent* by Fletcher Pratt, Blue Ribbon Books, 1939). A tabulation of word frequencies can be found in *Relative Frequency of English Speech Sounds* by G. Dewey, Harvard University Press, 1923.

The frequency of words appears to follow a law typical of large samples; namely if the symbols of the set are ranked in order of descending frequency, then the frequency of the $r$th symbol is given by

$$P(r) = \frac{C}{r} \tag{3.56}$$

where $C$ is a constant. Thus

$$\log P(r) = \log C - \log r$$

This relationship is known as Zipf's law and is shown in Fig. 3.20 for the words of the English language, where $C \sim 0.1$.

From the discussion it should be clear that in an English language message, transmittal of the letter X (which has low frequency) conveys much more information than transmittal of the letter E (which has high frequency). This relationship is reminiscent of the definition of *entropy* in statistical systems. Entropy is a macroscopic measure of the probability that a random arrangement (of the positions and momenta of the molecules) will lead to that state of the system. *High entropy* means *lack of information* about the specific arrangement of the molecules because many microstates can lead to that particular macroscopic state. When only a few microstates can contribute to a macroscopic state of the system

Table 3.1. *Frequency of occurrence of letters in English*

| | Letter | Frequency of occurrence in 1000 words | Frequency of occurrence in 1000 letters |
|---|---|---|---|
| 1. | E | 591 | 131.05 |
| 2. | T | 473 | 104.68 |
| 3. | A | 368 | 81.51 |
| 4. | O | 360 | 79.95 |
| 5. | N | 820 | 70.98 |
| 6. | R | 308 | 68.32 |
| 7. | I | 286 | 63.45 |
| 8. | S | 275 | 61.01 |
| 9. | H | 237 | 52.59 |
| 10. | D | 171 | 87.88 |
| 11. | L | 153 | 33.89 |
| 12. | F | 132 | 29.24 |
| 13. | C | 124 | 27.58 |
| 14. | M | 114 | 25.36 |
| 15. | U | 111 | 24.59 |
| 16. | G | 90 | 19.94 |
| 17. | Y | 89 | 19.82 |
| 18. | P | 89 | 19.82 |
| 19. | W | 68 | 15.39 |
| 29. | B | 65 | 14.40 |
| 21. | V | 41 | 9.19 |
| 22. | K | 19 | 4.20 |
| 23. | X | 7 | 1.66 |
| 24. | J | 6 | 1.32 |
| 25. | Q | 5 | 1.21 |
| 26. | Z | 3 | 0.77 |

the entropy is small, hence the expression that 'information is negative entropy'.

Entropy is defined through

$$s = k \ln W \tag{3.57}$$

where $W$ is the number of microscopic states and $k$ is Boltzmann's constant; the form of Eq. (3.57) is the only one that assures the additivity of the entropy when two systems $A$, $B$ are coalesced into a single system $C$

$$S_C = S_A + S_B = k \ln(W_A W_B) = k \ln W_C$$

By analogy we define the information content $h_i$ of a message i through the probability $p_i$ of its occurrence

$$h_i = -p_i \log_2 p_i \tag{3.58}$$

The minus sign is necessary since by definition $p_i < 1$. Suppose we are

Fig. 3.20. Word frequency v. word order for the English language obeys Zipf's law. (From E. W. Montroll and W. W. Badger, *Introduction to Quantitative Aspects of Social Phenomena*, Gordon and Breach Publishers, New York (1974) by permission.)

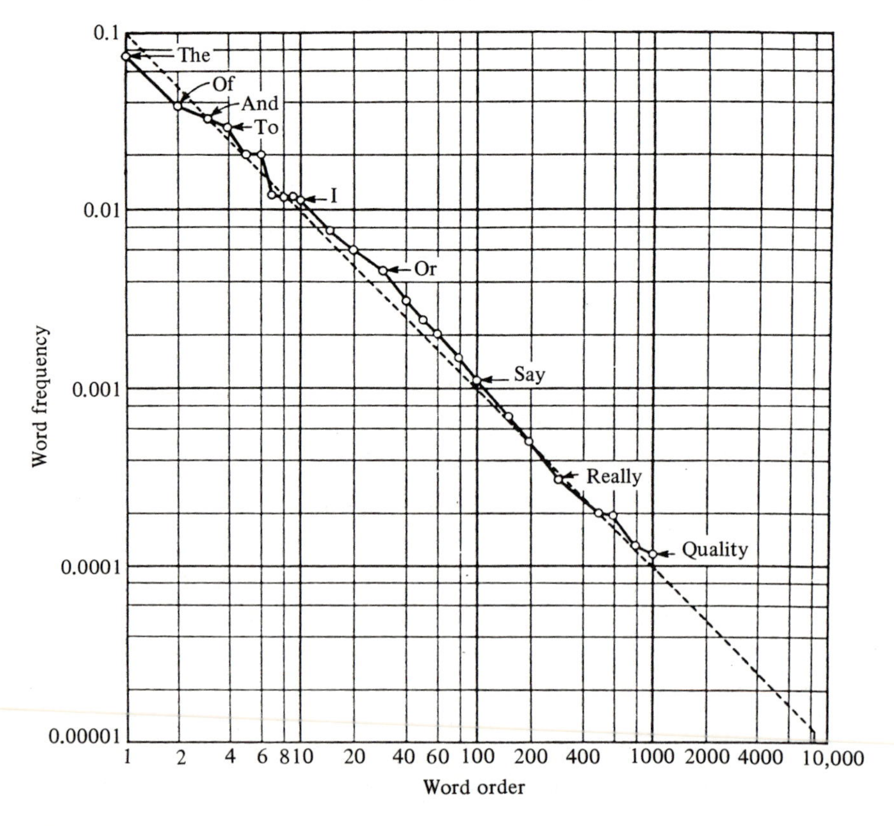

sending a message in binary form and that the two symbols '0' and '1' have equal probability of being transmitted. Therefore, each symbol has probability $\frac{1}{2}$ and according to Eq. (3.58) the corresponding information content is

$$\left.\begin{aligned} h_1 &= -p_1 \log_2 p_1 = \tfrac{1}{2} \\ h_0 &= -p_0 \log_2 p_0 = \tfrac{1}{2} \end{aligned}\right\} \qquad (3.59)$$

The information content of a message consisting of one binary digit is the sum of the information carried by the two symbols, or states

$$H(\text{binary digit}) = h_1 + h_0 = 1 \qquad (3.60)$$

The unit of information has been defined in terms of the *binary digit* as in Eq. (3.60) and is called a *bit*; this word has been derived from a contraction of bi(nary) (digi)t.

We can extend the argument to binary transmission where the two states have unequal probability, $p_0 \neq p_1$. If $N$ digits have been transmitted, state '0' will be received $Np_0$ times while state '1' will be received $Np_1$ times. The information content of this transmission of $N$ digits is

$$\begin{aligned} H_N &= H_{N0} + H_{N1} \\ &= -Np_0 \log_2 p_0 - Np_1 \log_2 p_1 = -N \sum_{i=0}^{1} p_i \log_2 p_i \end{aligned} \qquad (3.61)$$

The information per digit is defined through $H = H_N/N$; in general $H < 1$ and $H = 1$ if and only if $p_0 = p_1$. If a larger set of symbols is used in the message, the information content per message has the same form as for the binary case but with the sum extending over all symbols.

$$H = -\sum_{i=1}^{n} p_i \log_2 p_i \quad \text{(bits)} \qquad (3.62)$$

The information is maximal when all $p_i$ are equal to each other. If one $p_i = 1$ and all others are zero, the information content of a message is zero.

We can apply this formalism to the English language with its 26 letters. If all letters were equiprobable, ($p_i = \frac{1}{26}$), receipt of one letter would contain

$$H = -\sum_{i=1}^{26} \tfrac{1}{26} \log_2(\tfrac{1}{26}) = \log_2(26) = 4.7 \text{ bits} \qquad (3.63)$$

This is not surprising because with 4 binary digits we can encode only 16 symbols, whereas with 5 binary digits we can encode 32 symbols. Thus $H$ can also be interpreted as the *number of binary digits* needed to fully encode the set of symbols used to transmit the message. On the other hand we have seen that in the ASCII code, 7 bits are used to encode each letter of the alphabet (see Table 2.2). This may appear wasteful, but it provides redundancy. In reality in the English language the letters are not equiprobable and using the correct probabilities leads to

$$H = -\sum_{i=1}^{26} p_i \log p_i = 4.2 \text{ bits} \tag{3.63'}$$

This result implies that receipt of one letter in a message, such as in a teletype text, carries *less* than maximal information.

We can define the *relative* entropy of a set of symbols as the ratio of the information content of a single symbol to the maximum possible information content. As an example, for the letters of the English alphabet

$$\text{Relative entropy} = \frac{H}{H_{max}} = \frac{4.2}{4.7} = 0.9$$

We say that the *redundancy* of the letters in the English language is

$$\text{Redundancy} = 1 - (\text{Relative entropy}) = 0.1$$

If we take into account digram and trigram probabilities the information content is even smaller and the redundancy of the language is approximately 0.5. This means that in writing a text in English only half the letters can be freely chosen; the other half is imposed by the structure of the language.

It is possible to devise coding schemes that are maximally efficient and approach the theoretical limit given by Eq. (3.62). Such codes are known as Huffman codes but do not have the simplicity of less compact codes such as ASCII. Furthermore, redundancy in a coding system is useful for error recovery as we discuss in the next section. Shannon's work has made possible the quantitative evaluation of the information content of messages; this content, measured in bits, is independent of the method of coding or transmission.

## 3.8 Channel capacity

Since information is measured in bits, the *rate* at which information is transmitted will be measured in bits per second. We write

$$I = (\text{bits/s})$$

If we wish to transmit $R$ messages/s, each message having an information content of $H$ bits, then the transmission rate $I$ must be

$$I = HR \quad (\text{baud}) \tag{3.64}$$

Here we use the commonly used notation where 1 bit/s = 1 baud, this unit having been named in honor of the French engineer J. B. F. Baudot.

A communications channel has a capacity $C$ at which it can transmit information. If the bandwidth of the channel is $W$ and the number of symbols that are being transmitted is $m$, then the capacity is

$$C = 2W \log_2(m) \quad \text{(baud)} \tag{3.65}$$

In a binary channel, $m = 2$ and therefore $C = 2W$. This result is analogous to the sampling theorem (Eq. (3.20)) which states that an analog signal of harmonic content, (or bandwidth $W$), must be sampled at rate $f_s = 2W$. If the digitization is done at several levels, the channel capacity increases logarithmically. There are however good reasons for staying with binary transmission because of the lower probability for errors.

Eq. (3.65) is given for a noiseless channel. For a channel with signal to noise ratio $S/N$ the channel capacity is given by *Shannon's equation*

$$C = W \log_2(1 + S/N) \tag{3.66}$$

The above equation is the basis for the design of any communication channel. Note that when $S/N = 3$ the capacity is equal to that of a noiseless binary channel (Eq. (3.65)). For larger values of $S/N$ one can use multilevel encoding rather than straight binary so as to have the same capacity as an optimally chosen noiseless channel.

As an application of Shannon's equation we consider facsimile transmission. Let the desired transmission parameters be

| | |
|---|---|
| Transmission rate | $R = 1$ frame/minute |
| Size of frame | $10 \times 10$ inch |
| Horizontal line density | 100 lines/inch |
| Resolution of lines | 100 dots/inch |
| Intensity shadings for each dot | 32 |

We want to find the minimal required bandwidth if the channel has a signal to noise ratio $S/N = 20$ db.

First we calculate the information content of each frame, which has

$$[(10 \times 100) = 10^3 \text{ lines}] \times [(10 \times 100) = 10^3 \text{ dots/line}] = 10^6 \text{ dots}$$

Each dot can be any one of 32 symbols (all *a priori* equiprobable) so that according to Eq. (3.62)

$$H = 10^6 \log_2(32) = 5 \times 10^6$$

The message rate is $R = \frac{1}{60}\,\text{s}^{-1}$, and therefore the transmission rate is

$$I = 8.3 \times 10^4 \text{ baud}$$

A $S/N$ ratio of 20 db corresponds to

$$S/N = 10^{(20/10)} = 10^2$$

and thus the channel capacity from Eq. (3.66) is

$$C = W \log_2(1 + 10^2) \simeq 6.6\ \text{W}$$

If we equate the channel capacity to the desired transmission rate we find the minimum bandwidth required for the channel

$$W_{\min} = \frac{8.3 \times 10^4}{6.6} = 1.2 \times 10^4 \text{ Hz} \tag{3.67}$$

If we had used the noiseless channel formula with $m = 32$ we would have concluded

$$W_{\min} = 0.83 \times 10^4 \text{ Hz} \tag{3.67'}$$

which is below the Shannon limit. Even the result of Eq. (3.67) is too optimistic because it does not allow for any redundancy. Since the bandwidth of typical telephone lines is of order $2 \times 10^4$ Hz we conclude that the specifications of this example correspond to realistic parameters for facsimile transmission.

From Shannon's equation it appears as if one could maintain a given channel capacity by increasing the bandwidth at the expense of $S/N$, or vice versa. In the limit that the receiver noise is thermal, no further improvement in $S/N$ is possible, and for transmission over very long distances more often than not $S/N \ll 1$. It is in these cases that Shannon's equation is of great importance. We can then recast Eq. (3.66) in a form where the channel capacity depends only on the received power, $P_S$, and the noise temperature $T$ of the receiver. Since $S/N \ll 1$

$$\log_2\left(1 + \frac{S}{N}\right) = \frac{1}{\log_e(2)} \log_e\left(1 + \frac{S}{N}\right) \simeq 1.44 \frac{P_S}{P_N}$$

The noise power is assumed to be thermal (see Eq. (3.46))

$$P_N = kT\Delta f = kTW \tag{3.68}$$

where $\Delta f = W$ is the bandwidth of the receiver. Thus Eq. (3.66) becomes

$$C = W \log_2\left(1 + \frac{S}{N}\right) \rightarrow 1.44 \frac{P_S}{kT} \tag{3.69}$$

In this limit, increasing the bandwidth does not improve the channel capacity because it increases the noise power in the same proportion.

The limit of Eq. (3.69) is representative of communication with distant spacecraft where the transmitter power is limited and the distance over which the signal is transmitted is very long. As an application let us consider transmission of messages from a satellite orbiting Mars. We assume the following plausible parameters.

| | |
|---|---|
| Transmitter power | $P_t = 1$ Watt |
| Transmitter antenna gain | $G_t = 10^5$ |

(this corresponds to an antenna diameter of 6 m assuming microwave transmission, $\lambda \simeq 3$ cm)

| | |
|---|---|
| Receiving antenna diameter | $D_r = 29$ m |

Distance Mars–Earth $R \simeq 4.3 \times 10^{11}$ m
Receiver temperature $T_r = 300$ K

We first calculate the signal power received from the orbiter

$$P_S = (G_t P_t)\frac{A_r}{4\pi R^2} = G_t P_t \frac{\pi(D_r/2)^2}{4\pi R^2} = 1.4 \times 10^{-17}\ \text{Watt}$$

The channel capacity is calculated from Eq. (3.69) and we find

$$C = \frac{1.44}{kT} P_S = \frac{1.44}{(1.38 \times 10^{-23}) \times 300} 1.4 \times 10^{-17} \simeq 5 \times 10^3\ \text{baud}$$

The messages being transmitted could be TV pictures of Mars obtained at low resolution with a matrix of $512 \times 512$ dots (pixels) each, with 4 shadings (black and white picture). The information content of such a frame is

$$H = (512 \times 512) \times \log_2(4) = 5 \times 10^5\ \text{bits}$$

Thus the rate at which the pictures can be transmitted is

$$R = \frac{I}{H} = \frac{C}{H} = \frac{5 \times 10^3}{5 \times 10^5} = \frac{1}{100}\ \text{pictures/s}$$

This corresponds to 36 pictures/hour, a fairly realistic rate for the present state of technology.

## Exercises

### *Exercise 3.1*

A communications channel has a bandwidth $B = 10$ MHz and a signal to noise ratio of 15. The receiver operates at room temperature ($T = 300$ K) and the noise is only thermal.

(a) Find the maximum *rate* of information that the channel can transmit.
(b) Find the power at the receiver.

### *Exercise 3.2*

What are the relevant 'orders of magnitude' answers to the following questions

(a) The highest frequency electromagnetic wave that can be carried on two open wires.
(b) The frequency of visible light.
(c) The sampling frequency necessary to digitize an audio signal such as a telephone message.

*Exercise 3.3*

We wish to transmit television pictures consisting of

200 bits/line
200 lines/frame at the rate of 50 frames/second

(a) What is the minimum bandwidth required for the channel?
(b) Calculate the power due to thermal noise in the above bandwidth interval for $T = 300$ K.
(c) Assume that the transmitter emits with 0.2 W power isotropically and that the receiving antenna has an equivalent area of 0.4 $m^2$. What is the largest distance at which signals can be received with a signal to noise ratio of 10?

*Exercise 3.4*

(a) Search through 100 words to establish the frequency of the letters E, P, X in the English language. Establish the conditional probability that H follows T.
(b) Search through 1000 words in an English text and establish the frequency of the 10 most encountered words. Make a log–log plot of the word frequency, $f$, v. the word order, $n$ (most frequent word, second most frequent etc.) to show that Zipf's law, shown in Fig. 3.21, is indeed obeyed.

*Exercise 3.5*

Show that the effect of combining two Gaussian distributions of standard deviation $\sigma_1$ and $\sigma_2$ results in a Gaussian distribution of standard deviation

$$\sigma = (\sigma_1^2 + \sigma_2^2)^{1/2}$$

# 4

# GENERATION AND PROPAGATION OF ELECTROMAGNETIC WAVES

## 4.1 Maxwell's equations

All electromagnetic (em) phenomena are completely and uniquely determined by a set of differential equations, the famous Maxwell's equations. These equations predict the existence of em waves and govern their propagation and generation. We will write the equations for a region of space where no dielectric or permeable materials are present, and use the MKS system.

$$\nabla \cdot \mathbf{E} = \rho/\varepsilon_0 \tag{4.1}$$

$$\frac{1}{\mu_0}(\nabla \times \mathbf{B}) - \varepsilon_0 \frac{\partial \mathbf{E}}{\partial t} = \mathbf{J} \tag{4.2}$$

$$\nabla \cdot \mathbf{B} = 0 \tag{4.3}$$

$$\nabla \times \mathbf{E} + \frac{\partial \mathbf{B}}{\partial t} = 0 \tag{4.4}$$

Here **E** and **B** are the electric and magnetic fields which are vectors (**B** is really a pseudovector), $\rho$ and **J** are the electric charge density and electric current density respectively. The dielectric permittivity of the vacuum is designated by $\varepsilon_0$ and the magnetic permeability of the vacuum by $\mu_0$. Introduction of these quantities is required to ascertain consistency of units and dimensions and in the MKS system they have the values

$$\left.\begin{aligned} \varepsilon_0 &= 9.954 \times 10^{-12} \quad \text{coul/V-m} \\ \mu_0 &= 4\pi \times 10^{-7} \quad \text{V-s/A-m} \end{aligned}\right\} \tag{4.5}$$

Maxwell's equations exhibit a great degree of symmetry: Eqs. (4.1) and (4.2) are inhomogeneous and consist of one scalar and one vector equation; the driving terms are the charge density $\rho$ and the current density **J**. Eqs. (4.3) and (4.4) are analogous to Eqs. (4.1) and (4.2) except that they are homogeneous; this reflects the absence of magnetic monopoles in nature. Finally there is the very important difference in sign between Eqs. (4.2) and (4.4). This is what makes possible the existence of em waves in free space as we show below.

In free space there are no charges or currents so we set $\rho = 0$, $\mathbf{J} = 0$ in Eqs. (4.1, 2). We take the curl of Eq. (4.4) and use a relation from vector calculus

$$\nabla \times (\nabla \times \mathbf{E}) = -\nabla^2 \mathbf{E} + \nabla(\nabla \cdot \mathbf{E})$$

Since by Eq. (4.1), $\nabla \cdot \mathbf{E} = 0$ in free space, we have from Eq. (4.4)

$$\nabla \times (\nabla \times \mathbf{E}) = -\nabla^2 \mathbf{E} = -\nabla \times \frac{\partial \mathbf{B}}{\partial t} = -\frac{\partial}{\partial t}(\nabla \times \mathbf{B}) \tag{4.6}$$

where in the last step we interchanged the order of differentiation. We note that $(\nabla \times \mathbf{B})$ is given by Eq. (4.2), and in free space

$$\nabla \times \mathbf{B} = \mu_0 \varepsilon_0 \frac{\partial \mathbf{E}}{\partial t} \tag{4.6'}$$

By taking the time derivative of Eq. (4.6′) and equating it to Eq. (4.6) we obtain

$$\nabla^2 \mathbf{E} - \mu_0 \varepsilon_0 \frac{\partial^2 \mathbf{E}}{\partial t^2} = 0 \tag{4.7}$$

Eq. (4.7) is a wave equation where the velocity of propagation $c$ is

$$c^2 = \frac{1}{\mu_0 \varepsilon_0} \tag{4.8}$$

Using the values of $\varepsilon_0$, $\mu_0$ given in Eqs. (4.5) we find that $c = 3 \times 10^8$ m/s, the observed velocity of propagation of light in free space. Thus light is an electromagnetic wave.

The solutions of Eq. (4.7) are waves as can be seen by considering the simple case where $\mathbf{E}$ has only one component, say $E_x$, and varies only along one direction in space, say the $Z$-axis. Then Eq. (4.7) simplifies to

$$\frac{\partial^2 E_x}{\partial z^2} - \frac{1}{c^2}\frac{\partial^2 E_x}{\partial t^2} = 0 \tag{4.7'}$$

with solutions

$$E_x = f(z \pm ct)$$

where $f$ is any arbitrary doubly differentiable function of argument $(z \pm ct)$. The solution $E_x = f(z - ct)$ represents waves propagating to positive $z$ whereas $E_x = f(z + ct)$ propagates to negative $z$. The function $f(z \pm ct)$ can always be represented by a Fourier expansion, thus it suffices to study the propagation of harmonic waves, such as given by

$$E_x = E_0 \cos(\omega t - kz) \tag{4.9}$$

where

$$\omega/k = c \tag{4.10}$$

$k = 2\pi/\lambda$ is the wave vector.

An em wave has both a magnetic and an electric field, normal to one another, and both transverse to the direction of propagation. We show this for the simple example of a plane wave propagating along $z$ and having only an $E_x$ component, but the result is absolutely general. Under these assumptions only the $y$-component of the curl in Eq. (4.4) is different from zero and Eq. (4.4) reads

$$(\nabla \times \mathbf{E})_y = \frac{\partial E_x}{\partial z} = -\frac{\partial B_y}{\partial t}$$

Using Eq. (4.9) for $E_x$ we find

$$\frac{\partial E_x}{\partial z} = kE_0 \sin(\omega t - kz) = -\frac{\partial B_y}{\partial t}$$

$$B_y = \frac{k}{\omega} E_0 \cos(\omega t - kz) = \frac{1}{c} E_x, \qquad B_x = B_z = 0 \tag{4.9$'$}$$

Thus the plane wave of this example has the form shown in Fig. 4.1 (at $\omega t = \pi/2$).

The electromagnetic wave carries energy along its direction of propagation, and the energy flux, that is the energy crossing unit area in unit time, is given by the Poynting vector

$$\mathbf{S} = \mathbf{E} \times \mathbf{H} = \frac{1}{\mu_0} (\mathbf{E} \times \mathbf{B}) \tag{4.11}$$

Fig. 4.1. Electromagnetic plane wave propagating along the $Z$-axis and polarized along the $X$-axis.

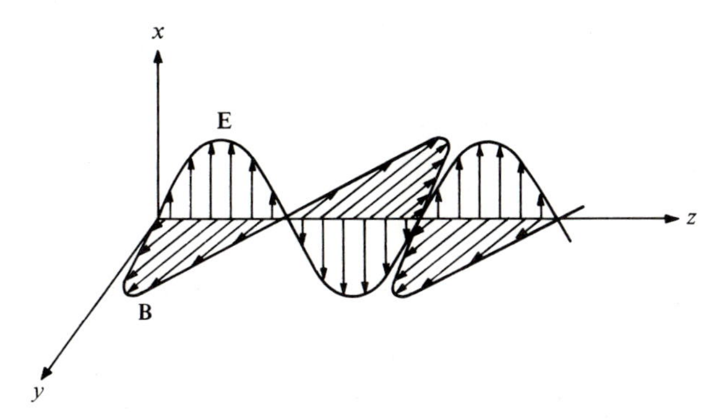

**S** points along the direction of propagation $\mathbf{k}/|\mathbf{k}|$. The time dependence of **S** is $\cos^2(\omega t - kz)$ and the net energy flux is obtained by time averaging **S**, which introduces a factor of $\frac{1}{2}$. Thus

$$\langle \mathbf{S} \rangle = \tfrac{1}{2}(\mathbf{S})_{\max} = \frac{1}{2\mu_0} E_0 B_0 \frac{\mathbf{k}}{|\mathbf{k}|} = \frac{1}{2}\left(\frac{\varepsilon_0}{\mu_0}\right)^{1/2} E_0^2 \frac{\mathbf{k}}{|\mathbf{k}|}$$

The energy density in a region where an electric and magnetic field are present is in general

$$u = \tfrac{1}{2}\varepsilon_0 |\mathbf{E}|^2 + \tfrac{1}{2}\frac{1}{\mu_0}|\mathbf{B}|^2 \tag{4.12}$$

Thus, the energy flux is related to the energy density in the wave through

$$|\mathbf{S}| = cu$$

as can be easily verified by comparing Eqs. (4.11) and (4.12) and using Eq. (4.9′).

The electric and magnetic fields are vectors, so that in principle they can have three independent components. In an em wave however both **E** and **B** must be transverse to the direction of propagation and this reduces their independent components to only *two*. To be concrete we will assume that the wave propagates along $z$ and we label the two components by $E_x$ and $E_y$ as in Fig. 4.2. The direction along which the $E$-field oscillates is called the direction of *polarization* of the wave. If the two components are in phase, the polarization vector remains fixed in space and we say that the wave is linearly polarized. If the two components are equal and 90° out of phase the polarization vector rotates – at frequency $\omega$ of course – and we say that the wave if circularly polarized. In general, the wave can be elliptically polarized. Even though an em wave always has definite polarization a source often emits em radiation containing waves

Fig. 4.2. The electric field vector in a plane transverse to the direction of propagation: (*a*) linear polarization, (*b*) circular polarization, (*c*) elliptical polarization.

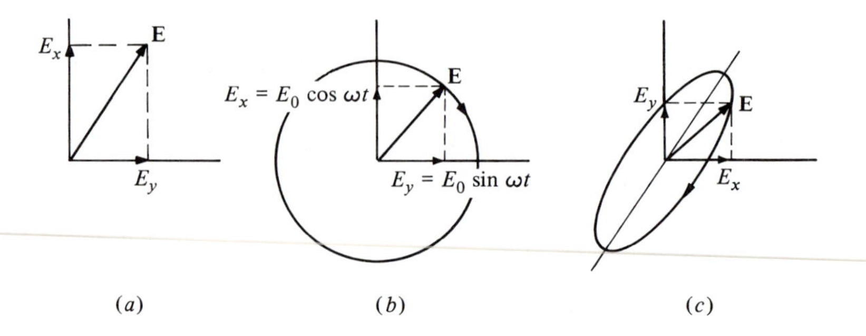

of all possible polarizations; in that case we say that the light, or radiation, is *unpolarized.*

## 4.2 Radiation and antennas

We now want to examine how em waves are generated. From Maxwell's inhomogeneous equations we see that the sources of the field are a time-dependent charge or current density. In fact uniformly moving electric charges do not radiate but accelerated charges do. The total power radiated by a charge $e$ subject to an acceleration $\dot{v}$ is given by Larmor's equation

$$P = \frac{2}{3} \frac{e^2}{4\pi\varepsilon_0} \frac{\dot{v}^2}{c^3} \tag{4.13}$$

Thus accelerated electrons in atoms emit light; microwave radiation is emitted from the motion of electrons in solid state devices. The motion of the conduction electrons in an antenna or the motion of free electrons in a klystron gives rise to radiation in the HF or VHF bands.

If an em wave is emitted isotropically from a source, the **E** and **B** fields must fall off with the distance $R$ from the source as $1/R$. Then, the Poynting vector **S** is proportional to $1/R^2$ and the total radiated power remains constant and independent of the distance from the source

$$P = \int \mathbf{S} \cdot d\mathbf{A} = 4\pi R^2 |\mathbf{S}| = 4\pi R^2 \frac{C}{R^2} = \text{constant} \tag{4.14}$$

We can try to compensate this reduction in flux with distance by focussing the radiation by means of directional antennas; for instance, in a plane wave propagating along $z$ the time-averaged Poynting vector is independent of $z$.

The simplest radiating system is an electric dipole whose moment oscillates in time. This is shown in Fig. 4.3($a$) and we assume that the dipole moment **p** has the time dependence

$$|\mathbf{p}| = p_0 \cos \omega t = ed \cos \omega t \tag{4.15}$$

We could produce such a moment if the two charges executed harmonic motion, their position being given by $z = \pm(d/2)\cos\omega t$. Thus their acceleration would be $\dot{v} = \ddot{z} = \pm\omega^2(d/2)\cos\omega t$, and according to the Larmor equation the total radiated power

$$P = \frac{2}{3} \frac{e^2}{4\pi\varepsilon_0} 2\left(\frac{d}{2}\right)^2 \frac{\omega^4}{c^3} = \frac{2}{3} \frac{1}{4\pi\varepsilon_0} \frac{\langle \ddot{p}^2 \rangle}{c^3} \tag{4.16}$$

where the brackets indicate a time average. The moving charges can be thought of as a current $I = I_0 \sin \omega t = 2e\omega \sin \omega t$ and using $(2\pi/\lambda = \omega)$ we can rewrite Eq. (4.16) as

$$P = \frac{1}{48\pi}\left(\frac{\mu_0}{\varepsilon_0}\right)^{1/2} I_0^2 \left(\frac{2\pi d}{\lambda}\right)^2 \tag{4.17}$$

This result ignores retardation effects and is strictly valid only when $(d/\lambda) \ll 1$. In that case however the antenna is inefficient and little power is radiated for a given current. Instead we must use antennas with dimensions of the order of the wavelength. Such a simple dipole antenna is shown in Fig. 4.3(*b*); it is fed at its center and the current distribution is assumed of the form

$$I = I_0 \cos\left(\frac{\pi z}{d}\right) \cos \omega t$$

with

$$\lambda = 2d \qquad \text{therefore} \qquad \omega = \frac{2\pi c}{\lambda} = \frac{\pi c}{d}$$

Such an arrangement is called a half-wave antenna. The *angular distribution* is very similar to that of a dipole antenna which in the limit $d/\lambda \ll 1$ is

$$\frac{dP}{d\Omega} = \frac{I_0^2}{128\pi^2}\left(\frac{\mu_0}{\varepsilon_0}\right)^{1/2}\left(\frac{2\pi d}{\lambda}\right)^2 \sin^2\theta \tag{4.18}$$

This is shown in Fig. 4.3(*c*); the dipole is assumed oriented along the $Z$-axis, and obviously the distribution is symmetric in azimuth. The energy

Fig. 4.3. Dipole radiation: (*a*) two equal and opposite charges separated by a distance form an electric dipole, (*b*) dipole half-wave antenna (the distribution of current flow at a particular instant of time is indicated by the dashed curve), (*c*) the angular distribution of the power emitted by the antenna shown in (*b*).

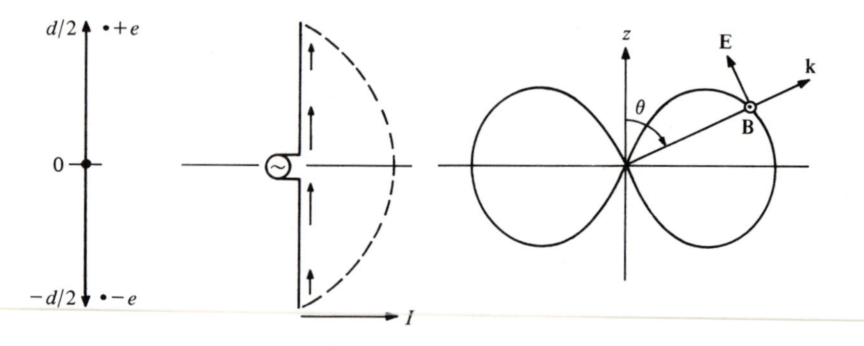

flux, i.e. the energy $\mathrm{d}E$ crossing the area $\mathrm{d}A$ in unit time, at a distance $R$ from the source is

$$\frac{\mathrm{d}P}{\mathrm{d}A}=\frac{1}{R^2}\frac{\mathrm{d}P}{\mathrm{d}\Omega}$$

Integration of Eq. (4.18) over all angles is easily carried out by noting that

$$\int \sin^2\theta\,\mathrm{d}\Omega=\frac{8\pi}{3}$$

and therefore Eq. (4.18) predicts a total power in agreement with Eq. (4.17) which was derived from Larmor's equation.

We note that in the MKS system that we are using, $(\mu_0/\varepsilon_0)^{1/2}$ has dimensions of an impedance and

$$Z_0=\left(\frac{\mu_0}{\varepsilon_0}\right)^{1/2}=377\ \Omega \tag{4.19}$$

is termed the *impedance of free space*. Then Eq. (4.17) can be written in the form

$$P=\tfrac{1}{2}R_{\mathrm{eff}}I_0^2 \tag{4.17$'$}$$

and an antenna is now characterized by its effective 'radiation' impedance. For instance, for the half-wave antenna where $\lambda=2d$, $R_{\mathrm{eff}}=50\ \Omega$; this result is not exact because $d/\lambda\ll 1$ is not fulfilled but it is quite close to the exact value $R_{\mathrm{eff}}=73\ \Omega$.

## 4.3 Directional antennas

We saw that the simple dipole antenna emits more radiation in the equatorial plane and much less at small polar angles. Thus it has some *directionality*. We can improve the directionality further by using an array of dipoles which are fed by signals with a definite phase relationship. As the simplest example we consider two dipole antennas positioned a distance $\lambda/4$ apart along the $Y$-axis, as shown in Fig. 4.4($a$). For an observer on the $Y$-axis at $y>0$, the signal from antenna 2 will arrive a quarter of a period later than from antenna 1. If however the source driving the antenna 2 has its phase advanced by $\pi/2$ the signals from both antennas will arrive in phase and interfere constructively. On the other hand, if the observer is located at $y<0$ he receives the signal from antenna 2 first and given its phase advance, the two signals arrive with a phase difference of 180° and therefore interfere destructively. The radiation pattern from such an arrangement in the plane normal to the dipoles is shown in Fig. 4.4($b$).

To express the previous argument analytically we consider the electric field on the $Y$-axis where at a point $y > 0$ each antenna contributes

$$E_1 = E_0 \cos\left(\omega t - ky + \frac{k\lambda}{8} + \phi_1\right)$$

$$E_2 = E_0 \cos\left(\omega t - ky - \frac{k\lambda}{8} + \phi_2\right)$$

The total field is given by $(E_1 + E_2)$; by using the trigonometric relation for the sum of the cosines, (and noting that $k\lambda = 2\pi$), we find

$$E = E_1 + E_2 = 2E_0 \cos\left(\omega t - ky + \frac{\phi_1 + \phi_2}{2}\right) \cos\left(\frac{\pi}{4} + \frac{\phi_1 - \phi_2}{2}\right)$$

The energy flux is proportional to $|E|^2$. If we take the time average, the first cosine contributes $\frac{1}{2}$ and

$$\langle S \rangle \propto 2E_0^2 \cos^2\left(\frac{\pi}{4} + \frac{\phi_1 - \phi_2}{2}\right)$$

Thus, if $\Delta\phi = \phi_1 - \phi_2 = -\pi/2$ maximum power is radiated toward $+y$; if $\Delta\phi = \pi/2$ no power is radiated towards $+y$ but maximum power is radiated towards $-y$. If $\Delta\phi = 0$ only one half of maximal power is radiated along the. $Y$-axis.

As more individual antennas are used in an array, the radiation pattern becomes narrower and the side lobes become smaller. The advantage of a *phased array* antenna is that one can change the direction in which it points by simply changing the phase of the elements. This is much more efficient than trying to mechanically rotate a massive antenna.

A very simple but common example of a phased array are the 'Yagi' antennas, used for household reception of TV, shown schematically in Fig. 4.5. If we take the frequency to be in the VHF band where $f = 500$ MHz, then $\lambda = c/f = 0.6$ m. Thus we can expect the individual

Fig. 4.4. Directional antenna: ($a$) two dipoles separated by $\lambda/4$, ($b$) the resulting radiation pattern in the plane normal to the dipoles.

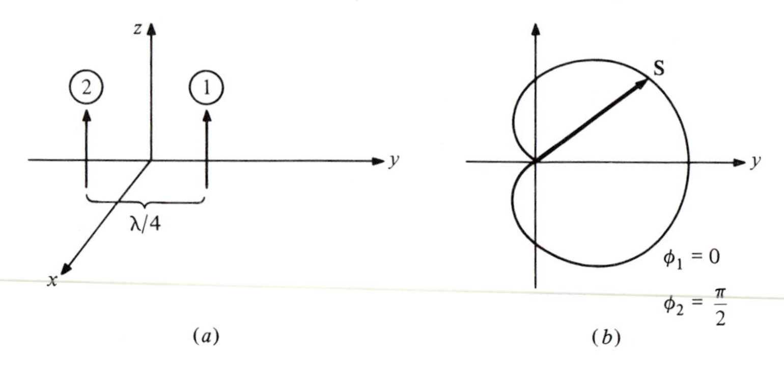

dipoles to be spaced $\lambda/4$ or about half a foot apart from each other. The length of the dipoles establishes the phase of the individually received signals. Note that television transmission has the polarization vector in the horizontal plane so that the array also must be horizontal.

It is a general law that the directionality of an antenna is related to its overall dimensions $L$ by $L\delta\theta \sim \lambda$. For instance, in the example of Fig. 4.4 we have approximately $L \sim \lambda$, and therefore we expect $\delta\theta \sim 1$ radian. Thus, large antennas are needed for long wavelengths as also indicated by Eq. (4.17). For microwaves, $\lambda$ is of order of centimeters and one can obtain excellent directionality with parabolic reflectors of a few meters diameter.

A parabola is the locus of all points that are equidistant from the directrix and the focus $F$ as shown in Fig. 4.6. If we choose the $Y$-axis as the directrix and place the focus, $F$, on the $X$-axis at the coordinate $x_F = 2a$, the equation of the parabola becomes $y^2 = 4(x - a)$. Given that the angle of reflection equals the angle of incidence it is easy to show that any ray originating at the focus is reflected in a direction parallel to the $X$-axis. A parabolic reflector is generated by rotating a parabola around the $X$-axis.

Fig. 4.5. A 'Yagi' antenna.

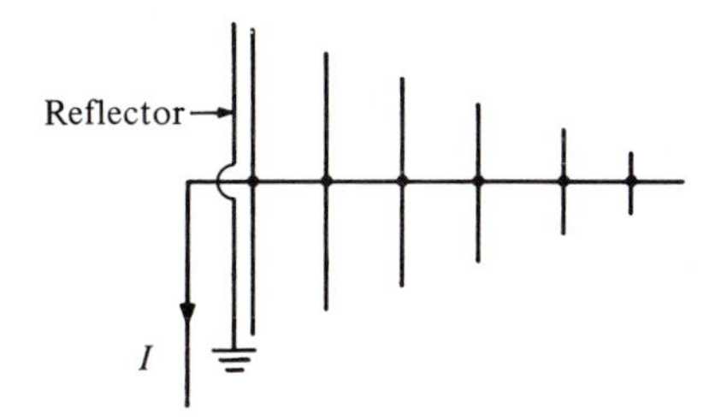


Fig. 4.6. A parabolic reflector focuses parallel light to a point; for large curvature the parabola can be approximated by a spherical surface of radius $R = 2a$.

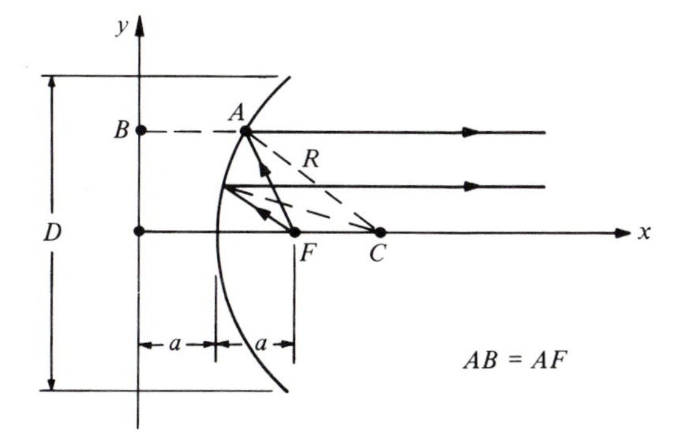

Thus the emerging pattern will always have azimuthal symmetry. As long as the diameter $D$ of the reflector does not exceed the focal distance $a$, the parabolic shape can be approximated by a spherical reflector of radius of curvature $R = 2a$.

The larger the diameter of the reflector, the more power will be directed into the $X$-direction. The ultimate limit in the directivity (or focussing) of the reflector comes from the wave nature of light. This is so because even an idealized source must have dimensions at least of order $\lambda$ (see for instance also the discussion following Eq. (4.17)). It can be shown that a ray originating at a distance $\delta x$ from the focus emerges at an angle $\delta\theta$ with respect to the horizontal, where

$$\delta\theta = \frac{\delta x}{2a} \tag{4.20}$$

If we set $2a \sim R \sim D/2$, and $\delta x \sim \lambda/2$, we obtain

$$\delta\theta = \lambda/D \tag{4.20$'$}$$

Eqs. (4.20) are a form of 'uncertainty relations' such as given by Eq. (3.6). They are equally applicable to optical lenses and are referred to as the *diffraction limit*.

We define the *antenna gain*, at angle $\theta$, $\phi$, as the ratio of the power per unit solid angle radiated at angle $\theta$, $\phi$ to the power that would have been radiated from an isotropic source, i.e. in the absence of the antenna. Namely

$$G(\theta, \phi) = \frac{\mathrm{d}P}{\mathrm{d}\Omega}(\theta, \phi)\frac{1}{(P_t/4\pi)} \tag{4.21}$$

where $P_t$ is the total power. Obviously the total power is unaffected by the antenna and

$$P_t = \int \frac{\mathrm{d}P}{\mathrm{d}\Omega}\,\mathrm{d}\Omega = \frac{P_t}{4\pi}\int G(\theta, \phi)\,\mathrm{d}\Omega$$

so that the antenna gain function is normalized as

$$\int G(\theta, \phi)\,\mathrm{d}\Omega = 4\pi \tag{4.21$'$}$$

Parabolic antennas or 'dishes' are symmetric in $\phi$ and we will assume that all the radiation is contained in the angular range $\delta\theta \ll 1$. With this simplified model we can integrate Eq. (4.21′)

$$4\pi = 2\pi \int_{-1}^{1} G(\theta)\,\mathrm{d}\cos\theta = 2\pi \int_{0}^{\pi} G\sin\theta\,\mathrm{d}\theta = \pi G\delta\theta^2$$

where $G$ is the gain in the forward cone, which we presumed constant.

Thus $G = 4/\delta\theta^2$; a better calculation gives for the gain in the forward direction, (which is the maximum gain), $G_0 = \pi^2/\delta\theta^2$. We can use $\delta\theta$ from Eq. (4.20′) and since the area of the dish is $\pi D^2/4 = A$, the expression for $G_0$ becomes

$$G_0 = \left(\frac{4\pi}{\lambda^2}\right)A \tag{4.22}$$

Thus the energy flux in the forward direction at a distance $R$ from the antenna will be

$$\frac{dP}{dA} = \frac{1}{R^2}\frac{dP}{d\Omega} = \frac{P_t}{4\pi R^2}G_0 = \frac{P_t}{R^2}\left(\frac{A}{\lambda^2}\right) \tag{4.23}$$

Eq. (4.23) shows the importance of a large area dish and of short wavelength.

*Example:* As an application of the above we will derive the power received by a *radar* operating with the following parameters

| | |
|---|---|
| Target distance | $R$ |
| Power in pulse | $P_t$ |
| Antenna forward gain | $G_0 = (4\pi/\lambda^2)A$ |
| Reflection cross section | $\sigma$ |

The energy flux at the target is given by Eq. (4.23) and the energy reflected from the target equals the incident flux multiplied by the reflection cross-section (an area). Thus

$$P_{\text{reflected}} = \frac{\sigma P_t G_0}{4\pi R^2}$$

The reflected power is radiated isotropically so that the reflected energy flux at the receiver is

$$\frac{dP_r}{dA} = \frac{1}{4\pi R^2}P_{\text{reflected}}$$

The received power is $P_r = A(dP_r/dA)$ where $A$ is the area of the dish. Thus

$$P_r = \frac{\sigma P_t G_0}{(4\pi R^2)^2}A = \frac{\sigma P_t}{4\pi R^4}\left(\frac{A}{\lambda}\right)^2 \tag{4.24}$$

where we have assumed optimal pointing. We see again the importance of short wavelength for radar; the received signal drops off as the fourth power of the target distance.

We can introduce typical numerical values

$\sigma \sim 10\ \text{m}^2$

$\lambda \sim 10\ \text{cm}$ (3 GHz)

$P_t \sim 1\ \text{kW}$

$G_0 \sim 10^3$ (this corresponds to $A \simeq 1\ \text{m}^2$)

To have good resolution the pulse width must be narrow, say $\delta t \sim 10^{-7}$ s which would imply a large bandwidth. However we can integrate the signal in time and therefore assume $\Delta f = 10^3$ Hz. If the receiver amplifier temperature is $T_N = 300$ K the noise power is (see Eq. (3.46))

$$P_N = kT\Delta f = 4 \times 10^{-18} \text{ W}$$

If we demand a $S/N \sim 3$ the received power must be in excess of $10^{-17}$ W which will set the range of the radar. From Eq. (4.24)

$$R^4 = \frac{\sigma P_t}{4\pi}\left(\frac{A}{\lambda}\right)^2 \frac{1}{P_r} = \frac{10 \times 10^3}{4\pi}\left(\frac{1}{0.1}\right)^2 \frac{1}{10^{-17}} \sim 10^{22}(\text{m}^4)$$

Thus $R \sim 300$ km which is a long distance. If we had demanded a signal to noise ratio $S/N \sim 250$, the range would have decreased to $R \sim 100$ km.

## 4.4 Reflection, refraction and absorption

So far we have considered the propagation of em waves in free space. The presence of matter affects the externally applied electric and magnetic fields and therefore the propagation properties of the wave as well. This comes about because matter contains electric charges and/or magnetic dipoles that are influenced by the external fields. For instance when an electric field is applied to a dielectric material, the dielectric will remain overall neutral but the electrons will be displaced from their equilibrium positions giving rise to a net dipole moment. The induced dipole moment per unit volume is called the polarization, **P**, of the material.* The polarization is related to the externally applied field **E**, through $\mathbf{P}/\varepsilon_0 = \chi_e \mathbf{E}$ where $\chi_e$ is the dielectric susceptibility of the material.

Maxwell's equations in the presence of matter retain their form (Eqs. (4.1)–(4.4)) if we replace the dielectric permittivity and magnetic permeability of the vacuum $\varepsilon_0$, $\mu_0$ by their values in the material

$$\left.\begin{aligned} \varepsilon &= K_e\varepsilon_0 = (1 + \chi_e)\varepsilon_0 \\ \mu &= K_m\mu_0 = (1 + \chi_m)\mu_0 \end{aligned}\right\} \qquad (4.25)$$

Here, $\chi_e$ and $\chi_m$ are the dielectric and magnetic susceptibilities of the material; $K_e$ is the *dielectric constant* and $K_m$ is the *relative magnetic permeability*.† When we introduce these values in Eq. (4.2) the velocity

* It is unfortunate that the same nomenclature is used as for the 'polarization vector' of an em wave; in both cases however we are trying to indicate a preferred direction in space.

† In the literature the dielectric constant is designated by $K$, and the relative magnetic permeability by $\mu_r$. We prefer $K_e$ and $K_m$ to emphasize the symmetry between electric and magnetic phenomena.

of propagation of the em wave in the material becomes

$$c' = \frac{1}{\sqrt{(\varepsilon\mu)}} = \frac{c}{\sqrt{(K_e K_m)}} = \frac{c}{n} \tag{4.26}$$

The ratio of the velocity of an em wave in free space to the velocity in the material is the *refractive index n* already introduced in Section 3.1. Eq. (4.26) expresses the refractive index in terms of the electromagnetic properties of the material.

When an em wave reaches the boundary (or interface) between two materials of different refractive index, part of the wave is *reflected* while the transmitted part changes its direction, it is *refracted*. These phenomena have been known since antiquity and are shown in Fig. 4.7 for the case $n_2 > n_1$. We define the angles of incidence $\theta_i$, reflection $\theta_r$, and transmission $\theta_t$ with respect to the normal to the interface. The transmitted and reflected rays lie in the plane defined by the normal and the incident ray, the *plane of incidence*. The angle of reflection equals the angle of incidence

$$\theta_i = \theta_r \tag{4.27}$$

whereas the angle of refraction is given by *Snell's law*

$$n_i \sin\theta_i = n_t \sin\theta_t \tag{4.28}$$

The laws of reflection and refraction can be derived by using the boundary conditions that must be obeyed by the em fields. Namely

$$\left.\begin{aligned} (E_t)_1 &= (E_t)_2 & \qquad \varepsilon_1(E_n)_1 &= \varepsilon_2(E_n)_2 \\ (B_t)_1/\mu_1 &= (B_t)_2/\mu_2 & \qquad (B_n)_1 &= (B_n)_2 \end{aligned}\right\} \tag{4.29}$$

where the subscripts 't' and 'n' refer to the components of the fields 'tangential' and 'normal' to the interface. The boundary conditions are a direct consequence of Maxwell's equations and we will use them later.

The laws of reflection and refraction can also be derived from a 'least-time' principle for the trajectory between two points $A$ and $B$ (Fermat's principle). This is shown in Fig. 4.8 where the velocity in medium 1

Fig. 4.7. Reflection and refraction of light at the interface between two media with different refractive indices.

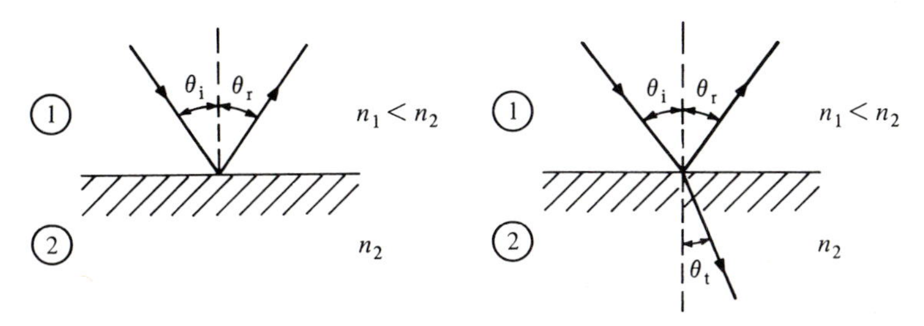

is $c_1 = c/n_1$ and in medium 2 it is $c_2 = c/n_2$. In terms of the coordinates shown in the figure the time of travel is

$$t = t_1 + t_2 = \frac{n_1}{c}(x_1^2 + y_1^2)^{1/2} + \frac{n_2}{c}[(x - x_1)^2 + y_2^2]^{1/2}$$

Minimizing $t$ with respect to $x_1$ yields

$$n_2 \frac{(x - x_1)}{[(x - x_1)^2 + y_2^2]^{1/2}} = n_2 \sin\theta_2 = n_1 \frac{x_1}{(x_1^2 + y_1^2)^{1/2}} = n_1 \sin\theta_1$$

The law of reflection can be obtained in a similar fashion if we assume that on its way from $A$ to $C$ the ray has to reach the interface at some point.

If the wave moves from a region of high refractive index to a region where the index is smaller, i.e. $n_2 < n_1$, Snell's law can be satisfied only when the angle of incidence $\theta_i$ is smaller than the critical angle $\theta_c$, where

$$\frac{n_1}{n_2}\sin\theta_c = 1$$

when $\theta_i > \theta_c$ there can be no transmitted wave and the incident wave is *totally* reflected.

We will now show how the refractive index can be obtained from simple considerations about the structure of matter. Furthermore we will consider gaseous materials in which case we can ignore the interaction between different atoms. We choose the polarization vector along the $X$-axis, and write for the electric field

$$E_x = E_0 e^{-i\omega t}$$

The atomic electrons will feel a force $F_x = eE_x$; we assume that the electrons are bound to the atom by a linear restoring force, and that their motion is damped. Then the equation of motion is of the form

$$\ddot{x} = 2\gamma\dot{x} + \omega_0^2 x = \frac{eE_0}{m} e^{-i\omega t} \tag{4.30}$$

Fig. 4.8. Illustration of Fermat's principle of 'least time' for refraction and reflection.

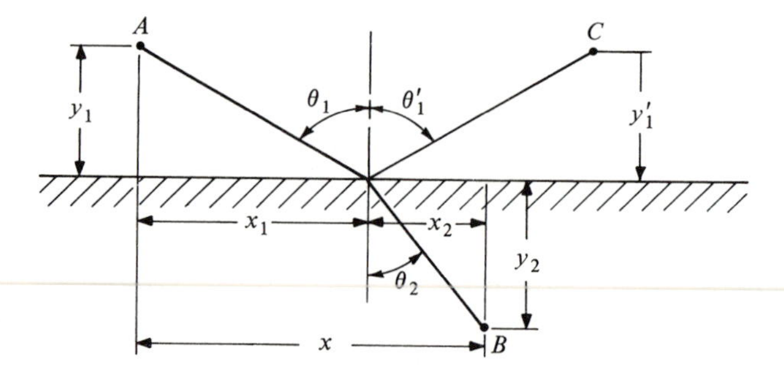

The solution of Eq. (4.30) is obtained by choosing $x = x_0 e^{-i\omega t}$, which leads to

$$x_0 = \frac{E_0(e/m)}{(\omega_0^2 - \omega^2) - i2\gamma\omega} \tag{4.31}$$

where $m$ is the mass of the electron and $\omega_0$ and $\gamma$ are parameters depending on properties of the particular atoms involved; $\omega_0$ is the angular frequency of strong absorption lines, and $\gamma$ their relative width.

If there are $N$ electrons per unit volume the polarization of the material is

$$P = Nex_0 = \left[\frac{Ne^2}{m}\frac{1}{(\omega_0^2 - \omega^2) - i2\gamma\omega}\right]E_0 \tag{4.32}$$

The electric susceptibility is given by

$$\chi_e = P/(\varepsilon_0 E_0)$$

and from Eq. (4.25)

$$\frac{\varepsilon}{\varepsilon_0} = 1 + \chi_e = 1 + \frac{P}{\varepsilon_0 E_0} \tag{4.33}$$

In most materials the magnetic susceptibility at optical frequencies is much smaller than $\gamma$, so that from Eq. (4.26) the refractive index becomes

$$n = \left(\frac{\varepsilon\mu}{\varepsilon_0\mu_0}\right)^{1/2} = \left(\frac{\varepsilon}{\varepsilon_0}\right)^{1/2} = \left(1 + \frac{P}{\varepsilon_0 E_0}\right)^{1/2}$$

Using Eq. (4.32) for the polarization we obtain

$$n = \left[1 + \frac{Ne^2}{m\varepsilon_0}\frac{1}{(\omega_0^2 - \omega^2) - i2\gamma\omega}\right]^{1/2} = \hat{n}(1 + i\kappa) \tag{4.34}$$

In this case, the refractive index has both a *real* part $\hat{n}$, and an *imaginary* part, $\hat{n}\kappa$. The imaginary part gives rise to absorption of the wave.

In general, in gases $\chi_e \ll 1$ and $n$ is close to 1 so that we can expand the radical in Eq. (4.34) to find the real and imaginary parts of $n$

$$\hat{n} = \mathrm{Re}(n) = 1 + \frac{Ne^2}{2m\varepsilon_0}\frac{(\omega_0^2 - \omega^2)}{(\omega_0^2 - \omega^2)^2 + 4\gamma^2\omega^2} \tag{4.35}$$

$$\hat{n}\kappa = \mathrm{Im}(n) \simeq \kappa = \frac{Ne^2}{2m\varepsilon_0}\frac{2\gamma\omega}{(\omega_0^2 - \omega^2)^2 + 4\gamma^2\omega^2} \tag{4.35'}$$

These functions are shown in Fig. 4.9. Near the resonance frequency $\omega \sim \omega_0$, the absorption becomes very strong and the medium can be opaque to the transmission of radiation at that particular wavelength. Similarly the real part of the refractive index can become less than 1 just above the resonance. This implies that the phase velocity $c'$ exceeds $c$; however the group velocity $v_g = d\omega/dk$ remains less than $c$ as required by special relativity. Because $(\hat{n} - 1)$ depends on frequency, it leads to

Fig. 4.9. Variation of the refractive index as a function of frequency in the region of a resonance. The real part is given by the curve labeled $(\hat{n} - 1)$ and the imaginary part by the curve labeled $\hat{n}\kappa$.

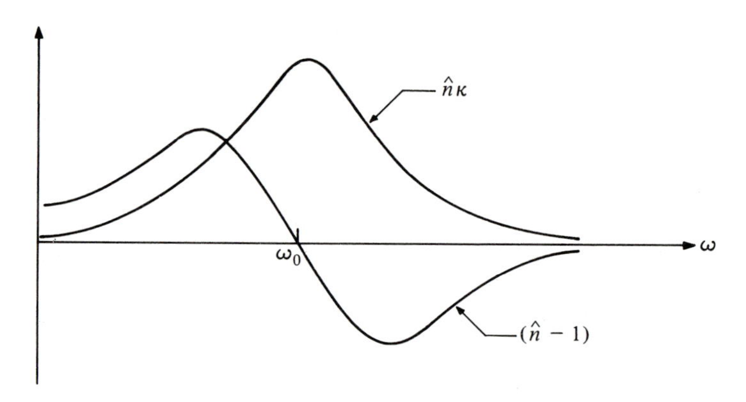


*dispersion*; thus different Fourier components propagate with different velocities, and short pulses become distorted when propagating through the medium.

For isolated atoms such as in gases the most important resonances are in the ultraviolet, so that for incident visible light we have $\omega \ll \omega_0$; furthermore the lines are not very broad, $2\gamma < \omega$. Thus Eqs. (4.35) simplify to

$$n \simeq \hat{n} = 1 + \frac{Ne^2}{2m\varepsilon_0\omega_0^2} \tag{4.36}$$

and $\kappa = 0$. The above expression is reasonably accurate for gases in spite of the simple model that was used. For instance for hydrogen gas, if we take $\omega/2\pi \simeq 3 \times 10^{15}$ (which corresponds to excitation from the ground state), and $N = 5 \times 10^{25}/\text{m}^3$ we find

$$n - 1 = 2.2 \times 10^{-4}$$

very close to the measured value $(n - 1) = 1.4 \times 10^{-4}$.

## 4.5 The ionosphere

One of the most interesting phenomena of early radio transmission was the reception of signals at large distances far beyond the line of sight between transmitter and receiver. Such transmission is due to the reflection of signals from the ionosphere; it is much more pronounced at night and is strongly affected by atmospheric conditions. We can explain the reflection of em waves from the ionosphere if we treat the latter as a partially ionized plasma containing free electrons. The em wave interacts with the plasma and we can describe this interaction in terms of a refractive index.

The plasma is overall electrically neutral; the ionized electrons are free to move whereas the much heavier atomic ions do not contribute to current flow. Maxwell's equations then take the form

$$\nabla \cdot \mathbf{B} = 0 \qquad \nabla \times \mathbf{E} + \frac{\partial \mathbf{B}}{\partial t} = 0$$

$$\nabla \cdot \mathbf{E} = 0 \qquad \nabla \times \mathbf{B} - \varepsilon\mu \frac{\partial \mathbf{E}}{\partial t} = \mu \mathbf{J} = \mu\sigma\mathbf{E}$$

In the last equation we used Ohm's law to express the current density in terms of the conductivity $\sigma$ of the plasma

$$\mathbf{J} = \sigma\mathbf{E} \tag{4.37}$$

By following the same steps that led to Eq. (4.7) the wave equation in this case becomes

$$\nabla^2\mathbf{E} - \mu\sigma \frac{\partial \mathbf{E}}{\partial t} - \mu\varepsilon \frac{\partial^2 \mathbf{E}}{\partial t^2} = 0 \tag{4.38}$$

where the term $-\mu\sigma(\partial\mathbf{E}/\partial t)$ introduces 'damping' in the propagation of the wave and thus affects the velocity of propagation. To solve Eq. (4.38) we will assume, as before, a plane wave solution

$$E = E_x = E_0 e^{-i(\omega t - kz)} \tag{4.39}$$

and we wish to establish the relation between $\omega$ and $k$, imposed by Eq. (4.38). Introducing Eq. (4.39) into Eq. (4.38) we obtain

$$(-k^2 + i\mu\sigma\omega + \omega^2\mu\varepsilon)E_x = 0$$

with solution

$$k^2 = \varepsilon\mu\omega^2\left[1 + i\frac{\sigma}{\varepsilon\omega}\right] \tag{4.40}$$

We can evaluate the conductivity $\sigma$ by using the model introduced in Eq. (4.30) but setting $\omega_0 = \gamma = 0$ since the electrons are not bound to atoms. Then

$$\frac{d}{dt}(\dot{x}) = \frac{e}{m} E_0 e^{-i\omega t}$$

and

$$\dot{x} = -\frac{e}{i\omega m} E_0 e^{-i\omega t}$$

Furthermore $\mathbf{J} = eN\mathbf{v}$ which in our case is $J_x = eN\dot{x} = \sigma E_0 e^{-i\omega t}$, with $N$

the density of free electrons. Thus the conductivity of the plasma is

$$\sigma = \mathrm{i}\,\frac{e^2 N}{\omega m} \tag{4.41}$$

The conductivity is imaginary (the current is 90° out of phase with respect to the applied electric field) and is frequency dependent. Introducing $\sigma$ from Eq. (4.41) in Eq. (4.40) and setting $\varepsilon \simeq \varepsilon_0$, $\mu \simeq \mu_0$ we find for the wave vector squared

$$k^2 = \frac{\omega^2}{c^2}\left[1 - \frac{4\pi N}{m\omega^2}\frac{e^2}{4\pi\varepsilon_0}\right] \tag{4.42}$$

which is the desired relation between $k$ and $\omega$.

It is convenient to introduce the *plasma frequency*, $\omega_p$, through

$$\omega_p^2 = \frac{4\pi N}{m}\frac{e^2}{4\pi\varepsilon_0}$$

so that

$$k = \frac{\omega}{c}\left(1 - \frac{\omega_p^2}{\omega^2}\right)^{1/2} \tag{4.43}$$

When $\omega < \omega_p$ the wave vector is imaginary, $k = \mathrm{i}\beta$ and Eq. (4.39) takes the form

$$E = E_x = E_0 \mathrm{e}^{-\mathrm{i}(\omega t - kz)} = E_0 \mathrm{e}^{-\mathrm{i}\omega t}\mathrm{e}^{-\beta z}$$

That is, the wave is attenuated in the plasma over a distance $l = 1/\beta$. Since the wave cannot propagate in the plasma it is reflected from it.* When $\omega > \omega_p$, the wave vector is real and the plasma is transparent to the wave. We can express Eq. (4.43) in terms of the refractive index $n = ck/\omega$

$$n = \left(1 - \frac{\omega_p^2}{\omega^2}\right)^{1/2} \tag{4.44}$$

Thus when $\omega > \omega_p$ the refractive index is real and $n < 1$. Even in that case the wave will undergo total reflection if the angle of incidence is larger than the critical angle $\theta_c$, where $\sin\theta_c = n$.

The ionosphere extends from approximately 50 to 300 km above the earth's surface. At such height the density is low as compared to that on the surface and one finds layers of ionized gas distributed roughly as indicated in Table 4.1. The ionization is produced by sunlight, in particular its UV component, and to a much lesser extent by cosmic rays and X-rays.

* For the same reason, high frequency em waves cannot propagate inside a metal and visible light is reflected from the surface of metals. In metals $N \sim 10^{23}$ but the conductivity is dominated by the collisions of the electrons with the lattice sites and thus Eq. (4.41) is not directly applicable.

The degree of ionization depends on the time of the day, and is maintained, in part, due to the earth's magnetic field. The effect of the field is to trap the electrons, just as protons are trapped by the earth's magnetic field to form the *Van Allen belts* which are located at a distance of ~1000 km from the surface.

We evaluate the plasma frequency for a free electron density $N_e = 10^5/\text{cm}^3$, to find

$$\omega_p^2 = 4\pi \frac{N_e}{m_e} \frac{e^2}{4\pi\varepsilon_0} = \frac{10^{11}}{0.9 \times 10^{-30}} \frac{(1.6 \times 10^{-19})^2}{(8.85 \times 10^{-12})}$$
$$= 3.2 \times 10^{14} (\text{rad/s})^2$$

or

$$\nu_p = \frac{\omega_p}{2\pi} = \frac{1.8 \times 10^7}{2\pi} = 2.8 \times 10^6 \text{ Hz}$$

Frequencies below $\nu_p$ cannot penetrate the ionosphere. At night, the free electron density decreases by a factor of about 100, so that $\nu_p \sim 300$ kHz which is typical of AM radio transmission. Signals at this frequency can now be reflected from high lying layers of the ionosphere and reach long distances as shown in Fig. 4.10. It should be appreciated that the

Table 4.1. *Ionosphere layers*

| Height (km) | Designation | $N_e$ (free electrons/cm$^3$) |
|---|---|---|
| 50–90 | D layer | $10^2$ |
| 90–130 | E layer | $10^5$ |
| 130–300 | F layer | $10^5$–$10^6$ |

Fig. 4.10. The layers of the ionosphere give rise to the reflection of electromagnetic waves.

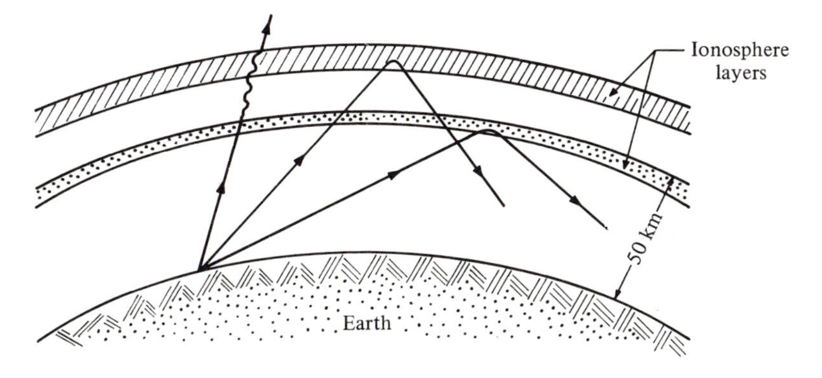

ionospheric layers do not present sharp interfaces but that the refractive index changes gradually as the electron density first increases and then drops as a function of height.

Propagation in the layers of the ionosphere is further complicated by the presence of the earth's magnetic field which causes the electrons to spiral at their cyclotron frequency

$$\omega_B = \frac{eB}{m}$$

This modifies the refractive index from the expression given by Eq. (4.44) to

$$n = \left(1 - \frac{\omega_p^2}{(\omega \pm \omega_B)\omega}\right)^{1/2} \tag{4.44'}$$

The earth's field is weak, $B \simeq 0.5$ gauss $= 5 \times 10^{-5}$ T and results in a cyclotron frequency for electrons

$$\omega_B = \frac{(1.6 \times 10^{-19}) \times (5 \times 10^{-5})}{0.9 \times 10^{-30}} \sim 10^7 \text{ rad/s}$$

This frequency is in the MHz range, and thus the earth's field has a significant influence on the propagation of radio transmission.

## 4.6 Satellite communications

Reflection from the ionosphere provided the principal means for over the horizon communications until the introduction of communication satellites in the 1960s. It is now possible to transmit a signal from an earth station to an orbiting satellite; the signal is amplified and retransmitted from the satellite, thus providing coverage over very large distances with high frequency carriers. For even longer range communications such as world-wide coverage, links are established between satellites in different orbits.

Communication satellites can be placed in polar, inclined or equatorial orbits, as shown in Fig. 4.11(*a*). The most common positioning for communication purposes is in geostationary orbits. For a satellite to remain fixed over the same earth location, it must be in equatorial orbit and rotate with a 24 hour period. This fixes the orbit radius since for a circular orbit

$$\omega^2 = \left(\frac{v}{r}\right)^2 = \frac{GM_\oplus}{r^3} \tag{4.45}$$

where $G$ is Newton's constant, $M_\oplus$ the mass of the earth and $\omega, r$ the angular velocity and radius of the orbit. We can rewrite Eq. (4.45) in terms of $g$, the acceleration of gravity on the earth's surface and the earth's radius $R_\oplus$

$$\omega^2 = g\frac{R_\oplus^2}{r^3} \qquad \text{or} \qquad r = \left[\frac{gR_\oplus^2}{\omega_\oplus^2}\right]^{1/3} \tag{4.45'}$$

Using $g = 9.8$ m/s$^2$, $R_\oplus = 6.37 \times 10^6$ m and $\omega_\oplus = 2\pi/(24 \text{ hours}) = 7.2 \times 10^{-5}$ rad/s we find

$$r = 4.23 \times 10^7 \text{ m}$$

Thus the height of the equatorial orbit is $h = r - R_\oplus$; the exact value is

$$h = 35\,889 \text{ km}$$

Even when placed in a geostationary (also called geosynchronous) orbit, the satellite does not remain perfectly fixed with respect to the earth, but wobbles about its intended reference point.

We can calculate the coverage provided by a satellite located at a height $h$ above the earth with the help of Fig. 4.11($b$). The satellite is at $S$ and the observer at $A$; $\phi_i$ is the angle of inclination of the satellite above the horizon of the observer and $\phi_E$ is the angle subtended by the observer at the center of the earth with respect to the satellite direction. The coverage angle $\beta$ as seen from the satellite is

$$\beta = \frac{\pi}{2} - \phi_i - \phi_E$$

Since

$$\frac{\sin\beta}{R_\oplus} = \frac{\sin(\pi/2 + \phi_i)}{h + R_\oplus}$$

Fig. 4.11. ($a$) Various possible satellite orbits, ($b$) earth coverage by a satellite located at a height $h$ above the earth.

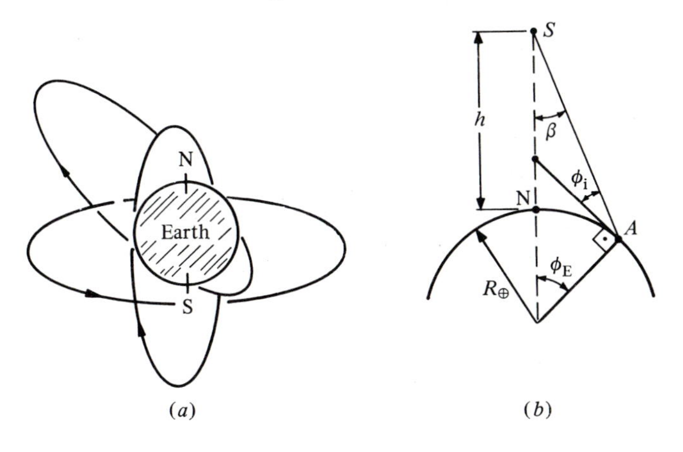

We obtain

$$\frac{\cos(\phi_i + \phi_E)}{R_\oplus} = \frac{\cos \phi_i}{h + R_\oplus}$$

$$\phi_E + \phi_i = \text{arc cos}\left[\frac{R_\oplus}{h + R_\oplus} \cos \phi_i\right]$$

For a geostationary satellite $h = 35\,900$ km and choosing $\phi_i > 20°$, one finds $\phi_E + \phi_i < 82°$ or $\phi_E < 62°$. This covers a large part of a hemisphere but not regions near the poles. In terms of solid angle

$$\Delta\Omega = 2\pi(1 - \cos \phi_E) = 2\pi \times (0.53)$$

Since at most $2\pi$ can be covered (half of the earth is invisible even from infinite distance and at zero inclination), the satellite can be viewed from 53% of the maximum area. In practice, communication satellites have directional antennas which increase the power of the transmitted signal at the expense of coverage.

We briefly analyse the power requirements for satellite communications and will consider separately the *uplink* from the *downlink*. If $P_t$ is the total power of the transmitter and $G_0$ the maximum antenna gain, the power received by a geostationary satellite is

$$P_s = \frac{P_t}{4\pi(h)^2} G_0 A_s \tag{4.46}$$

where $A_s$ is the area of the satellite antenna. We assume $P_t = 10^3$ W, $G_0 = 10^4$ which is easy to achieve for a transmitter on the earth; also $A_s = 1\ \text{m}^2$ and $h \simeq 35\,900$ km. Then $P_s \simeq 6 \times 10^{-10}$ W. We can assume a

Fig. 4.12. Locations of the geostationary communication satellites as of 1985.

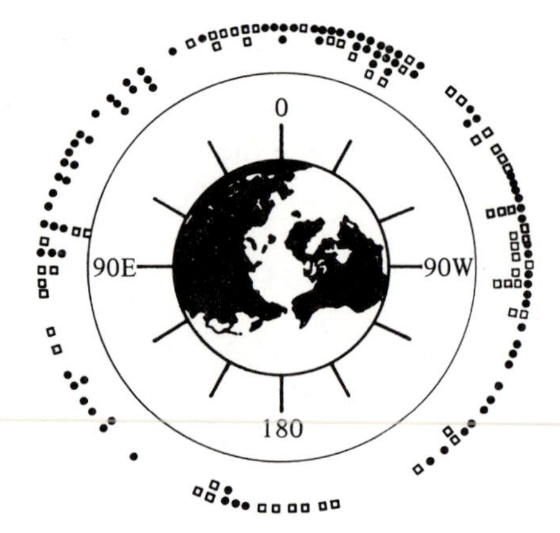

noise temperature $T_N = 10^3$ K for the receiver in the satellite and a bandwidth $\Delta f = 3 \times 10^7$ Hz; thus the noise power is (see Eq. (3.46))

$$P_N = kT\Delta f \simeq 4 \times 10^{-13}\ \text{W}$$

Then the $S/N$ at the satellite's receiver is in excess of 1000 and therefore poses no problems.

It is more difficult to achieve a good $S/N$ in the downlink because satellite power and antenna size are restricted. We assume $P_t = 1$ kW as before, but $G_0 = 10^2$; this still allows coverage on the earth of an area of radius $R \sim 4500$ miles, that is in excess of the entire continental U.S. We also take the receiving antenna of typical home TV dishes with an area of 6 m$^2$; the received power then is $P_r = 4 \times 10^{-11}$ W, and given the same bandwidth and receiver noise we find $S/N = 100$, which is adequate for home reception.

There are already a large number of communication satellites in equatorial orbit and more are planned. These are indicated in Fig. 4.12 (courtesy of the *New York Times*, September 15, 1985). A problem of interference arises when the angular separation between orbiting satellites is smaller than the beamwidth of the earth antenna and if they operate in the same frequency band. Thus the number of satellites that can be placed in equatorial orbit is limited, and this limit is being rapidly approached.

## 4.7 Waveguides and transmission lines

It is possible to transmit em energy by using a conducting medium, such as a pair of 'electrical wires': this is the standard way by which electrical power is distributed. For communications we use high frequency em waves which do not penetrate deep into a conductor but are reflected from its surface. Thus for short wavelengths, $\lambda < 10$ cm, it is more efficient to contain the em radiation inside a metallic structure, a *waveguide*, just as water is contained inside a garden hose. In general the wavelength has to be shorter than the dimensions of the guide if the wave is to propagate without attenuation. We will see in the following section that even visible light can be trapped in a waveguide but in that case the walls are dielectric rather than conducting.

The behavior of the em fields at an interface is governed by the boundary conditions given by Eqs. (4.29). Inside a perfect conductor the electric field vanishes; since the tangential component, $E_t$, must be continuous at the boundary (see the first of Eqs. (4.29)) it follows that $E_t$ for the *external* field must also vanish at the boundary. Thus, if an em wave is normally incident on a conductor, as in Fig. 4.13(*a*), the electric field of the reflected wave must be reversed with respect to the electric field of the incident wave. As a result the incident and reflected waves combine to give rise to a *standing wave* as in (*b*) of the figure. Standing waves are apparent in many physical phenomena, as for instance when we wiggle a rope that is tied at one end.

Mathematically, we can write for the two waves

$$E_{\mathrm{i}} = E_0 \mathrm{e}^{-\mathrm{i}(\omega t - kz)}$$

$$E_{\mathrm{r}} = -E_0 \mathrm{e}^{-\mathrm{i}(\omega t + kz)}$$

where $E_{\mathrm{i}}$ propagates to positive $z$, whereas the reflected wave, $E_{\mathrm{r}}$, propagates to negative $z$ and has reversed amplitude at $z = 0$. Then

$$E = E_{\mathrm{i}} + E_{\mathrm{r}} = E_0 \mathrm{e}^{-\mathrm{i}\omega t}[\mathrm{e}^{ikz} - \mathrm{e}^{-ikz}] = 2\mathrm{i}E_0 \mathrm{e}^{-\mathrm{i}\omega t} \sin kz \qquad (4.47)$$

which is a standing wave of wavelength $\lambda = 2\pi/k$. A standing wave does *not* transmit energy along the $Z$-axis in contrast to traveling waves which carry energy along their direction of propagation.

Next we consider a wave which is *obliquely* incident onto the conducting surface as shown in Fig. 4.14(*a*). The electric field is polarized parallel to the interface and the angle of incidence is $\theta$. We define the wavevector $\mathbf{k}$ along the direction of propagation and can resolve it into two components $k_x$ and $k_z$, indicated in (*b*) of the figure.

$$|\mathbf{k}| = \frac{2\pi}{\lambda} = \frac{\omega}{c} = (k_x^2 + k_z^2)^{1/2} \qquad (4.48)$$

Fig. 4.13. Generation of standing waves by reflection from a conductor: (*a*) the two traveling waves at three different instants of time, (*b*) the resulting sum of the two waves is a standing wave.

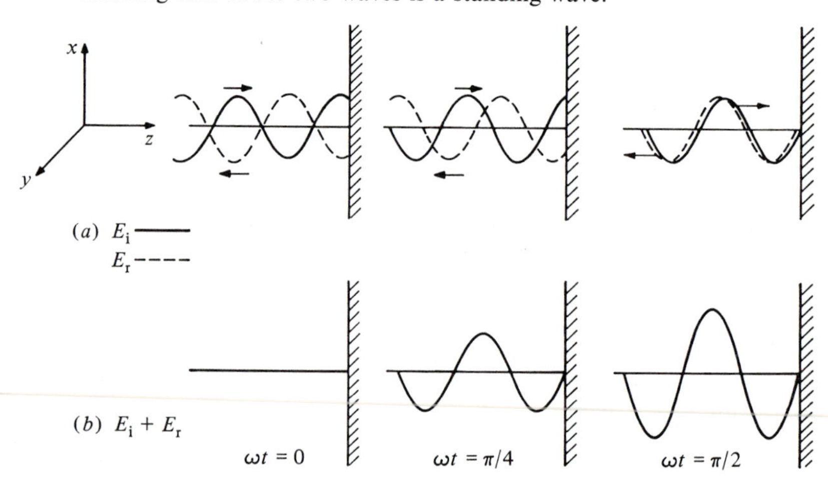

Since Maxwell's equations are linear we can treat independently the wave that is incident normal to the interface $k_n = k_x = k \cos \theta$, and the wave that is propagating parallel to the interface, where $k_p = k_z = k \sin \theta$. The parallel component is unaffected by the boundary, but the normal component is reflected and must always have a *node* at the interface, that is the electric field amplitude must be zero. We now place a second conducting surface, parallel to the first one at a distance $b$, as shown in Fig. 4.15($a$). Since the normal component must have a node at that surface as well, the wavevector must obey

$$k_n b = m\pi$$

where $m$ is an integer. Using $k_n = k \cos \theta = (2\pi/\lambda) \cos \theta$, we find that propagation can occur only at angles $\theta$ that satisfy the condition

$$\lambda = \frac{2b \cos \theta}{m} \qquad m = 1, 2, 3, \ldots \tag{4.49}$$

If $\lambda > 2b$ Eq. (4.49) cannot be satisfied and the wave *will not* propagate.

The different values of the integer $m$ correspond to different distributions of the amplitude of the standing wave in the $x$-direction. This is shown in Fig. 4.15 and we speak of different *modes*. Furthermore we have assumed

Fig. 4.14. Reflection from a metallic boundary: ($a$) the angle of incidence is $\theta$ and the wave is polarized along the $Y$-axis (out of the plane of the paper), ($b$) resolution of the wavevector into two components along the $X$- and $Z$-axes.

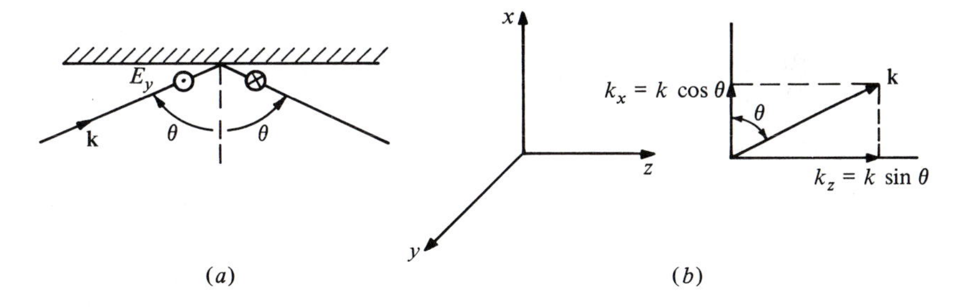


Fig. 4.15. Propagation between two metallic surfaces results in a standing wave in the plane normal to the direction of propagation: ($a$) lowest mode, ($b$) next higher mode.

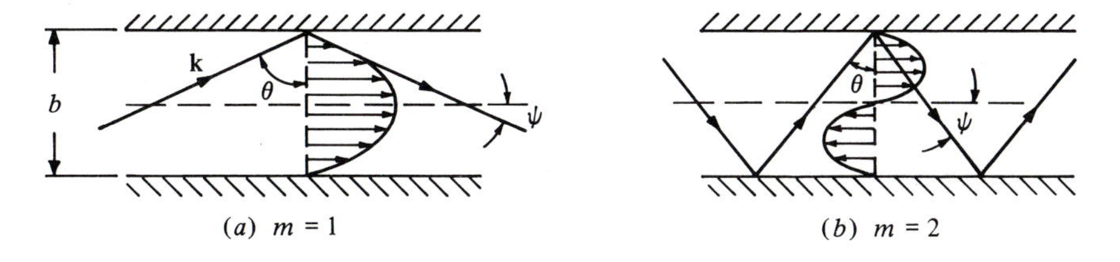

that the electric field is along the $y$-direction, parallel to the interface; thus there is no electric field component along the $z$-axis and the modes are called transverse electric (TE-modes). We can think of the energy propagating between the two conducting surfaces by a zig-zag path as the wave bounces off the two conducting surfaces. The higher the mode, $m$, the more bounces occur for a given length of propagation.

It is interesting to examine the field distribution along the $z$-direction (the direction of propagation) as well. We have $k_p = k \sin\theta$, so that the 'wavelength' $\lambda_p$ which determines the periodicity along $z$, and is defined by $k_p = 2\pi/\lambda_p$ has the value

$$\lambda_p = \frac{2\pi}{k_p} = \frac{2\pi}{k \sin\theta} = \frac{\lambda}{\sin\theta}$$

Thus the phase velocity $v_p$ is

$$v_p = \nu\lambda_p = \frac{\nu\lambda}{\sin\theta} = \frac{c}{\sin\theta} \tag{4.50}$$

which is larger than $c$. However, the velocity at which energy propagates down the $z$-axis, the group velocity $v_g$, is determined by the zig-zag pattern and is less than $c$. We easily find from Fig. 4.15 that

$$v_g = c \sin\theta \tag{4.50'}$$

when $\theta \to 0$ the group velocity $v_g \to 0$ and there is no propagation in that particular mode. The wavelength at which this occurs is called the cut-off wavelength $\lambda_0$ and is given (see Eq. (4.49)) by

$$\lambda_0 = \frac{2b}{m} \tag{4.51}$$

So far we have considered propagation between two parallel conducting surfaces. If we add two more surfaces to form an enclosure, we obtain a *waveguide*. The field amplitudes must satisfy the boundary conditions at all four 'walls' and this gives rise to modes characterized by two indices, $n$ and $m$. The field patterns for the lowest mode in a *rectangular* waveguide, the $TE_{0,1}$ mode, are shown in Fig. 4.16. Cylindrical waveguides obey the same principles but are not used as much as rectangular guides.

*The coaxial line:* At frequencies below the microwave band, it is not convenient to use waveguides since they become bulky while transmission in a conductor still involves too much attenuation. In this case the em fields are contained between *two* conducting surfaces (that can be at different potentials) and we speak of a *transmission line*. The transmission line can be thought of as the transition between bulk conductors and waveguides.

The most widely used type of transmission line is the coaxial line where the conducting surfaces are two coaxial cylinders. This is shown in Fig. 4.17; the inner radius is $r_{in} = a$ and the outer radius $r_{out} = b$ and the region between the two conductors is filled with a dielectric ($\sigma = 0$) of permittivity $\varepsilon$ and permeability $\mu$. The simplest mode of propagation is TEM (transverse electric and magnetic) where the electric field is radial and the magnetic field tangential as indicated in the figure.

To find the propagation properties of an em wave in this structure we will directly solve Maxwell's equations (Eqs. (4.1)–(4.4)) but in cylindrical coordinates. We are interested only in the region between the two conductors, where there is no current flow, $J = 0$ and thus Eqs. (4.2) and (4.4) become

$$\nabla \times \mathbf{E} = -\frac{\partial \mathbf{B}}{\partial t}, \qquad \nabla \times \mathbf{B} = \mu\varepsilon \frac{\partial \mathbf{E}}{\partial t} \tag{4.52}$$

In cylindrical coordinates the components of the curl of a vector **A** are

Fig. 4.16. Field line patterns for the $TE_{01}$ mode propagating in a rectangular waveguide. The graphs show a 'snapshot' of the **E** and **B** fields at a particular instant of time.

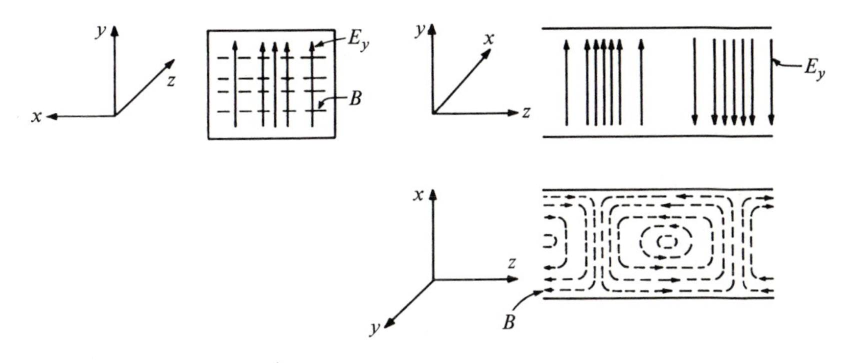


Fig. 4.17. Coaxial transmission line.

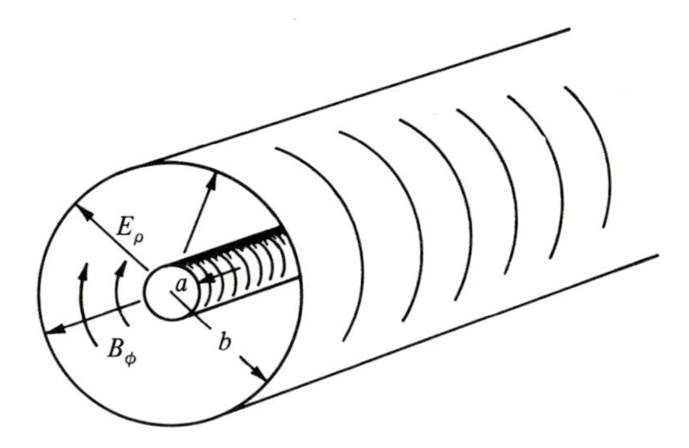

given in general by

$$\left.\begin{aligned}(\nabla \times \mathbf{A})_\rho &= \frac{1}{\rho}\frac{\partial A_z}{\partial \phi} - \frac{1}{\rho}\frac{\partial}{\partial z}(\rho A_\phi)\\(\nabla \times \mathbf{A})_\phi &= \frac{\partial A_\rho}{\partial z} - \frac{\partial A_z}{\partial \rho}\\(\nabla \times \mathbf{A})_z &= \frac{1}{\rho}\frac{\partial}{\partial \rho}(\rho A_\phi) - \frac{1}{\rho}\frac{\partial A_\rho}{\partial \phi}\end{aligned}\right\} \tag{4.53}$$

where $\rho$ is the radial coordinate.

We will choose the fields as sketched in Fig. 4.17, that is

$$\left.\begin{aligned}B_\phi &= B_0 e^{-i(\omega t - kz)}\\E_\rho &= E_0 e^{-i(\omega t - kz)}\end{aligned}\right\} \tag{4.54}$$

and all other components are set to zero. The above fields represent a wave traveling in the positive $z$-direction, and are transverse to the direction of propagation and to one another. Then the first of Eqs. (4.52) gives

$$(\nabla \times \mathbf{E})_\phi = \frac{\partial E_\rho}{\partial z} = ikE_\rho = -\frac{\partial B_\phi}{\partial t} = i\omega B_\phi$$

The second of Eqs. (4.52) gives

$$(\nabla \times \mathbf{B})_\rho = -\frac{1}{\rho}\frac{\partial}{\partial z}(\rho B_\phi) = -\frac{\partial B_\phi}{\partial z} = -ikB_\phi$$

$$= \mu\varepsilon \frac{\partial E_\rho}{\partial t} = -\mu\varepsilon i\omega E_\rho$$

We summarize the two results

$$\left.\begin{aligned}E_\rho &= \frac{\omega}{k} B_\phi\\E_\rho &= \frac{1}{\mu\varepsilon}\frac{k}{\omega} B_\phi\end{aligned}\right\} \tag{4.55}$$

For these equations to hold we must have

$$k^2 = \varepsilon\mu\omega^2 = \frac{\omega^2}{c^2} K_e K_m \tag{4.55$'$}$$

Thus the wave propagates in the coaxial line with phase velocity

$$v_p = c' = \frac{\omega}{k} = \frac{c}{(K_e K_m)^{1/2}} \tag{4.56}$$

Note that there is no cut-off frequency in this case, and all frequencies, even dc current, can propagate in a coaxial line. The key feature of this geometrical arrangement is that the high frequency fields are shielded by the outer conductor.

We can relate the fields $E_\rho$, $B_\phi$ to the current $I$ flowing in the conductor and to the voltage difference $V$ between the two conductors. From Ampère's law we have

$$2\pi\rho\frac{B_\phi}{\mu} = I \tag{4.57}$$

Using Eqs. (4.55) we can express $E_\rho$ in terms of the current $I$

$$\frac{\omega}{k}\varepsilon E_\rho = \left(\frac{\varepsilon}{\mu}\right)^{1/2} E_\rho = \frac{I}{2\pi\rho} \tag{4.57'}$$

where in the second step we set $\omega/k = 1/(\varepsilon\mu)^{1/2}$ according to Eq. (4.55′).

From Eq. (4.57) we see that the current $i$ must have the same space–time dependence as the electric field. Thus, recalling Eq. (4.54) we write

$$I = I_0 e^{-i(\omega t - kz)}$$

The voltage can be found by integrating the radial electric field from its value at the inner conductor to that at the outer conductor

$$V = \int_a^b E_\rho \, d\rho = \frac{1}{2\pi}\left(\frac{\mu}{\varepsilon}\right)^{1/2} I \int_a^b \frac{d\rho}{\rho} = \frac{I}{2\pi}\left(\frac{\mu}{\varepsilon}\right)^{1/2} \ln\left(\frac{b}{a}\right)$$

The *impedance*, $Z$, of the coaxial line is given by the ratio $V/I$,

$$Z = \frac{V}{I} = \frac{1}{2\pi}\left(\frac{\mu}{\varepsilon}\right)^{1/2} \ln\left(\frac{b}{a}\right) = \frac{1}{2\pi} Z_0 \left(\frac{K_m}{K_e}\right)^{1/2} \ln\left(\frac{b}{a}\right) \tag{4.58}$$

where we have used $Z_0$ to designate the impedance of free space

$$Z_0 = \left(\frac{\mu_0}{\varepsilon_0}\right)^{1/2} = \left(\frac{4\pi \times 10^{-7}(\text{V-s/A-m})}{8.85 \times 10^{-12}(\text{coul/V-m})}\right)^{1/2} = 377 \text{ Ohm}$$

already introduced in Section 4.2 (see Eq. (4.19)).

As an example we can consider a coaxial cable with conductor ratios $b/a = 4$ and filled with polystyrene which has $K_e \sim 2.5$, $K_m \sim 1.0$. Then the impedance is

$$Z = \frac{377}{2\pi}\frac{1}{\sqrt{2.5}}\ln(4) = 53\ \Omega$$

which is a typical value for the impedance used in r.f. systems.

## 4.8 Fiber optics

We saw in the previous section that high frequency em waves can propagate in a guide with conducting walls. It is also possible for an em wave to propagate in a guide with dielectric walls. While the principle of dielectric wave guides was known since 1910, only recent technological advances have made such guides practical. Dielectric waveguides are used in the visible and consist of very thin fibers whose refractive index is precisely controlled; light can be trapped inside such a fiber and will propagate with very little attenuation.

The optical fibre is constructed of layers of dielectric with differing refracting index (smaller values of $n$ at larger radii) as shown in Fig. 4.18($a$). For simplicity we will consider first a flat dielectric slab of index $n_1$ bounded by two plane dielectrics of index $n_2 < n_1$ as shown in ($b$) of the figure. An em wave is incident onto the interface, at angle $\theta$. We know from Snell's law (Eq. (4.28)) that if $\theta > \theta_c$ there can be no transmitted ray and all the energy is reflected back into medium 1. The critical angle $\theta_c$ is found by setting $\theta_2 = 90°$ (or $\sin\theta_2 = 1$) in Eq. (4.28) so that

$$\sin\theta_c = \frac{n_2}{n_1} \tag{4.59}$$

The complement of $\theta$ is the propagation angle $\psi = 90° - \theta$. As an example, if we have a lucite rod ($n_1 = 1.5$) bounded by air ($n_2 = 1.0$), the critical angle is $\sin\theta_c = 0.67$ and $\theta_c = 58°$, so that $\psi_c = 42°$; rays with propagation angles $\psi < \psi_c$ will be trapped inside the lucite.

In optical fibers the refractive index is $n_1 \sim 1.5$ roughly the same as for lucite, but the boundary layer has a refractive index $n_2$ very close to $n_1$, within a few percent. Thus it is convenient to define the difference between refractive indices through

$$\Delta = \frac{n_1^2 - n_2^2}{2n_1^2} \simeq \frac{n_1 - n_2}{n_1} \tag{4.60}$$

We can express $\psi_c$ in terms of $\Delta$, the fractional change in the refractive

Fig. 4.18. Optical fiber consisting of two layers of dielectric with different refractive index: ($a$) the radii and indices are as indicated, ($b$) simple model of the fiber assumes only the presence of two plane boundaries.

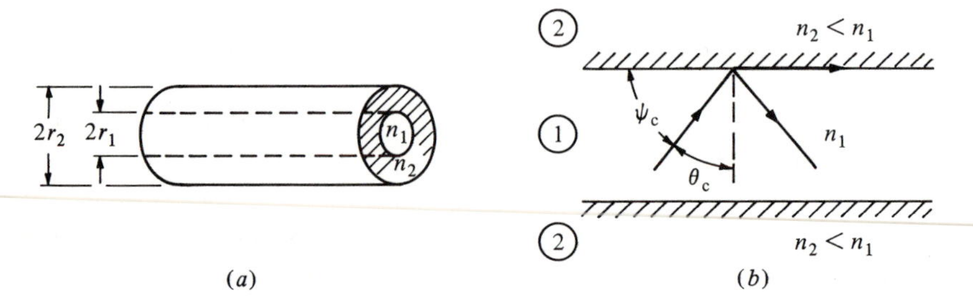

index

$$\psi_c = \cos^{-1}\left(\frac{n_2}{n_1}\right) = \sin^{-1}\left\{\left[1 - \frac{n_2^2}{n_1^2}\right]^{1/2}\right\} = \sin^{-1}(2\Delta)^{1/2}$$

Thus

$$\sin\psi_c = (2\Delta)^{1/2} \simeq \psi_c \tag{4.61}$$

where in the last step we assumed that $\Delta \ll 1$. For a ray to have $\psi < \psi_c$ *inside* the fiber, the entrance angle $\psi_{ext}$ must satisfy $\sin\psi_{est} < n_1 \sin\psi_c$ as shown in Fig. 4.19. As an example we consider a fiber with $\Delta = 0.01$. Then

$$\psi_c \simeq \sqrt{0.02} = 0.14 \text{ rad} = 8^\circ \qquad \text{and} \qquad n_1\psi_c \simeq 12^\circ$$

The limiting entrance angle $\psi_a \simeq n_1\psi_c$ is referred to as the numerical aperture of the fiber; in the present example N.A. $= \psi_a = 0.21$.

Since the em wave will be reflected at the top *and* the bottom interface, a standing wave is established in the transverse direction, exactly as for a metallic wall waveguide (see Fig. 4.15). Thus the angle of incidence must satisfy the condition

$$2b\cos\theta = (N+1)\frac{\lambda_0}{n_1} \tag{4.62}$$

where $\lambda_0$ is the free-space wavelength and $N$ can be zero or an integer, $N = 0, 1, 2, 3, \ldots$ For a fiber of circular cross-section of radius $a$, Eq. (4.62) remains valid if we set $b \simeq 2a$. Thus we can write for the modes propagating in an optical fiber

$$4an_1k_0 \sin\psi = 2\pi(N+1) \qquad N = 0, 1, 2, \ldots \tag{4.63}$$

As an application we introduce into Eq. (4.63) typical parameters for an optical fiber, where we have chosen

$$\lambda_0 = 850 \text{ nm}$$
$$a = 40\ \mu\text{m}$$
$$n_1 = 1.5$$
$$\Delta = 0.01$$

Fig. 4.19. Definition of the limiting entrance and cut-off angles for propagation of light in an optical fiber.

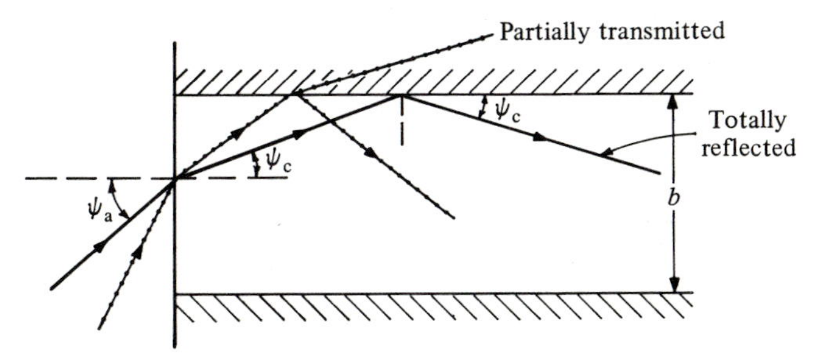

The propagation angle for the lowest mode ($N = 0$) is found to be

$$\psi_0 \sim \sin \psi_0 = \frac{\lambda_0}{4an_1} = \frac{850 \times 10^{-3}}{4 \times 40 \times 1.5} = 3.5 \times 10^{-3} \text{ rad}$$

or $\psi_0 \simeq 0.2°$. Thus the ray propagates in a direction very close to the fiber axis. Next we calculate the highest mode that can be supported by this fiber. For this we can insert Eq. (4.61) into (4.63) to obtain

$$N_{\max} + 1 = \frac{4an_1}{\lambda_0} (2\Delta)^{1/2} = 40$$

The result of the previous example shows that a signal injected into a fiber can propagate in several modes. This is a drawback because each mode has a different group velocity and thus the signal is dispersed. The group velocity is given by

$$v_g = v \cos \psi = \frac{c}{n_1} \cos \psi$$

and depends on the mode, through $\cos \psi$. For the $N$th mode we have

$$v_{gN} \simeq \frac{c}{n_1} [1 - \psi_N^2]^{1/2} \simeq \frac{c}{n_1} \left[1 - \frac{1}{2} \left\{\frac{\lambda_0 (N+1)}{4an_1}\right\}^2\right]$$

$$= \frac{c}{n_1} \left[1 - \Delta \left(\frac{N+1}{N_{\max} + 1}\right)^2\right] \tag{4.64}$$

The effects of dispersion can be compensated for by constructing fibers with a graded index, as shown in Fig. 4.20(*a*). Modes with small propagation angles $\psi_N$ are confined to the center of the fiber, whereas as $N$ (and thus $\psi_N$) increases, the rays can reach larger radii as shown in (*b*) of the figure. Clearly when $N$ (and thus $\psi_N$) is large, the path length is

Fig. 4.20. Graded index optical fiber: (*a*) radial profile of the refractive index, (*b*) propagation of rays entering at different angles (note the compensation of dispersion).

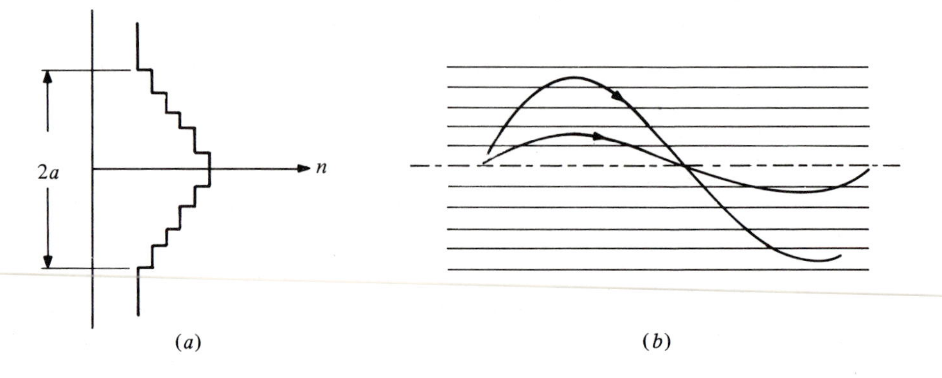

longer and if the refractive index was homogeneous, the group velocity of the mode would be smaller as indicated by Eq. (4.64). However, the index $n$ is a function of radius and large angle rays propagate in a region of smaller refractive index, that is where their velocity is higher. As a result the longer path length is compensated by the higher velocity and all modes propagate along the fiber axis with very similar velocity. The change in the refractive index is usually proportional to the square of the radius (parabolic)

$$n^2(r) = [n^2(r=0)](1 - gr^2)$$

In that case, as can be seen from Fig. 4.20($b$) the fiber has focussing properties and a short section of fiber can be used as a lens.

Attenuation in commercial fibers is of order of 1 db/km, namely a decrease in intensity of 25% in one km. In communication applications, fibers are driven by solid state lasers and the signal is detected by photodiodes. The bandwidth of an optical fiber channel is extremely large as compared to VHF or microwave carriers. If we assume that the optical carrier is modulated to 1% of its frequency, this would result in a bandwidth $W \sim 10^{13}$ Hz, that is in the terahertz range. In practice the limitation on the usable bandwidth does not come from the properties of the channel but from the availability of devices for modulating and demodulating the carrier. Such devices are based on various electrooptic effects and can operate at best in the GHz range.

## 4.9 The laser

The laser is a source of intense, very highly collimated and monochromatic, coherent em radiation. The radiation is usually in the visible but lasers operate effectively in the infrared and at other frequencies as well. The laser was first proposed by C. H. Townes and A. L. Schawlow in 1958 and its name is an acronym for '*L*ight *A*mplification by *S*timulated *E*mission of *R*adiation'; this is a precise description of the process involved in a laser. The radiation is emitted when atoms, or molecules or electrons in a solid, make a transition to a state of lower energy; because the atoms are *stimulated* to radiate, the resulting radiation has the same direction, frequency and phase as the radiation already present in the laser cavity. It is this property that makes the laser such a unique source of em radiation. Needless to say that practical applications of the laser appear today in all areas of technology, of medicine, as well as in every scientific field.

As we know, atoms have discrete energy levels and transitions can take place between these levels. In the absence of external excitations the atom is found in its lowest level, the ground state. A transition from a state of energy $E_2$ to a state of lower energy $E_1$ is accompanied by the emission of a photon of angular frequency $\omega$, where

$$\hbar\omega = E_2 - E_1 \tag{4.65}$$

Conversely, an atom in the state with energy $E_1$ can absorb a photon of frequency $\omega$ and make a transition to the state of energy $E_2$ provided the condition of Eq. (4.65), $E_2 - E_1 = \hbar\omega$ is satisfied. These two processes are indicated graphically in Fig. 4.21; *spontaneous emission* in (*a*), and *absorption* in (*b*).

If the atom is in the state $E_2$ and radiation of frequency $\omega$, where $\hbar\omega = E_2 - E_1$ is incident on it, the atom makes a transition to the state of energy $E_1$ and emits a photon of frequency $\omega$. Thus if initially only one photon of frequency $\omega$ was present, after the interaction of the em radiation with the atom, two photons of that frequency are present. This process is called *stimulated emission* and is indicated in (*c*) of the figure. Einstein was the first to show that stimulated emission must take place in order to assure the equilibrium between an assembly of atoms and the electromagnetic field. He also showed that for the same radiation intensity the probability of absorption equals the probability of stimulated emission.

We will designate the *probability per unit time* for each of the processes shown in Fig. 4.21 as follows:

| | |
|---|---|
| $A$ | probability/unit time for spontaneous emission |
| $W_{12} = B(\mathrm{d}u/\mathrm{d}\omega)$ | probability/unit time for absorption |
| $W_{21} = B(\mathrm{d}u/\mathrm{d}\omega)$ | probability/unit time for stimulated emission |

Here $\mathrm{d}u/\mathrm{d}\omega$ is the energy density of the em radiation, per unit frequency interval;* the energy density in an interval $\mathrm{d}\omega$, near the frequency $\omega_0$ is

Fig. 4.21. Electromagnetic transitions in a two-level system: (*a*) spontaneous emission, (*b*) absorption, (*c*) stimulated emission.

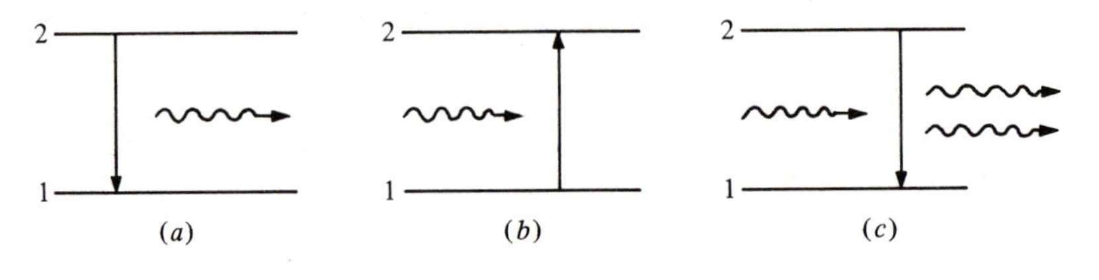


* Often $u(\omega)$ is used to designate the energy density per unit frequency; we are using the more cumbersome expression $\mathrm{d}u/\mathrm{d}\omega$ to remain consistent with the definition of energy density already given by Eq. (4.12).

given by

$$u(\omega_0, \mathrm{d}\omega) = \left.\frac{\mathrm{d}u}{\mathrm{d}\omega}\right|_{\omega_0} \mathrm{d}\omega \tag{4.66}$$

The coefficients $A$ and $B$ will be determined later. Note however that the above definitions imply $W_{21} = W_{12}$.

We now consider an assembly of atoms with a density of $N_t$ atoms/cm$^3$, where $N_1$ is the density of atoms in state 1 and $N_2$ the density in state 2, and $N_1 + N_2 = N_t$ (see Fig. 4.21). The energy difference of the two states is $E_2 - E_1 = \hbar\omega$ and radiation at frequency $\omega$ is incident on the assembly; the energy density of the radiation $(\mathrm{d}u/\mathrm{d}\omega)\,\mathrm{d}\omega$ corresponds to a photon density of $N_\gamma$ photons/cm$^3$. Because of the three processes indicated in Fig. 4.21 photons will be created as well as absorbed; the rate of change of the photon density will be

$$\frac{\mathrm{d}N_\gamma}{\mathrm{d}t} = AN_2 - W_{12}N_1 + W_{21}N_2 \tag{4.67}$$

In most applications $W_{21}N_2 \gg AN_2$ so that Eq. (4.67) simplifies to

$$\frac{\mathrm{d}N_\gamma}{\mathrm{d}t} = B\left(\frac{\mathrm{d}u}{\mathrm{d}\omega}\right)(N_2 - N_1) \tag{4.67$'$}$$

If $N_2 > N_1$ the number of photons increases and the assembly of atoms acts as an *amplifier* of the incident radiation. If $N_2 < N_1$ the number of photons decreases, and the radiation is absorbed by the atoms.

Under conditions of thermal equilibrium the population of levels of energy $E$ is governed by the Boltzmann distribution

$$N(E) = N_t \mathrm{e}^{-E/kT}$$

where $T$ is the temperature and $k$ is Boltzmann's constant. Thus

$$\frac{N_2}{N_1} = \frac{N_t \mathrm{e}^{-E_2/kT}}{N_t \mathrm{e}^{-E_1/kT}} = \mathrm{e}^{-(E_2 - E_1)/kT} \tag{4.68}$$

Since $E_2 > E_1$, we find that $(N_2/N_1) < 1$. An assembly of atoms in thermal equilibrium will absorb radiation. In order to amplify the incident radiation we must create a *population inversion* in the assembly of atoms. This can be achieved in several atomic, molecular or condensed matter systems which therefore can be made to lase.

To achieve population inversion, the atoms are *pumped* from their ground state to a state of higher energy from where the level 2 is populated. The simplest scheme is that of the three-level laser shown in Fig. 4.22($a$).* The action of the pump is to excite the state 3, which spontaneously

* It is evident that in a two-level system the pump would depopulate the upper level leading only to an equalization of the population of the two states but not to inversion.

decays by a fast transition to state 2; the population of $N_2$ can then exceed $N_1$ and a lasing transition can occur between levels 2 and 1. To sustain the lasing action, the pump must remove atoms from the ground state faster than the lasing transition populates that state. The four-level laser shown in (*b*) of the figure is better suited to continuous laser operation. State 1, the lower level of the laser transition decays fast to the ground state 4; thus $N_1$ is always much less than $N_2$. The upper state 2 is populated by a fast spontaneous transition from the pumped state 3; the upper state should have a fairly long lifetime so that spontaneous decay does not compete with the stimulated emission.

We will now consider these effects quantitatively. The probability rate for spontaneous emission is given by

$$A = \frac{4}{3}\frac{\omega^3}{c^2}\left(\frac{e^2}{4\pi\varepsilon_0}\frac{1}{\hbar c}\right)|\langle x\rangle_{12}|^2 \tag{4.69}$$

$A$ has units of inverse time and $\hbar$ is Planck's constant divided by $2\pi$

$$\hbar = \frac{h}{2\pi} = 1.05 \times 10^{-34}\ \text{J-s} \tag{4.70}$$

We also note the appearance of the dimensionless combination

$$\frac{e^2}{4\pi\varepsilon_0}\frac{1}{\hbar c} = \alpha \simeq \frac{1}{137} \tag{4.70$'$}$$

which is called the *fine structure constant*, $\alpha$. The quantity $\langle x\rangle_{21} = \langle x\rangle_{12}^*$ is the 'matrix element' of $x$ between the states 1 and 2; it is necessary to know the quantum mechanical wave functions of the two states 1, 2 in order to calculate $\langle x\rangle_{12}$ explicitly. However $\langle x\rangle$ has dimensions of length and is of the order of atomic dimension $\langle x\rangle \sim 10^{-8}$ cm. If we use this value and $\omega = 3 \times 10^{15}$ rad/s in Eq. (4.69) we find $A \simeq 10^7\ \text{s}^{-1}$, that is, atomic states should have lifetimes in the range of $10^{-7}$ s; this agrees with observation.

Fig. 4.22. The principle of laser operation: (*a*) three-level laser, (*b*) four-level laser.

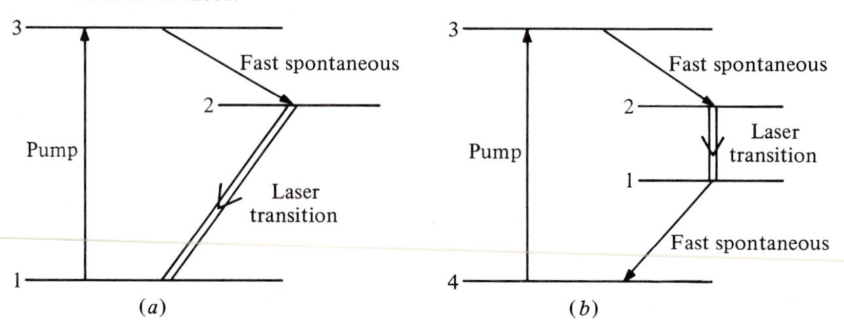

The absorption probability $W_{12}$ will depend on the intensity of the em radiation as indicated by the definition $W_{12} = B(\mathrm{d}u/\mathrm{d}\omega)$, but also on how close the frequency $\omega$ of the radiation is to $\omega_0$, the energy difference of the two states (divided by $\hbar$); $\omega_0 = (E_2 - E_1)/\hbar$. We will adopt the point of view that the incident radiation is monochromatic at frequency $\omega$, and that the atomic levels have an energy width $\Delta\omega_0$. Absorption is maximal when $\omega = \omega_0$ and falls off quickly when $|\omega - \omega_0| \geqslant \Delta\omega_0/2$; this can be described quantitatively by introducing a *line shape function* $g(\omega - \omega_0)$ as shown in Fig. 4.23. The line shape function is symmetric about $\omega_0$ and normalized to unity through

$$\int_{-\infty}^{\infty} g(\omega - \omega_0)\,\mathrm{d}\omega = 1 \tag{4.71}$$

Thus $g(\omega - \omega_0)$ has dimensions of inverse frequency.

We can now express $W_{12}$ in terms of the *intensity* of the incident radiation $I$, that is the energy crossing unit area in unit time. The intensity is related to the energy density $u$ of the radiation through

$$I = uc \tag{4.72}$$

and the corresponding flux of photons is

$$F = \frac{I}{\hbar\omega} = \frac{uc}{\hbar\omega} \tag{4.72$'$}$$

The absorption rate equals the stimulated emission rate and they are given by

$$W_{12} = W_{21} = \frac{4\pi^2}{3\hbar}\left(\frac{e^2}{4\pi\varepsilon_0}\frac{1}{\hbar c}\right)|\langle x \rangle_{12}|^2 g(\omega - \omega_0) I \tag{4.73}$$

This equations contains the same matrix element as for spontaneous emission (Eq. (4.69)); it also depends on the line-shape function of the atomic levels introduced in Eq. (4.71). It is convenient to write Eq. (4.73) in terms of the photon flux $F$ and an *absorption cross-section* $\sigma$, as

$$W_{12} = \sigma F \tag{4.73$'$}$$

Fig. 4.23. Typical line-shape function characterizing the transitions between energy levels.

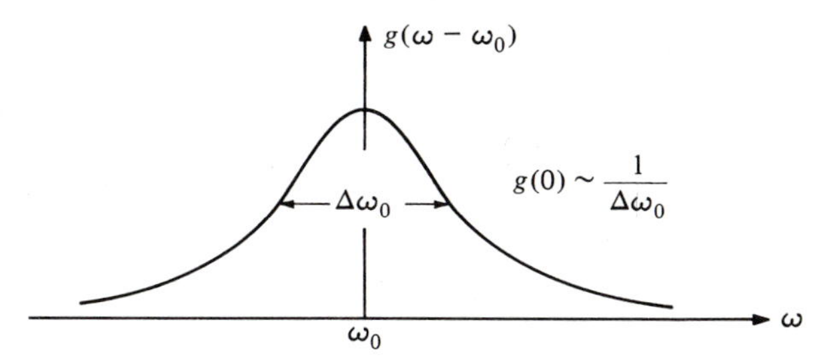

where $\sigma$ depends only on the atomic system and on the frequency $\omega$. From Eqs. (4.73) the cross-section is found to be given by

$$\sigma = \frac{4\pi^2}{3}\left(\frac{e^2}{4\pi\varepsilon_0}\frac{1}{\hbar c}\right)|\langle x\rangle_{12}|^2 \omega g(\omega - \omega_0) \tag{4.74}$$

The cross-section has the dimensions of area and if the density of atoms is $N$ (atoms/cm$^3$) the product $\mathrm{d}P = \sigma N\, \mathrm{d}z$ gives the probability that a photon will be absorbed in a path of length d$z$. For atomic systems $\omega g(\omega - \omega_0) \sim \omega/\Delta\omega_0 \sim 10^5$ and for excited states $\langle x\rangle \sim 10^{-9}$ to $10^{-10}$ cm; thus $\sigma$ ranges from $10^{-14}$ to $10^{-16}$ cm$^2$.

The rate at which the photon flux changes when a beam of photons traverses a path length d$z$ in the atomic medium can be obtained from Eqs. (4.67). As before we ignore spontaneous transitions so that

$$\mathrm{d}F = W_{12}(N_2 - N_1)\,\mathrm{d}z = \sigma F(N_2 - N_1)\,\mathrm{d}z$$

or

$$\frac{\mathrm{d}F}{F} = \sigma(N_2 - N_1)\,\mathrm{d}z \tag{4.75}$$

For a finite length $z$,

$$\Delta F = F(z) - F(0) = [\mathrm{e}^{\sigma(N_2 - N_1)z} - 1]F(0) \tag{4.75$'$}$$

Thus, as also found in Eq. (4.67′), if $N_2 > N_1$ the photon flux will increase and the medium will act as an amplifier. If we are able to feed back through the medium a fraction of the amplified radiation, the amplifier will behave as an oscillator provided the *gain* exceeds the losses in the system.

To achieve feedback the radiation is trapped between two mirrors on either side of the medium. On each reflection a small fraction, $\beta$, of the incident beam is absorbed, and another small fraction $T$, is transmitted through the mirror; furthermore for each pass through the cavity, a small fraction $\eta$ of the beam is lost due to a variety of causes. If the fractional decrease in the flux, for one pass, is $\delta F/F$, then

$$1 - \frac{\delta F}{F} = (1 - \beta)(1 - T)(1 - \eta) \equiv \mathrm{e}^{-\gamma} \tag{4.76}$$

In general $\beta$, $T$ and $\eta$ are much less than one and so is $\gamma$. Introducing the losses in Eq. (4.75′) we obtain

$$\Delta F = \{\mathrm{e}^{[\sigma(N_2 - N_1)l - \gamma]} - 1\}F \tag{4.77}$$

where $l$ is the length of the active medium. If the exponent is positive, that is if

$$(N_2 - N_1) \geqslant \frac{\gamma}{l\sigma} \tag{4.78}$$

the system will lase. The value of $(\gamma/l\sigma)$ defines the *threshold* value of the population inversion density.

We can write Eq. (4.77) in more familiar form by noting that $\mathrm{d}F/F = \mathrm{d}I/I$ and that the time for one traversal is $L/c$, where $L$ is the length of the cavity, thus

$$\frac{\mathrm{d}I}{\mathrm{d}t} = \frac{cI}{L}\{\mathrm{e}^{[\sigma(N_2 - N_1)l - \gamma]} - 1\} \tag{4.79}$$

Near threshold the exponent is close to zero and Eq. (4.79) is approximated by

$$\frac{\mathrm{d}I}{\mathrm{d}t} \simeq \frac{cI}{L}[\sigma(N_2 - N_1)l - \gamma] \tag{4.79'}$$

As an example, we consider an optical system where the total losses in one pass are $\gamma = 0.03$, the absorption cross-section is $\sigma = 10^{-16}\ \mathrm{cm}^2$ and the active medium length $l = 30$ cm. Then the threshold inversion density is

$$(N_2 - N_1)_{\mathrm{thr}} = \frac{\gamma}{\sigma l} = \frac{0.03}{30 \times 10^{-16}} = 10^{13}\ \mathrm{atoms/cm}^3$$

This is a very small fraction of all the atoms contained in $1\ \mathrm{cm}^3$ and this is why it is possible readily to observe lasing in so many different atomic systems. We recall that even in a gas, the atomic density is in excess of $10^{19}/\mathrm{cm}^3$. The factor $\sigma(N_2 - N_1)$ is referred to as the gain per unit path. Typically gas lasers operate with gains of 0.2 db/m whereas molecular systems can have gains as high as 40 db/m.

In Eqs. (4.77) or (4.79) we have an expression for the rate of change of the photon flux. For a complete description of the system we must also have expressions for the rate of change of the population densities $N_2$ and $N_1$. For simplicity we will consider a four-level lasing system so that we can set $N_1 = 0$. Then

$$\left.\begin{aligned} \frac{\mathrm{d}N_2}{\mathrm{d}t} &= W_\mathrm{p}N_\mathrm{g} - \sigma(cN_\gamma)N_2 - AN_2 \\ \frac{\mathrm{d}N_\gamma}{\mathrm{d}t} &= \sigma(cN_\gamma)[N_2 - N_1] - \frac{\gamma c}{l}N_\gamma \end{aligned}\right\} \tag{4.80}$$

Here $W_\mathrm{p}$ is the probability rate of the pump, and $N_\mathrm{g} = N_\mathrm{t} - N_2$ is the density of atoms in the ground state. We have replaced the flux $F$ by $F = cN_\gamma$, where $N_\gamma$ is the photon density in the cavity. The two coupled equations must be solved for $N_2$ and $N_\gamma$ given appropriate initial conditions; we can usually set $N_1 = 0$ in the second equation.

Under equilibrium conditions, $\mathrm{d}N_2/\mathrm{d}t = \mathrm{d}N_\gamma/\mathrm{d}t = 0$ and ignoring the spontaneous term and also setting $N_1 = 0$ the solution of Eqs. (4.80) is simply

$$N_2 = \frac{\gamma}{l\sigma} \qquad N_\gamma = \frac{W_p N_g}{c\sigma N_2}$$

or

$$N_\gamma = \frac{W_p N_g}{\gamma} \frac{l}{c} \tag{4.81}$$

The power extracted from the laser is given by the photon flux in the cavity $F = cn_\gamma$, multiplied by the transmission loss $T$, the cross-sectional area of the beam $A$ and the photon energy $\hbar\omega$. Thus

$$P_{\text{trans}} = W_p N_g (lA) \left(\frac{T}{\gamma}\right) (\hbar\omega) \tag{4.82}$$

As a simple example consider $\hbar\omega = 2$ eV, $N_g = 10^{19}/\text{cm}^3$, $A = 0.03$ cm$^2$, $l = 30$ cm and $(T/\gamma) = \frac{1}{3}$. If the pump rate is $W_p = 1/\text{s}$ then the transmitted power is 1 Watt, which corresponds to a laser of significant power.

## 4.10 Properties of laser radiation

We have seen that in the laser the radiation must be trapped in a cavity so as to interact effectively with the active medium. In the visible, the cavity is formed by two mirrors which also serve to focus the radiation as shown schematically in Fig. 4.24. The reflectivity of the mirrors is $R = 1 - \beta$ where $\beta$ is the *reflection loss*. The number of reflections $N_{1/e}$ that can be achieved before the light intensity is decreased to 1/e of its initial value is obtained from

$$R^N = \frac{1}{\text{e}} \qquad \text{or} \qquad N(\ln R) = -1$$

But

$$\ln R = \ln(1 - \beta) \simeq -\beta \qquad \text{for } \beta \ll 1$$

Thus

$$N_{1/e} = \frac{1}{\beta} = \frac{1}{1 - R} \tag{4.83}$$

If the cavity length is $d$, the mean path-length of a photon in the cavity is $L_\gamma = \text{d}N_{1/e} = \text{d}/(1 - R)$. Typical reflectivities for laser mirrors are in excess of $R \geqslant 0.99$.

The optical cavity, as any cavity, can be characterized by a quality factor, $Q$, where

$$Q = 2\pi \frac{\text{energy stored in system}}{\text{energy lost/cycle}} = \frac{2\pi}{T} \frac{U}{\mathrm{d}U/\mathrm{d}t} = \omega \frac{U}{\mathrm{d}U/\mathrm{d}t} = \frac{\omega}{\Delta\omega} \tag{4.84}$$

The resonant frequency of the cavity is $\omega$ and $\Delta\omega$ is the width of the resonance. Fig. 4.25 shows two examples of systems with high and low $Q$ respectively. By definition $U/(\mathrm{d}U/\mathrm{d}t)$ is the 'lifetime', $\tau$, of the energy stored in the cavity; thus it follows from Eq. (4.84) that

$$Q = \omega\tau \tag{4.84'}$$

To calculate the $Q$ of the optical cavity we note that the lifetime of a photon in the cavity is given by

$$\tau = \frac{L_\gamma}{c} = \frac{d}{c(1-R)}$$

and therefore

$$Q = \frac{2\pi\nu d}{c(1-R)} = 2\pi \frac{d}{\lambda} \frac{1}{(1-R)} \tag{4.85}$$

We see that the $Q$ (which is dimensionless) can also be interpreted as giving the number of oscillations that the em field undergoes before the stored energy decays to 1/e of its initial value.

In the cavity the radiation travels in both directions and consequently a pattern of standing waves will be established, with *nodes* at the mirror surfaces. Thus not all wavelengths can be supported in the cavity, but

Fig. 4.24. Confocal cavity resonator.

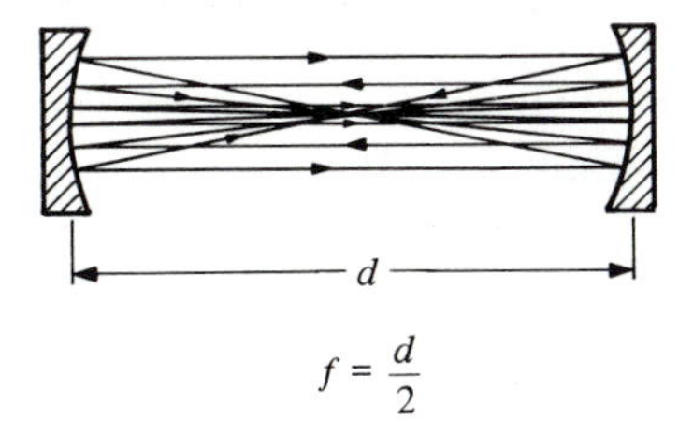


Fig. 4.25. Definition of the quality factor of an optical cavity in terms of the resulting line-width.

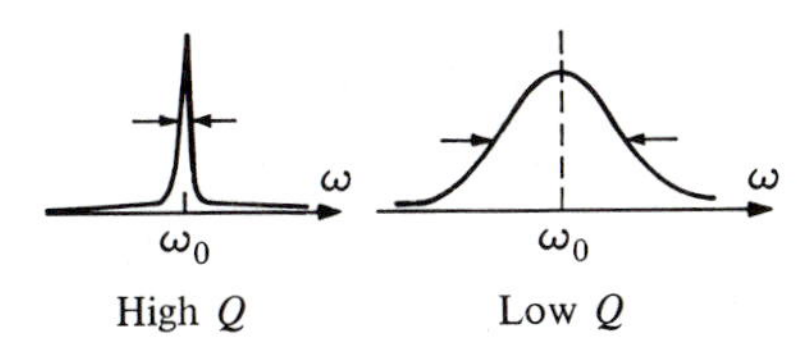

only those for which an integral number of wavelengths fits in the spacing between mirrors which is given by $d$ (see Fig. 4.26). Hence

$$d = q\frac{\lambda}{2} \qquad q = 1, 2, 3, \ldots \tag{4.86}$$

Typically, $d$ is of order of a meter whereas $\lambda \sim 5 \times 10^{-7}$ m. Thus $d/\lambda \sim 10^6$ and therefore $q$ is a very large number. Different values of $q$ correspond to different *modes* of oscillation. The spacing between modes $\Delta\nu_{\text{min}}$ is given by

$$\Delta\nu_{\text{min}} = \frac{c}{\lambda_{q+1}} - \frac{c}{\lambda_q} = \frac{c}{2d}[q + 1 - q] = \frac{c}{2d} \tag{4.87}$$

For $d = 1$ m, we find $\Delta\nu_{\text{min}} = 1.5 \times 10^8$ Hz.

The line width for the individual modes is determined by the $Q$ of the cavity. Using Eqs. (4.84) and (4.85) we write

$$\frac{\nu}{\Delta\nu} = Q = 2\pi\frac{d}{\lambda}\frac{1}{(1-R)}$$

or

$$\Delta\nu_s = \frac{c}{2\pi d}(1 - R)$$

For $d = 1$ m and $R = 0.99$ one obtains $\Delta\nu_s = 5 \times 10^5$ Hz which is indeed a narrow line; recall that in the visible $\nu \simeq 5 \times 10^{14}$ Hz. Finally we ask how many modes will be contained in the laser beam. This depends on the line shape function for the transition (see Fig. 4.23). For atomic systems the line width is often given in terms of *wave numbers*; in these units the line width is typically $\Delta\tilde{\nu} = 1\ \text{cm}^{-1}$. The line width in hertz is obtained by multiplying the wave number by the speed of light. Thus

$$\Delta\nu_g = c\Delta\tilde{\nu} = (3 \times 10^{10}\ \text{cm/s}) \times (1\ \text{cm}^{-1}) = 3 \times 10^{10}\ \text{Hz}$$

which is due in large part to Doppler broadening. We can now reconstruct the spectrum of the radiation emitted from a laser, which will be of the general form of Fig. 4.27. However, the system may lase simultaneously at more than one optical line.

In any laser mode the electric and magnetic fields are very nearly

Fig. 4.26. Standing waves in an optical cavity.

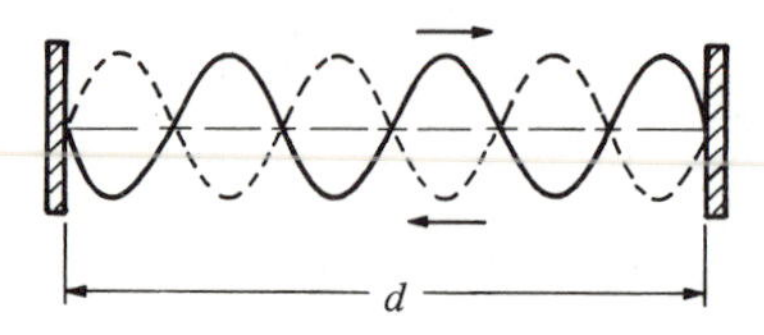

perpendicular to the cavity axis; thus we speak of TEM modes. In addition to the condition given by Eq. (4.86) a mode is also specified by the distribution of the em field in the transverse plane. This situation is analogous to the field distribution in waveguides or optical fibers. The distribution in the plane transverse to the direction of propagation is sketched in Fig. 4.28 for the lowest and next to lowest mode.

Lasers can also be used to produce very short pulses of intense radiation. If all the radiation stored in the cavity could be extracted in one pass, a short pulse of high peak power would result. This is achieved by suddenly changing the $Q$ of the cavity, and one speaks of a *Q-switched* laser. The pulse duration is of order

$$\Delta t = \frac{d}{c} \simeq \frac{1\,\mathrm{m}}{3 \times 10^{8}} \sim 3 \times 10^{-9}\ \mathrm{s}$$

Much narrower pulses can be obtained by taking advantage of the many modes contained in the spectrum. If these modes can be made to oscillate with the same phase, then they are equivalent to the discrete frequency amplitudes of a Fourier series. Correspondingly, in the time domain, the

Fig. 4.27. Typical frequency spectrum of the radiation emitted by a laser.

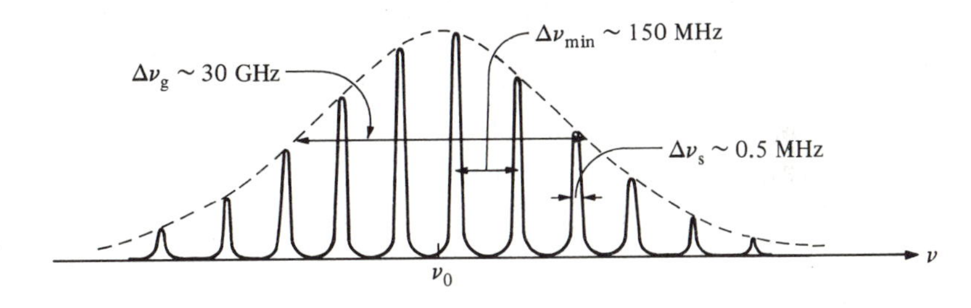


Fig. 4.28. Spatial distribution of the two lowest modes of laser radiation.

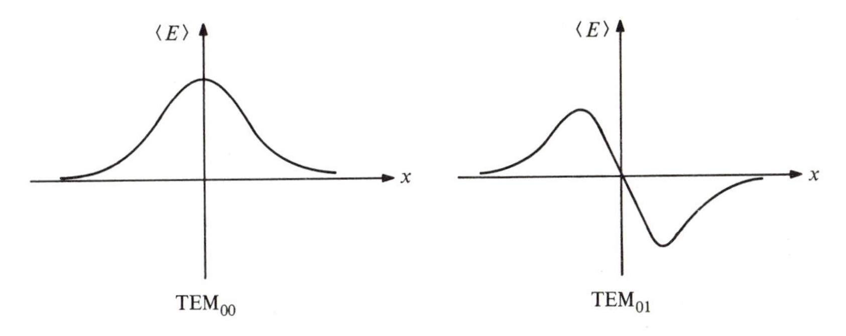

pulse will have a width typical of

$$\Delta t \sim \frac{1}{\Delta \nu_g} \sim \frac{1}{30\ \text{GHz}} \sim 30 \times 10^{-12}\ \text{s}$$

Pulses as narrow as $10 \times 10^{-15}$ s can be obtained by special techniques of 'compression' in the time domain.

The minimum width of the laser beam usually occurs in the middle of the cavity and is given by

$$w_s = \left(\frac{L\lambda}{\pi}\right)^{1/2} \tag{4.88}$$

where $L$ is the focal length of the cavity mirror. $w_s$ is the 'waist' of the beam and one can recognize the similarities of Eq. (4.88) to the diffraction limit of Eq. (4.20′). For typical values of $\lambda = 600$ nm and $L = 0.5$ m one finds

$$w_s \simeq 3 \times 10^{-4}\ \text{m} = 0.3\ \text{mm}$$

One can take advantage of the small spot size to make the transmitted beam highly parallel. Using a lens of the same focal length as the cavity mirrors the resulting beam divergence is

$$\theta_d = \frac{w_s}{L} = \left(\frac{\lambda}{L\pi}\right)^{1/2}$$

For the parameters defined above, $\theta_d \simeq 6 \times 10^{-4}$.

Wave phenomena are characterized by their degree of *coherence*, which measures how well the phase relationship is maintained at different space–time points. Consider for instance two points $P_1$ and $P_2$ and let the relative phase of the wave amplitude at these points be $\phi_{12}$ at some time $t = 0$. If the relative phase, at $P_1$ and $P_2$, remains equal to $\phi_{12}$ at all times the wave has perfect *spatial coherence*. Usually spatial coherence deteriorates as the separation between $P_1$ and $P_2$ increases. We define the *coherence length* $l_c$ as the distance over which the phase relationship is maintained.

We can also define the *temporal coherence* of the wave by considering the phase difference between times $t_1$ and $(t_1 + \tau)$ at a fixed point $P$. If this phase difference $\phi_\tau$ is maintained for all later times $t_2$ and $(t_2 + \tau)$, the wave has perfect temporal coherence. The *coherence time* $\tau_c$ measures how long the wave is emitted without 'glitches' or other changes in its phase. If the frequency of the source has a width $\Delta\nu$, then clearly, the coherence time cannot exceed $\tau_0 \simeq 1/\Delta\nu$. In this case the coherence length is also limited to $l_c \simeq c/(2\pi\Delta\nu)$.

The *idealized* 'plane wave' that we use in calculating wave phenomena has by definition perfect spatial and temporal coherence. Ordinary sources of visible light are very far from meeting these conditions and one needs to select a very small collimated beam to have a spatially coherent wave. In contrast, laser beams exhibit a degree of coherence several orders of magnitude higher than that of ordinary sources. Obviously, a non-coherent beam cannot exhibit interference phenomena.

There are two final properties of laser radiation that we mention: brightness and polarization. *Brightness* is defined as the power radiated into unit solid angle by unit surface of the source. Because of their high directionality and small beam size, even low power lasers have a brightness that exceeds that of other sources in the visible by orders of magnitude. Laser beams are usually linearly polarized with good accuracy. This is accomplished by orienting the windows on the discharge tube at the 'Brewster angle' with respect to the cavity axis. The polarization of the transmitted beam is then in the plane defined by the cavity and the normal to the window.

## Exercises

### *Exercise 4.1*

(a) Show that a parabolic reflector generates a parallel beam of light when a *point* source is placed at its focus.
(b) Find the position of the focus.
(c) Show *qualitatively* that when em radiation of wavelength $\lambda$ is focussed in terms of a reflector of aperture $\pi R^2$ the angular aperture of the antenna is

$$\Delta\theta \simeq \lambda/R$$

(d) Find the dimensions of the Arecibo radiotelescope and calculate its 'directionality' $\Delta\theta$ for frequencies near $f = 1\,\text{GHz}$.

### *Exercise 4.2*

Consider a fiber optics channel operating at a wavelength $\lambda = 600\,\text{nm}$. The channel will support 1% changes in wavelength.
(a) Find the bandwidth of the channel.
(b) Given a $S/N$ ratio of 10 find the capacity of the channel.
(c) Assuming that telephone communication needs a sampling rate $W = 10\,\text{kHz}$, how many simultaneous telephone conversations can the channel support?

### *Exercise 4.3*

An optical fiber of radius $a = 10\,\mu\text{m}$ operates in the visible ($\lambda = 600\,\text{nm}$).

The refractive index is $n = 1.5$ and the difference in the refractive index between the core and the cladding is

$$\Delta = \frac{n_1^2 - n_2^2}{2n_1^2} = 0.02$$

(a) Find the limiting entrance angle.
(b) Find the maximum mode number that the fiber will support.

***Exercise 4.4***

In a laser cavity the spot size in the $TEM_{00}$ mode is characterized by a radius

$$w_s = (\lambda L/\pi)^{1/2}$$

where $L$ is the distance between the mirrors.

(a) Give a *qualitative* derivation of this result based on the uncertainty relation for wave phenomena.
(b) Consider a $CO_2$ laser (where $\lambda = 10.6\ \mu m$) of length $L = 2$ m and calculate the spot size.
(c) For the above laser calculate the frequency difference between two adjacent $TEM_{00}$ modes. Given that the FWHM of the $CO_2$ laser line is 50 MHz, find how many $TEM_{00}$ modes fall within this width.

# PART C

# NUCLEAR ENERGY

The first sustained nuclear chain reaction was achieved by Enrico Fermi and his collaborators at the University of Chicago on December 2, 1942. Since then, nuclear energy has been one of the dominant factors in our society. Fission reactors are widely used to supplement the generation of electric power and, when it is realized, controlled thermonuclear fusion promises to be an inexhaustible source of energy for mankind. Yet the most decisive and terrifying aspect of nuclear energy so far has been in the production of weapons of unprecedented destructive power. These weapons if used in the large quantities presently available can alter the ecology of the planet and completely destroy, or at the least, radically change human and animal life from what we know it to be today.

In Chapter 5 we begin by discussing the units of energy, the various levels of energy consumption and supply, and the global balance of energy on the earth. The earth receives its energy from the sun, which generates energy by nuclear fusion. Next we review the facts associated with nuclear forces in order to discuss the release of energy in fission and fusion processes. We also discuss radioactivity, its detection and its effect on living organisms. One section is devoted to nuclear reactors and another section to the principles of controlled nuclear fusion. For completeness we also consider solar energy, which even though still economically impractical is an inexhaustible source of clean energy.

Chapter 6 addresses the use of nuclear energy in weapons and the effects produced by high yield nuclear explosions. We discuss the existing weapons arsenals and the efforts to control the spread and proliferation of nuclear weapons by treaty. This brings up the consideration of delivery vehicles and of intelligence gathering by reconnaissance satellites. Finally we examine proposed defensive systems, such as the 'Strategic Defense Initiative', and the limitations imposed on them by the laws of physics.

# 5

# SOURCES OF ENERGY

## 5.1 Introduction

When a force acts on a body and displaces it, the force does work and increases the energy of the body. For instance, a rock released in the earth's gravitational field gains kinetic energy as it falls, but it looses potential energy. While work and energy are equivalent definitions of the same physical quantity we think of energy as the 'capacity to do work'. Energy can manifest itself in various forms, but overall energy is always *conserved* in any physical process. Examples of different forms of energy are chemical energy, the rest-mass energy $E = mc^2$ of a body of mass $m$, or the random thermal energy of a material body.

Energy is measured, in the MKS system, in Joules. *Power* is energy produced (or consumed) per unit time and its unit is the Watt: 1 Watt = 1 Joule/s. *Energy flux* is the energy crossing unit area in unit time and is measured in Watts/$m^2$. Other commonly used units of energy are the calorie and the btu (British thermal unit). The relevant definitions and conversion factors are summarized below.

1 Joule = 1 N-m
1 calorie = 4.186 J
1 btu = 1055 J
1 Watt = 1 J/s

Even though energy is conserved, it is the conversion of energy from one form to another that is the basis of life, and also of modern technology. For instance, in an automobile engine chemical energy is first transformed to thermal energy; the thermal energy is converted to the mechanical energy that drives the automobile; finally the mechanical energy is again transformed to heat energy through friction, air resistance, etc. Through these processes the automobile and its passengers have reached their destination, which was the desired goal.

A crucial fact in the conversion of energy from one form to another is that some processes are *reversible*, while others are *irreversible*. At the microscopic scale of atomic and nuclear phenomena, all processes are presumed to be reversible. However even for systems of modest complexity, and in particular at the level of human experience, most processes are *irreversible*. This fact is contained in the second law of thermodynamics which states that in any process the total *entropy* of the system will remain the same or will increase. In reversible processes the total entropy of the system is unchanged; in irreversible processes the total entropy increases. When fuel is burned in an automobile engine the process is irreversible. The price we pay for the transportation is the consumption of the fuel, that is of an ordered form of energy which is converted to disordered energy, with a resultant increase in entropy. Thus, when we speak of the energy needs of the world we imply the need for *ordered* forms of energy.

Thermal energy can be converted to mechanical energy but only if two heat reservoirs are available. Let the reservoirs be at temperatures $T_1$ and $T_2$ with $T_2 > T_1$. To convert some of the thermal energy of the reservoir at $T_2$ to mechanical energy, we must extract from it energy $Q_2$ (in the form of heat) and deliver energy $Q_1$ to the reservoir at $T_1$; the difference between $Q_2$ and $Q_1$ has been converted to mechanical energy (or work) $W$

$$W = Q_2 - Q_1 \tag{5.1}$$

The efficiency, $\eta$, for this conversion process is defined by $\eta = W/Q_2 = (Q_2 - Q_1)/Q_2$. The highest possible efficiency is achieved when the thermal engine is reversible, in which case it can be easily proven that

$$\frac{Q_2 - Q_1}{Q_2} = \frac{T_2 - T_1}{T_2} \tag{5.2}$$

Eq. (5.2) shows that when $T_2 \gg T_1$ the conversion efficiency will be high. As an example, the temperature of superheated steam (at 3000 psi) is $T_2 = 700°\text{F}$ ($T_2 = 673$ K), and as in most practical applications, $T_1$ is fixed at the ambient temperature, $T_1 \simeq 300$ K. Thus the efficiency of a typical steam power plant cannot exceed

$$\eta = 1 - (300/673) = 0.55$$

In practice the efficiency is in the range of 40%, close to the upper limit. Eqs. (5.1) and (5.2) show that if we have sources of thermal energy at temperature much above the ambient level, they can be used to satisfy our needs for ordered energy: for instance, by producing electrical power. Hereafter we will simply refer to 'energy consumption' without reference to the fact that we really mean ordered energy.

Energy is used by living organisms, the intake being in the form of food, and allows them to accomplish biological functions as well as mechanical work. The main uses of energy, however, in our civilization are for heating, for industrial production, and for transportation. The energy needed or released in some typical processes is listed in Table 5.1, and is also shown in Fig. 5.1 on a logarithmic scale. Note that throughout this and the following chapter we always refer to metric tons, 1 ton $= 10^3$ kg; we use the abbreviations kt and Mt for $10^6$ and $10^9$ kg respectively.

For early man the main use of energy was for heating, and the source was the burning of wood. At present the principal sources of energy are fossil fuels: coal, oil and natural gas. Nuclear energy is an important component of electric power generation in the developed countries; it provides about 13% of the total electrical power in the U.S. Mechanical sources of energy such as hydroelectric power and wind power have been known and exploited for a long time but supply only a small fraction (less than 5%) of the total energy consumption. In recent years the use of geothermal, solar and even tidal energy has been researched and pilot plants have been constructed; nevertheless the exploitation of these sources is at present not economically advantageous and they are used only in very special cases. Fossil fuels, but also uranium and other fissile materials, are all depletable resources and the existing reserves will be exhausted in

Table 5.1. *Energy use and energy release in some typical processes*

| Process | Energy or energy per day |
|---|---|
| Energy flow from sun | $3.2 \times 10^{31}$ J/day |
| Energy flow to earth | $1.5 \times 10^{22}$ J/day |
| Total world human energy use | $9.0 \times 10^{17}$ J/day |
| Total U.S. human energy use | $2.0 \times 10^{17}$ J/day |
| Electrical output, large power plant (1 GW) | $9.0 \times 10^{13}$ J/day |
| Fission energy of 1 kilogram Uranium-235 | $8.0 \times 10^{13}$ J |
| Combustion energy of 1 barrel of oil | $6.0 \times 10^{9}$ J |
| U.S. energy use per capita | $9.0 \times 10^{8}$ J/day |
| World energy use per capita | $2.0 \times 10^{8}$ J/day |
| Food energy use per capita (2000 kcal/day) | $9.0 \times 10^{6}$ J/day |
| *Relative use of energy* (*U.S. 1970*) | |
| Heating | 0.22 |
| Refrigeration | 0.05 |
| Industrial | 0.48 |
| Transportation | 0.25 |

a short time (100–200 years at the present rate of consumption). Thus the realization of an *alternate energy source* such as controlled fusion appears to be an imperative for the survival of our civilization in its present form.

The development of our industrial society has been made possible by the availability of energy, and this in turn has created a greater demand for energy. In fact, the energy consumption globally has been growing exponentially

$$P(t) = P(0)e^{\alpha t} \tag{5.3}$$

so that the time in which $P(t)$ doubles is given by $\tau_D = (\ln_e 2)/\alpha$. Until 1977 the doubling time was $\tau_D = 29$ years but the rate of growth seems to be slowing down so that at present $\tau_D \simeq 50$ years. This is still too rapid a rate of growth in energy demand for the limited resources of our planet especially as the developing nations become industrialized demanding a greater share of the global energy use. Thus conservation of resources is a necessary component of any long term energy policy.

Fig. 5.1. Comparative orders of magnitude of possible energy production and of energy consumption world wide.

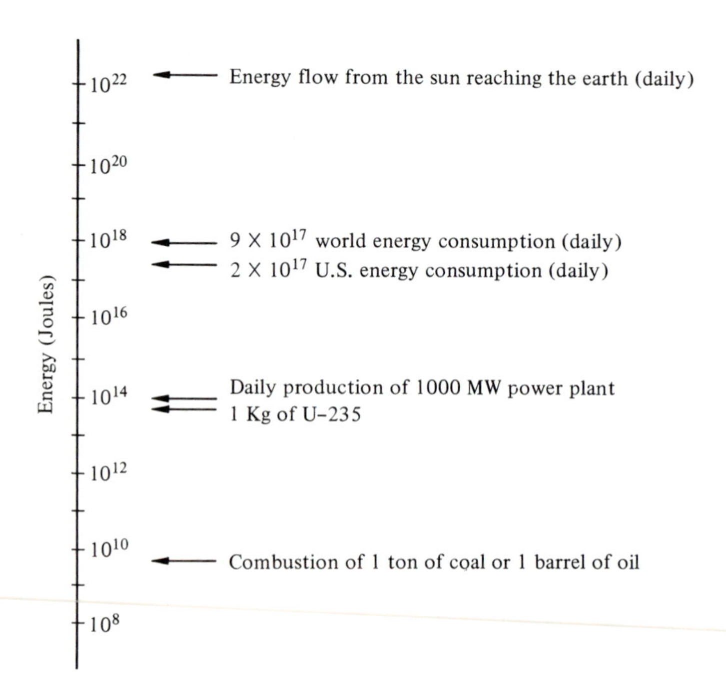

## 5.2 The terrestrial energy balance

Life on earth is crucially dependent on the average ambient temperature; a planet's temperature is a function of its distance from the sun and is determined by balancing the energy received from the sun with the total energy radiated by the planet. Clearly, no convection or conduction of heat is possible in the interplanetary space and thus energy transfer to and from the earth proceeds by radiation of em energy, as shown schematically in Fig. 5.2. The energy received from the sun is near the visible part of the spectrum while the radiated energy is in the infrared.

To discuss the balance in radiant energy we recall (see Chapter 4) that excited atoms emit radiation. If we have an assembly (i.e. a large number) of atoms that are maintained in a cavity of temperature $T$, and remain in equilibrium with the radiation, the spectrum of the radiation in the cavity is given by Planck's equation. The energy density per unit angular frequency ($\omega = 2\pi\nu$) interval is

$$\frac{\mathrm{d}u}{\mathrm{d}\omega} = \frac{\hbar\omega^3}{\pi^2 c^3} \frac{1}{\mathrm{e}^{\hbar\omega/kT} - 1} \tag{5.4}$$

The Planck distribution is plotted as a function of frequency in Fig. 5.3 and has a peak which shifts to higher frequencies as the temperature increases; the maximum in Eq. (5.4) occurs when $(h\nu/kT) = 2.831$.

The energy radiated into unit solid angle by unit normal area of the cavity in unit time can be shown to be

$$\frac{\mathrm{d}I(\omega)}{\mathrm{d}\Omega} = \frac{c}{4\pi} \frac{\mathrm{d}u}{\mathrm{d}\omega} \tag{5.4'}$$

Then the energy radiated over all angles is found from (5.4′) by integrating

Fig. 5.2. Illumination of the earth by the sun's radiant energy.

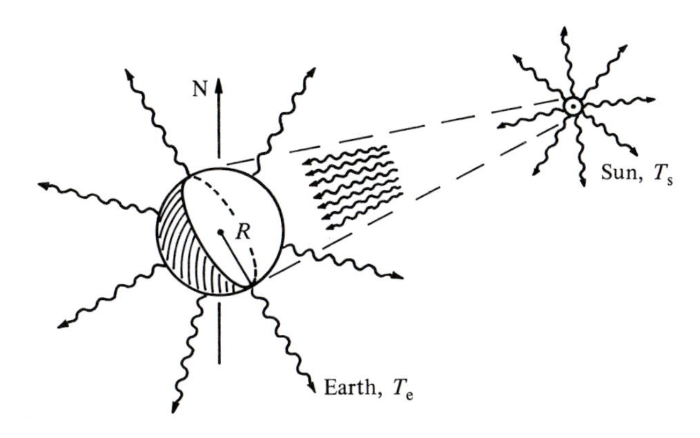

it over $\cos\theta\, d\Omega$ from $\theta = 0$ to $\theta = \pi/2$ so that

$$I(\omega) = \frac{c}{4\pi}\int_0^{\pi/2} \frac{du}{d\omega} \cos\theta \, d\cos\theta \int_0^{2\pi} d\phi = \frac{c}{4}\frac{du}{d\omega} \tag{5.4''}$$

Finally integration of Eq. (5.4″) over all frequencies yields the total radiated energy per unit surface of the emitter, per unit time

$$I_T = \int_0^{\infty} \frac{c}{4}\frac{du}{d\omega}\, d\omega = \left(\frac{\pi^2 k^4}{60\hbar^3 c^2}\right) T^4 \tag{5.5}$$

which is seen to depend on the *fourth power* of the temperature. The $T^4$ dependence of the total radiated power was known before Planck's discovery of his equation and is called the *Stefan–Boltzmann* law. Using $S_r$ (instead of $I_T$) for the energy radiated per unit area and unit time we write

$$S_r = \varepsilon \sigma T^4 \tag{5.6}$$

Here $\sigma$ is Stefan's constant as given by the expression in parentheses in Eq. (5.5); it has the numerical value

$$\sigma = 5.67 \times 10^{-8} \text{ Joule/m}^2\text{-s-(K)}^4$$

The coefficient $\varepsilon$, where $0 < \varepsilon < 1$, characterizes the ability of a physical system to radiate; it is called the *emissivity*. When $\varepsilon = 1$ the system radiates according to Planck's law and it is called a *black body*; this is the maximum possible rate of radiation. At the other limit, $\varepsilon = 0$ characterizes a system that does not radiate at all.

A physical system may also absorb em radiation, or it may reflect it. If energy flux $S_i$ is incident on the system, the absorbed energy flux is $S_a$ where

$$S_a = \frac{\text{energy absorbed}}{\text{unit normal area} - \text{unit time}} = aS_i$$

Fig. 5.3. The Planck distribution plotted as a function of frequency.

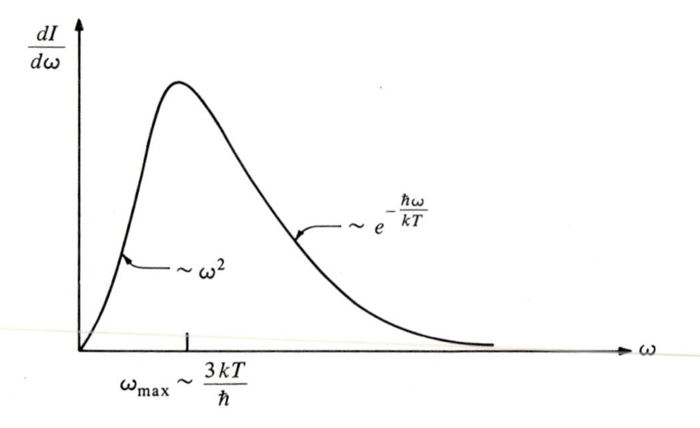

and $a$ is called the *absorbtivity* of the system. In general the emissivity and the absorbtivity are functions of the frequency and the coefficients should be written as $\varepsilon(\nu)$ and $a(\nu)$. However it is a very important and general result, known as Kirchoff's law, that

$$\varepsilon(\nu) = a(\nu) \tag{5.7}$$

This follows from conditions of equilibrium between the system and the surrounding em radiation. For instance, painting a surface white decreases the absorption of radiant heat, but also decreases its ability to radiate at that frequency.

From the Planck distribution we can understand the spectrum of radiation emitted by the sun and by the earth. The sun's *surface* is at a temperature $T_s \sim 5800$ K and therefore the peak of the Planck distribution is at $h\nu \sim 1.4$ eV, or $\lambda \simeq 900$ nm. This corresponds to an energy slightly below the visible window but a large fraction of the sun's spectrum falls into that window. Conversely the temperature of the earth's surface is $T = 300$ K and thus the spectrum of its radiated energy peaks at $h\nu = 0.07$ eV or $\lambda \sim 17$ $\mu$m, which is well into the infrared.* The energy flux reaching the earth from the sun is

$$S = 1.36 \times 10^3 \text{ Watts/m}^2 \tag{5.8}$$

and this value is known as the *solar constant*. We can use it to calculate the total energy radiated by the sun, what we call its luminosity $\mathscr{L}_\odot$; since the earth–sun distance is $R_S = 1.5 \times 10^{11}$ m, we find $\mathscr{L}_\odot = 4\pi R_S^2 S = 3.8 \times 10^{26}$ Watts.

The balance of energy for the earth is expressed by writing

$$\left.\frac{dE}{dt}\right|_{\text{absorbed}} = \left.\frac{dE}{dt}\right|_{\text{radiated}}$$

Since at any time, only half of the earth faces the sun, the normal area for absorption is $A_N = \pi R^2$ where $R = R_\oplus$ is the radius of the earth (see Fig. 5.2). Thus

$$\left.\frac{dE}{dt}\right|_{\text{absorbed}} = Sa(\pi R^2) \tag{5.9}$$

The rate at which energy is radiated is given by Stefan's law (Eq. (5.6)).

$$\left.\frac{dE}{dt}\right|_{\text{radiated}} = \varepsilon\sigma T^4(4\pi R^2) \tag{5.9$'$}$$

where the radiating area is the total surface of the earth $A = 4\pi R^2$. Equating the rates given by Eqs. (5.9) and if as a first approximation we

* Since $dI/d\lambda = -(c/\lambda^2)(dI/d\nu)$, the peak of the distribution $dI/d\lambda$ is at $(\hbar c/kT\lambda) \simeq 5$; this is at a different frequency than the peak of $dI/d\nu$, which occasionally can lead to confusion.

assume that (averaged over wavelength) $\varepsilon = a$, we find that the temperature of the earth's surface should be

$$T = \left(\frac{S}{4\sigma}\right)^{1/4} = 279\ \mathrm{K} = 6°\mathrm{C} \tag{5.10}$$

While this result is in reasonable agreement with the range of temperatures found on the earth, the assumption $\varepsilon = a$ is too crude to use in any detailed model.

To improve the calculation we will include (a) the effect of latitude, (b) the reflection of the incident energy by the cloud cover, and (c) the absorption and re-radiation of energy by the layers of the upper atmosphere. The first of these effects is indicated in Fig. 5.4(*a*) and we assume that the radiation is always incident parallel to the plane of the equator; in reality, of course, the angle of incidence depends on the position of the earth in its orbit around the sun (on the seasons). We consider a strip at latitude $\phi$ and of width $w = R\Delta\phi$. The total area of the strip is $A = w2\pi(R \cos \phi)$. The area normal to the direction of the incident radiation is $A_\perp = w2R \cos^2 \phi$ where we have included a factor of $\cos \phi$ for the inclination of the surface and replaced $2\pi R$ by $2R$ for the normal projection. Proceeding as in Eqs. (5.9), we set $SA_\perp = \sigma T^4 A$ to obtain

$$T(\phi) = \left[\frac{S}{\sigma}\frac{\cos \phi}{\pi}\right]^{1/4} = (296\ \mathrm{K})(\cos \phi)^{1/4} \tag{5.10'}$$

Fig. 5.4. (*a*) The effect of latitude on the flux received per unit area of the earth's surface. (*b*) Reflection ($\alpha_r$) and absorption ($\alpha_{in}$) of the radiation incident on the earth ($\alpha_s$). $\alpha_e$ represent the radiation emitted by the earth; it is partially transmitted ($\alpha_c$) and partially reflected ($\alpha_c$) by the clouds and upper atmosphere.

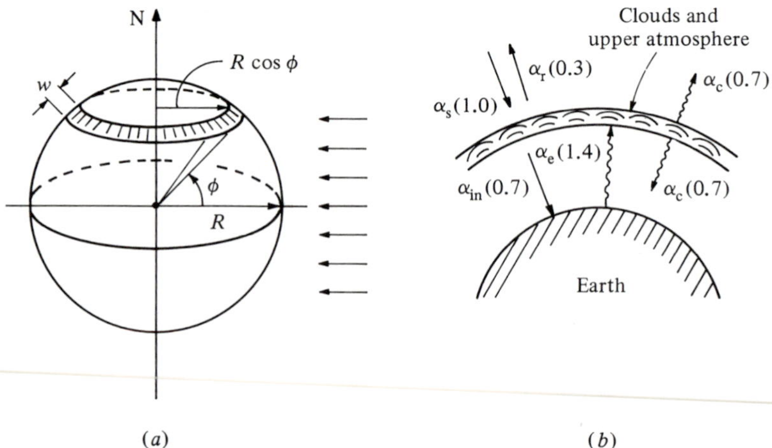

For instance if the equator is at $T = 23°C$ then at a latitude $\phi = 44°$ the weather would be freezing. Of course for a realistic model we must take into account the conduction of the earth's surface and convection through the motion of water and air masses.

Since we live on the earth's surface we are primarily interested in the surface temperature and thus in the radiation received and emitted by the surface. This is greatly affected by the reflection and absorption properties of the clouds and of the upper layers of the atmosphere. For instance, the thin ozone layer in the upper atmosphere strongly absorbs the ultraviolet providing the necessary protection to living beings; $CO_2$ layers strongly absorb in the infrared giving rise to the 'greenhouse effect'. On the average, half of the earth is covered by clouds which have a reflectivity of order 50% for visible, so that the solar radiation reaching the surface is typically $\sim 0.7$ of the solar flux. Because of the presence of such absorbing layers and since the incident and emitted radiation are at different wavelengths we cannot any more assume that $\langle a \rangle = \langle \varepsilon \rangle$.

We can construct a simple model taking the above factors into consideration, as shown in Fig. 5.4(*b*), where the energy balance involves the following components:

| | |
|---|---|
| $\alpha_s$ | incident radiation |
| $\alpha_r$ | reflected radiation |
| $\alpha_{in}$ | radiation reaching the earth's surface |
| $\alpha_e$ | total power radiated by the earth's surface; all the energy is assumed to be completely absorbed by the upper atmosphere |
| $\alpha_c$ | total power radiated by the upper atmosphere; the radiation is assumed isotropic |

Equilibrium implies the relations

$$\alpha_{in} = \alpha_s - \alpha_r, \qquad \alpha_s = \alpha_r + \alpha_c \qquad \text{and} \qquad \alpha_e = \alpha_{in} + \alpha_c$$

from where we immediately find that

$$\alpha_c = \alpha_{in} \qquad \text{and} \qquad \alpha_e = 2\alpha_{in}$$

The radiation incident on the earth is $\alpha_s = S\pi R^2$ and since only 0.7 of the incident radiation reaches the surface we write

$$\alpha_{in} = 0.7\alpha_s = 0.7S\pi R^2 \qquad \text{and} \qquad \alpha_e = 1.4S\pi R^2$$

The radiated power $\alpha_e$ is determined by the temperature of the earth's surface

$$\alpha_e = \left.\frac{dE}{dt}\right|_{rad} = \sigma T^4(4\pi R^2)$$

Finally, equating this result to $\alpha_e = 1.4S\pi R^2$ we find $T = 303\ K = 30°C$.

This is in good agreement with observation and shows how the absorbing layers of the upper atmosphere can significantly affect the surface temperature.

Even from our simple calculations we can appreciate how complex the energy balance of the earth is. Moreover, the equilibrium is not necessarily stable and may be *unstable* to an increase in atmospheric pollutants or major changes in the ecology. For instance, an increase in $CO_2$ would increase the 'greenhouse effect' and warm up the earth; this in turn would create more evaporation, increase the cloud cover and the density of the upper layers, further accentuating the greenhouse effect. Conversely, if the snow cap increases, it would result in more reflection from the earth's surface and thus lower the surface temperature; this would extend the snow cover even further, and so on. It is for these reasons that effects such as a 'nuclear winter' can have a permanent and catastrophic effect on the earth's climate.

## 5.3 The atomic nucleus

The atomic nucleus is an assembly of *protons* ($p$) and *neutrons* ($n$) which are bound (held together) by the nuclear force. The neutron, discovered by J. Chadwick in 1932, has the same mass and shares many of the properties of the proton except that it is electrically neutral; the proton and neutron are collectively called *nucleons*. The typical dimension of nuclei is $r_N \sim 10^{-13}$ cm, to be compared with that of atomic systems which have radii $r_a \sim 10^{-8}$ cm. The number of protons in the nucleus is designated by $Z$, which therefore also indicates its charge; the number of neutrons is designated by $N$ and $A = Z + N$ is the atomic number. In nature stable nuclei are found only with charge less than $Z = 92$ and atomic number less than $A = 238$. Nuclei with values of $Z$ as high as 106 have been produced artificially but they are unstable.

Since positive charges repel one another, a force stronger than the electromagnetic force must be present in order to hold the nucleus together; this force must have a *short range* because we do *not* feel its effects on a macroscopic scale. We speak of the *nuclear force* and it has been found that it has the same strength for the $pp$, $pn$ and $nn$ systems. We can model such a nucleon–nucleon force by a square potential well of radius $a \sim 10^{-13}$ cm, as shown in Fig. 5.5. The depth of the well is of order 20 MeV, a million times deeper than the energy of the ground state of the hydrogen atom, and is just barely sufficient to bind a proton and a neutron into a deuteron.

To simplify the calculations related to nuclear phenomena it is customary to express momentum, mass and energy in MeV and length in *fermis*.

$$1 \text{ fermi} = 1 \text{ F} = 10^{-13} \text{ cm} = 1 \text{ fm}$$

According to special relativity the energy $E$ of a particle of mass $m$ is given by Einstein's celebrated relation

$$E = mc^2 \tag{5.11}$$

Thus we express the mass of a particle, or nucleus, by its equivalent energy in MeV. For instance the mass of the proton is $m_p = 1.673 \times 10^{-27}$ kg; instead it is easier to use

$$\left.\begin{aligned} m_p c^2 &= 938.28 \text{ MeV} && \text{proton mass} \\ m_n c^2 &= 939.57 \text{ MeV} && \text{neutron mass} \\ m_e c^2 &= 0.511 \text{ MeV} && \text{electron mass} \end{aligned}\right\} \tag{5.11$'$}$$

Similarly, we express the momentum $p$ of a particle in MeV by giving the corresponding value of $cp$, which has units of energy. Finally the combination of the constants ($\hbar c$) appears often in calculations; it has dimensions of energy–time and in our units

$$\hbar c \simeq 200 \text{ MeV-F} \tag{5.12}$$

As an exercise, we estimate the depth of the potential well of Fig. 5.5 from simple arguments based on the *uncertainty principle*. This principle, which forms the foundation of quantum mechanics, states that we cannot measure simultaneously two complementary variables with arbitrary precision. For instance, if $\Delta p_x$ and $\Delta x$ are the uncertainties in the measurement of momentum and position, then

$$\Delta p_x \Delta x \geqslant \hbar$$

Thus, if we take $\Delta x$ to be of the order of the dimensions of the nuclear potential well $\Delta x \sim a \sim 1$ F, we find $c\Delta p_x \geqslant 200$ MeV; we now assume that

Fig. 5.5. A potential well of finite width and depth can be used as a model for the nuclear force.

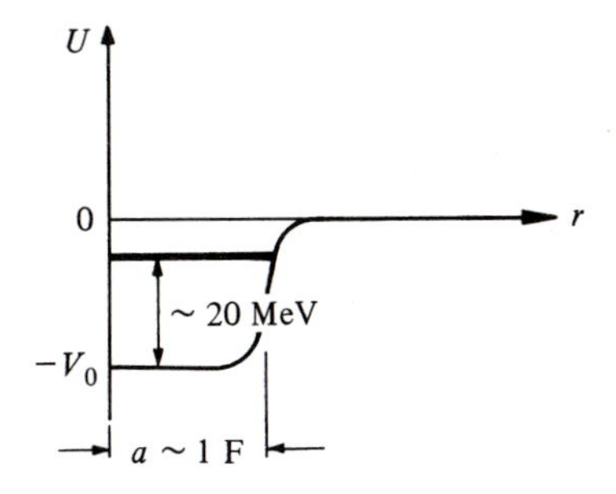

$cp \sim c\Delta p_x$ so that the kinetic energy is

$$\text{K.E.} = \frac{p^2}{2m} = \frac{(c\Delta p_x)^2}{2mc^2} \gtrsim 20 \text{ MeV}$$

where it is the nucleon (proton or neutron) mass that appears in the denominator. Since a nucleus is bound, the total energy must be negative; thus $U \leqslant -20$ MeV which is of the correct order of magnitude.

The existence of the nuclear force is the reason why a system of positive charges can be bound, but it is not sufficient to explain the structure of nuclei. We must take into account the Pauli exclusion principle according to which two protons (or two neutrons) with the same spin projection cannot occupy the same state. As we add protons and neutrons to a nucleus the width and depth of the effective potential well increases, but these particles cannot occupy filled energy levels and must therefore have increasingly higher energies. At a certain point the kinetic energy of any additional nucleon exceeds the depth of the potential well and the particle cannot be bound. It is for this reason that no nuclei exist with more than $\sim 250$ nucleons. We recall that the structure of atoms and of the periodic table are also a consequence of the exclusion principle.

The interplay between the repulsive Coulomb force and the exclusion principle favors nuclei where $N \sim Z$, as observed for light nuclei. However for high $Z$, nuclei have $N > Z$ because the short range nuclear force is less effective in balancing the em repulsion. In Fig. 5.6 the values of $Z$ and $N$ for the stable and *radioactive* nuclei are shown. By radioactive, we mean nuclei that spontaneously decay into another nucleus. Nuclei with the same value of $Z$ but different values of $N$ are called *isotopes* because they have the same atomic (chemical) properties. We designate a nucleus by the atomic symbol of its corresponding chemical element with a left subscript for $Z$, a right subscript for $N$ and a left superscript for $A$. For instance

$$^{238}_{92}\text{U}_{146}$$

is pronounced 'Uranium two-thirty-eight' and represents the nucleus of the Uranium ($Z = 92$) isotope with atomic number $A = 238$. Since $N = Z - A$ the right subscript is redundant and we will omit it. Other examples are

$^1_1$H the proton, which we will also designate by $p$
$^2_1$H the deuteron, also designated by $d$
$^4_2$He the nucleus of Helium-4, also known as an 'alpha-particle' and designated by $\alpha$
$^{16}_{8}$O Oxygen-16, etc.

Nuclei with an odd atomic number have half-integer spin; for instance the proton and neutron have spin $\frac{1}{2}$, $^{17}_{8}O$ has spin $\frac{5}{2}$ etc. When the atomic number is even the nucleus can have spin 0 and this is generally the case, but it can also have integer spin; for instance the deuteron has spin 1, whereas the $\alpha$-particle has spin 0. The $\alpha$-particle consists of two protons

Fig. 5.6. Chart of the 'nuclides' as a function of the number of protons, $Z$, and neutrons, $N = A - Z$; no nuclei can be formed far away from the stability region where $Z \sim N$. (From F. Bitter, *Currents, Fields and Particles*, John Wiley (1956).)

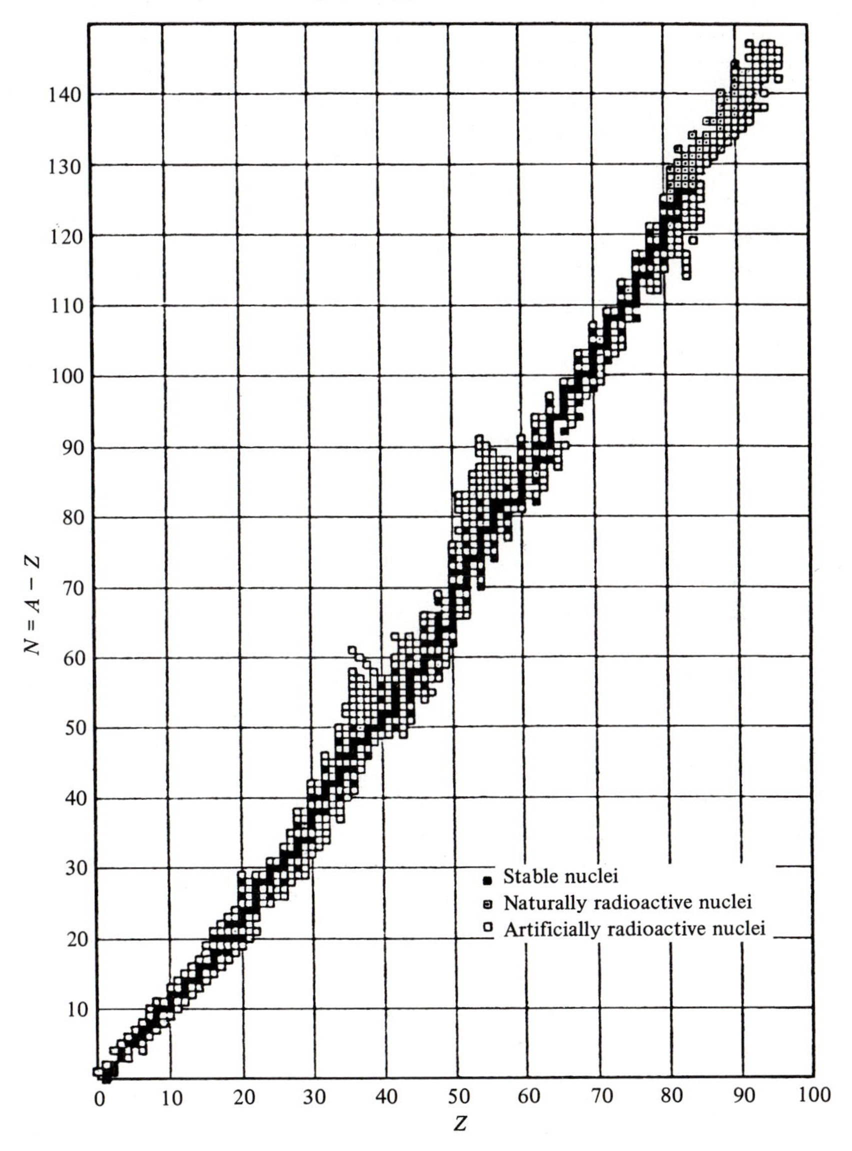

with their spins antiparallel and two neutrons also with antiparallel spins; because of this configuration it is an extremely tightly bound system.

Ordinarily a nucleus is found in its ground state, the state of lowest energy. However, following a nuclear collision a nucleus can be in an excited state; it can then return to its ground state by the emission of a photon as shown in Fig. 5.7(*a*). This process is analogous to radiative transitions in atoms except that the difference between energy levels is of the order of MeV; photons in that energy range, which exceeds the energy of X-rays, are referred to as $\gamma$-rays. Radiative transitions in nuclei are usually prompt, with lifetimes in the range of $10^{-12}$ s.

An unstable nucleus can also decay by emitting an electron. In this case, given a parent nucleus $A, Z$, the daughter nucleus has the same $A$-value but the $Z$-value has increased to $Z+1$. We speak of beta-decay, as shown schematically in Fig. 5.7(*b*). Positron emission is also possible in which case the $Z$-value of the daugher nucleus is $Z-1$. These processes are due to the *weak interaction* which changes a neutron into a proton as in Eq. (5.13) and vice versa as in Eq. (5.13′). For instance, since the neutron – proton mass difference exceeds the electron mass (see Eqs. (5.11′)) a *free* neutron can decay according to

$$n \rightarrow p + e^- + \bar{\nu}_e \tag{5.13}$$

This process takes place with a lifetime of 15 minutes. The inverse process

$$p \rightarrow n + e^+ + \nu_e \tag{5.13'}$$

cannot occur for a free proton, but is allowed when the proton is inside the nucleus, since a rearrangement of the nucleus inside the well can provide the necessary energy. The particle designated by $\nu_e$ is the *electron neutrino* and is a neutral particle of *zero mass* ($\bar{\nu}_e$ is the electron antineutrino); it escapes from the nucleus but carries away energy and

Fig. 5.7. Examples of nuclear decay: (*a*) de-excitation by emission of a photon, (*b*) beta decay (emission of an electron and of an antineutrino), (*c*) decay by the emission of an $\alpha$-particle (helium nucleus).

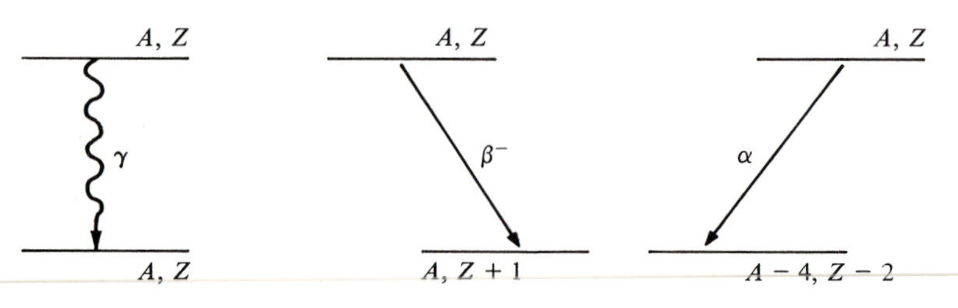

momentum. Nuclear reactors are copious sources of antineutrinos. The lifetime for $\beta$-decay varies for different transitions but can be of order of hours, days or even longer.

Heavy unstable nuclei often decay by the emission of an $\alpha$-particle. In this case a parent nucleus $A, Z$ decays to a daughter with $A-4, Z-2$ as shown in Fig. 5.7($c$). That this process is possible is due to the great stability of the $\alpha$-particle and to the quantum mechanical tunnel-effect. The lifetime for $\alpha$-decay is very long and can be of order of thousands or millions of years. The naming of these decays as $\alpha, \beta, \gamma$ radiations dates from the first discovery of natural radioactivity in 1900 by Becquerel and the Curies. Under very special circumstances a heavy nucleus can decay by *fission* into two lighter nuclei. This process is the source of energy in nuclear reactors and in the smaller nuclear weapons.

## 5.4 Nuclear binding, fission and fusion

We saw in the previous section that the nucleus is a *bound* system of protons and neutrons. This means that if one wishes to separate all the constituents of the nucleus energy must be *supplied*; the necessary amount of energy is called the *binding energy* of that nucleus. By convention the binding energy is a negative number, being defined as the difference between the rest mass of the nucleus and the sum of the rest masses of all its constituent nucleons

$$\text{B.E.} = {}^{A}_{Z}m - [Zm_{\text{p}} + (A-Z)m_{\text{n}}]$$

The heavier the nucleus the larger the binding energy. In fact, the binding energy per nucleon, B.E./$A$ has approximately the value of $-8$ MeV for most nuclei. Thus the binding energy decreases the mass of a nucleus from that of its free constituents by about 0.8%.

As examples we calculate B.E./$A$ for the deuteron and for the helium nucleus. The deuteron has a rest mass $m_{\text{d}}c^2 = 1875.63$ MeV; the mass of its two constituents, the proton and neutron was given in Eqs. (5.11′). Thus the binding energy for the deuteron is

$$(\text{B.E.})_{\text{d}} = m_{\text{d}}c^2 - (m_{\text{p}} + m_{\text{n}})c^2 = 1875.63 - 1877.85 = -2.22 \text{ MeV}$$

and we find B.E./$A = -1.1$ MeV. The mass of the ${}^{4}_{2}\text{He}$ nucleus is $m_\alpha c^2 = 3727.41$ MeV and therefore

$$(\text{B.E.})_\alpha = m_\alpha c^2 - (2m_{\text{p}} + 2m_{\text{n}})c^2 = -28.29 \text{ MeV}$$

which results in B.E./$A = -7.1$ MeV indicating that the ${}^{4}_{2}\text{He}$ nucleus is a tightly bound system as we have already remarked.

The values of B.E./$A$ for all nuclei are shown in Fig. 5.8 as a function of $A$; they vary slowly with $A$ except for the lightest elements. The minimum is in the region of the iron nucleus, $A \sim 56$, and this is why Fe is the most abundant element in burned out stars. It is the small deviation from flatness in the curve of B.E./$A$ that makes possible the release of energy in nuclear fission or fusion processes. For instance a nucleus in the region $A \sim 200$ has B.E./$A \sim -8$ MeV; if the nucleus breaks up into two nuclei in the region $A \sim 60$ where B.E./$A \sim -8.5$ MeV, the final system is *more tightly* bound, and the difference in binding will appear as kinetic energy of the fission products. Similarly, if two light nuclei fuse into a heavier nucleus, for instance $d + d \rightarrow {}^4_2\text{He} + \gamma$, the final system is more tightly bound and again the energy difference appears as kinetic and/or radiative energy.

We see that in order to release nuclear energy we must either induce the *fission* of a heavy nucleus or the *fusion* of two light nuclei. As long as the final system is more tightly bound than the initial state, energy will be released. This is equivalent to stating that the rest mass of the final system must be less than the rest mass of the initial state. That is, some of the rest mass is converted into kinetic energy as predicted by Einstein (see Eq. (5.11)), where

$$\Delta E = (m_{\mathrm{i}} - m_{\mathrm{f}})c^2 = (\Delta m)c^2 \tag{5.14}$$

To calculate the energy released in a nuclear reaction from Eq. (5.14) we need to know the masses of all nuclei with adequate precision. For practical reasons it is convenient to give the mass of the corresponding *neutral atom* in terms of the *atomic mass unit* (amu). One amu is $\frac{1}{12}$th of the mass of the carbon-12 atom, where

$$1\ \text{amu} = (931.5016 \pm 0.0026)\ \text{MeV}/c^2 = m_{\mathrm{a}}$$

Fig. 5.8. The binding energy per nucleon as a function of atomic number.

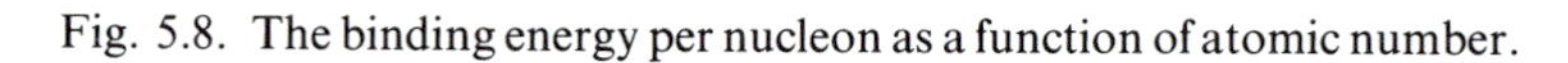

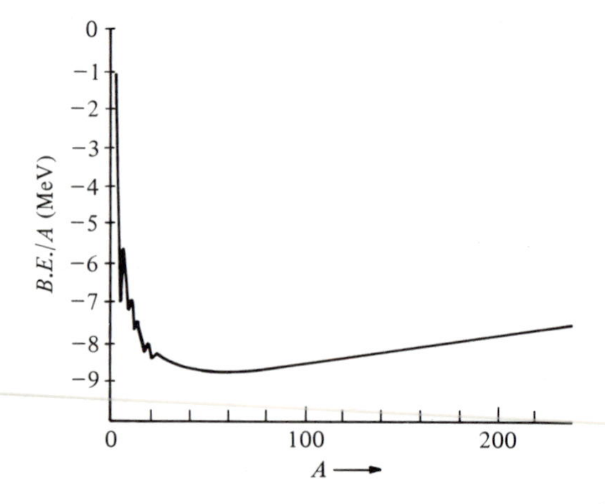

Furthermore the *mass excess* is defined through

$$(M - A) = {}^{A}_{Z}M - Am_{\mathrm{a}}$$

where ${}^{A}_{Z}M$ is the mass of the neutral atom with nucleus of charge $Z$ and atomic number $A$. The mass excess has been tabulated for all stable isotopes, and is usually given in MeV/$c^2$.* Thus the mass of a nucleus can be found from

$$^{A}_{Z}m = Am_{\mathrm{a}} + (M - A) - Zm_{\mathrm{e}} \tag{5.15}$$

In Eq. (5.15) the rest energy of the electrons has been included whereas the electron binding energy can be safely ignored.

As an example we calculate the energy released in the fission of ${}^{235}_{92}\mathrm{U}$ when it absorbs a slow neutron

$${}^{1}_{0}n + {}^{235}_{92}\mathrm{U} \rightarrow {}^{144}_{56}\mathrm{Ba} + {}^{89}_{36}\mathrm{Kr} + 3({}^{1}_{0}n) \tag{5.16}$$

The number of protons and the number of neutrons are the same on both sides of the equation since these particles cannot be created or destroyed during the reaction and no weak interactions such as in Eqs. (5.13) take place; thus there is also no change in the number of electrons nor are any neutrinos produced. The mass excess on the left-hand side is

$$\begin{aligned}(M - A)\,{}^{235}_{92}\mathrm{U} &= 40.916\\ (M - A)\quad {}^{1}_{0}n &= \underline{\phantom{4}8.071}\\ &\phantom{=}\ 48.987 \quad \mathrm{MeV}/c^2\end{aligned}$$

whereas on the right-hand side

$$\begin{aligned}(M - A)\,{}^{144}_{56}\mathrm{Ba} &\simeq -\ \ 80.0\\ (M - A)\ {}^{89}_{36}\mathrm{Kr} &= -\ \ 76.79\\ 3 \times (M - A){}^{1}_{0}n &= \underline{\phantom{-\ \ }24.21}\\ &\phantom{=}\ -132.58 \quad \mathrm{MeV}/c^2\end{aligned}$$

We see that the final products are more tightly bound, and the energy released in this process is

$$\Delta E = (m_{\mathrm{i}} - m_{\mathrm{f}})c^2 = 48.99 - (-132.58) = 181.57\ \mathrm{MeV}$$

The ${}^{235}_{92}\mathrm{U}$ nucleus fissions into two nuclei of almost equal mass, but not always those indicated in Eq. (5.16). In each fission, *on the average*, 200 MeV of energy are released and 2.4 neutrons are produced. The fission products are usually unstable nuclei, and are therefore highly radioactive.

Heavy nuclei are, in general, subject to fission when bombarded by neutrons. The explanation of this phenomenon is given in terms of the 'liquid drop' model sketched in Fig. 5.9. Since the neutron has no electric

* See for instance *Introduction to Modern Physics* by J. D. McGervey, Academic Press, New York, 1983.

charge it can easily penetrate the nucleus. The addition of the neutron increases the energy of the nucleus by $\sim 8$ MeV (the average B.E./$A$) and to dissipate that extra energy, the nucleus begins to oscillate, the oscillation amplitude grows, and eventually the nucleus fissions. The cross-section for fission varies drastically for the various nuclei; for $^{235}_{92}\mathrm{U}$ the cross-section increases for low energy or 'slow' neutrons and reaches $\sigma_f \sim 10^{-21}\ \mathrm{cm}^2$ for $E_n \sim 0.01$ eV. Neutrons of such low energy are referred to as *thermal* neutrons. In contrast $^{238}_{92}\mathrm{U}$ absorbs 'fast' neutrons ($E_n \gtrsim 1$ MeV) to form the isotope $^{239}_{92}\mathrm{U}$ which beta-decays to neptunium rather than fission. This reaction is indicated below, the $^{239}_{93}\mathrm{Np}$ beta-decaying with a 2-day lifetime to plutonium.

$$^{1}_{0}n + {}^{238}_{92}\mathrm{U} \longrightarrow {}^{239}_{92}\mathrm{U} \xrightarrow[\text{20 min}]{} {}^{239}_{93}\mathrm{Np} + e^- + \bar{\nu}$$

$$^{239}_{93}\mathrm{Np} \xrightarrow[\text{2 days}]{} {}^{239}_{94}\mathrm{Pu} + e^- + \bar{\nu}_e \qquad (5.17)$$

The artificial isotope $^{239}_{94}\mathrm{Pu}$ has a very large fission cross-section and is used extensively as fuel for weapons; it has a lifetime of 24 000 years.

The fact that more than one neutron is produced for every absorbed neutron opens up the possibility of a 'chain reaction' which would grow exponentially. However, the neutrons from the reaction of Eq. (5.16) are produced with kinetic energy in the range of 1–2 MeV. Before these neutrons can be absorbed to induce fission they must be slowed down to thermal energies. This is achieved by using a *moderating* material consisting of light nuclei, such as water or graphite. When a neutron scatters elastically by 180° from a nucleus of mass $M$, the kinetic energy after the collision is

$$(\mathrm{K.E.})_{\text{after}} = \left| \frac{M - m_n}{M + m_n} \right|^2 (\mathrm{K.E.})_{\text{before}}$$

Thus the efficiency of elastic collisions for slowing down neutrons improves greatly as $M \to m_n$, and this is why light materials are effective moderators.

The rate of growth of the chain reaction depends on the probability that a neutron will induce a fission reaction as compared to its probability

Fig. 5.9. Schematic representation of nuclear fission in the liquid drop model.

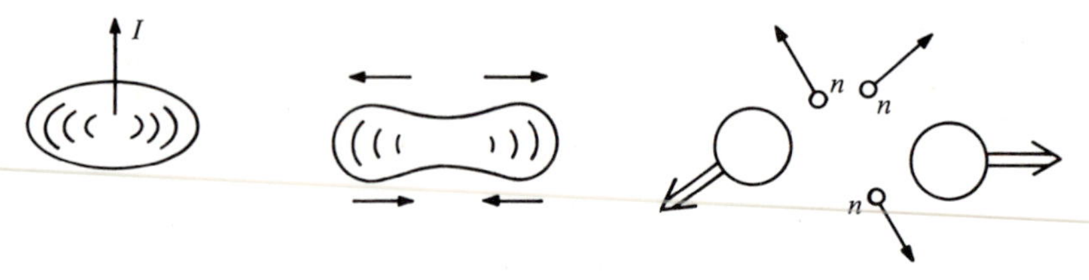

for escaping the volume of the fissile material or of being absorbed in a reaction that produces no further neutrons. We can introduce the concept of a *reproduction factor* $k$ which is defined as the average number of neutrons produced for every absorbed neutron in a system of very large dimensions. For instance for a graphite *natural* uranium lattice $k \sim 1.04$ to 1.07. Clearly if $k < 1$ no chain reaction is possible. If the probability that a neutron will escape is $P_e$, then the *multiplication factor* $K = k(1 - P_e)$. For a reactor operating at constant power we must have $K = 1$; this is achieved by controlling the reaction rate through the introduction of materials, such as cadmium, that absorb neutrons. For a weapon we want $K > 1$ so that the reaction will proceed very fast and give rise to an explosion.

Natural uranium is composed 99.3% of $^{238}U$ which absorbs neutrons and only 0.7% of the fissile $^{235}U$. Thus to increase the reproduction factor, commercial reactors use uranium that has been enriched in $^{235}U$ up to 3%. However with a suitable moderator reactors using natural uranium can operate satisfactorily. In contrast highly enriched uranium will undergo an explosive reaction if it is assembled in large enough quantities so as to exceed its 'critical mass'.

As an example, let us calculate the mean free path for a neutron in pure metallic $^{235}U$. In this case there is no moderator present and we take the fission cross-section (for fast neutrons $E > 10$ keV), $\sigma_F \simeq 5 \times 10^{-24}\ \text{cm}^2$. Since the density of uranium is 18 g/cm$^3$, the (number) density of nuclei is $n_F \simeq 5 \times 10^{22}$ nuclei/cm$^3$ and the mean free path is

$$l = 1/n_F \sigma_F \cong 4\ \text{cm} \tag{5.18}$$

It is reported that the *critical mass* for pure $^{235}U$ corresponds to a sphere of radius $r \sim 8$ cm in good agreement with our simple estimate for the mean free path. Such a sphere has a mass of $\sim 45$ kg and a chain reaction will proceed explosively if that amount of $^{235}U$ is tightly assembled. For 3% enriched uranium, the critical mass is of order of several tons.

The control of fusion reactions is much more difficult than for fission. This is because the initial particles must have high energy in order to overcome the Coulomb repulsion. In weapons this is achieved by triggering a fission explosive which produces enough energy to ignite the fusion reaction. Methods for maintaining a fusion reaction under controlled conditions will be discussed in Section 5.7.

## 5.5 Nuclear reactors

Nuclear reactors are devices in which a fission chain reaction can be sustained in equilibrium and the generated heat is converted to mechanical energy, usually by a steam-driven turbine. The fuel is enriched uranium and is encased in hermetically sealed zirconium tubes;* these are the fuel elements and are assembled in the *core* of the reactor. Most commercial U.S. reactors use water under high pressure as a moderator in which the fuel elements are immersed. The water acts as a primary coolant and generates steam in a heat exchanger. A schematic is shown in Fig. 5.10. The secondary loop uses the steam to drive the turbine and a second heat exchanger condenses the exhaust steam. The heat exchanger in the secondary loop uses cooling water from nearby rivers or lakes or air cooling towers. In that respect the reactor has the same properties as any large fossil fuel electric power generating plant.

Under normal operating conditions the power generated by the reactor is kept constant. This implies that the number of fissions per unit time is constant, and thus also the neutron flux in the reactor core remains constant. This means that for every neutron that is absorbed (either in a fission-producing reaction or otherwise) or leaves the core, exactly one neutron must be produced; no more and no less. The ratio of neutrons

Fig. 5.10. Schematic representation of the components of power generating reactors: (*a*) pressurized water reactor, (*b*) boiling water reactor.

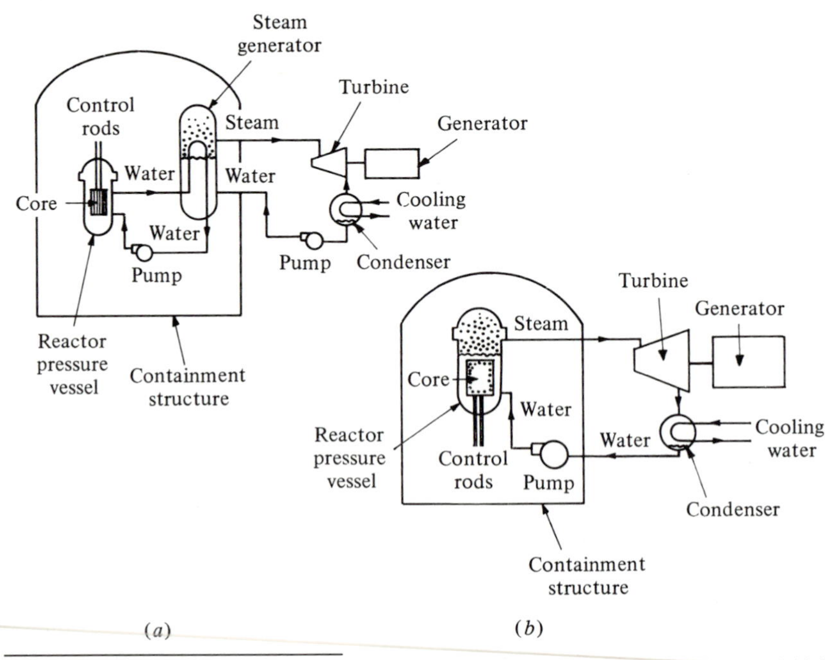


* Zirconium has a low neutron absorption cross-section and can withstand high temperatures.

produced to neutrons lost is the *multiplication factor* $K$, already introduced in the previous section. Neutrons are produced by fission of the fuel but also to a much lesser extent by the decay of unstable isotopes. This is balanced by the loss of neutrons which occurs because of (a) capture in the fuel that leads to fission, (b) capture in $^{238}$U or in the reactor structure that does not lead to fission and (c) escape from the core. Thus when

$K < 1$ the power is falling and the reactor is sub-critical
$K = 1$ the power is constant and the reactor is critical
$K > 1$ the power is rising and the reactor is super-critical

The multiplication factor depends on various parameters of the reactor and is controlled by adjusting the cadmium or boron control rods. To increase the power level the control rods are retracted making $K$ slightly larger than 1, and the power increases. When the desired power level is reached the control rods are inserted deeper into the core to make $K = 1$. The power is directly proportional to the neutron flux which is carefully monitored and used to provide the necessary feedback for stable operation.

The power generated in the reactor is given by

$$P = \phi_n n_F \sigma_F V Q \tag{5.19}$$

Here, $\phi_n$ is the neutron flux in the core, $\phi_n = n_t v$ with $v_n$ the velocity of the thermal neutrons and $n$ their number density. The fission cross-section is $\sigma_F$ and $n_F$ is the number density of fissile nuclei; $V$ is the volume of the core and $Q$ is the energy released per fission. Assuming an enrichment of 3%, the number density of $^{235}$U nuclei is

$$n_F = 0.03 \times \bar{n} = 0.03 \times (5 \times 10^{22}) = 1.5 \times 10^{21} \text{ cm}^{-3}$$

where $\bar{n}$ is the number density of uranium nuclei averaged over the dimensions of the core. For the fission cross-section in the presence of a moderator we will take

$$\sigma_F \simeq 10^{-22} \text{ cm}^2$$

and for the neutron flux we choose the realistic value

$$\phi_n = 10^{14} \text{ cm}^{-2} \text{ s}^{-1}$$

Thus for a core of two cubic meters, $V = 2 \times 10^6$ cm$^3$, recalling that the energy released per fission is of order $Q = 200$ MeV $= (2 \times 10^8) \times (1.6 \times 10^{-19})$ J, we find

$$P = 10^{14} \times (1.5 \times 10^{21}) \times 10^{-22} \times (2 \times 10^6) \times (3.2 \times 10^{-11})$$
$$\simeq 10^9 \text{ Watts}$$

Namely, we can expect a thermal power of 1 GW. If we assume a 40% efficiency for conversion to electrical power, the reactor can provide $\sim$400 MW which is equivalent to the rating of large fossil fuel power plants.

The mass of the core is $M \sim (20\ \text{g/cm}^3) \times (2 \times 10^6\ \text{cm}^3) = 40$ tons and the number of fissile nuclei is $N_F = n_F \times V = 3 \times 10^{27}$. The fission rate on the other hand is $R_F = P/Q \simeq 3 \times 10^{19}$/s; thus the total fuel would be exhausted in $10^8$ s, or in 3 years. Therefore it is necessary to refuel nuclear reactors approximately once a year. Refueling consists in removing the partially spent fuel elements and replacing them with new fuel elements containing uranium enriched to the desired level. The used fuel elements are then reprocessed to extract the remaining $^{235}$U but also the $^{239}$Pu that has been produced by the bombardment of the $^{238}$U. Refueling, transportation of the used fuel elements and reprocessing are exacting and potentially hazardous operations.

The stability of nuclear reactors is due in part to a detail in the chain of neutron production. The time it takes for a neutron to be *thermalized* and induce a fission is of order of 1 ms ($10^{-3}$ s). Thus if the reactor was adjusted so that $K = 1.001$, then in one second the reactor power would approximately double. This shows that the feedback and motion of the control rods would have to be extremely fast. However a small fraction of the neutrons in the core do not come directly from fission but are due to the decay of fission products, as for instance in the process

$$^{87}\text{Br} \rightarrow {}^{87}\text{Kr} + e^- + \bar{\nu} \rightarrow {}^{86}\text{Kr} + {}^1_0 n$$

which have a long lifetime. About 1.1% of the neutron flux is due to these secondary neutrons which are produced with an average delay of 14 seconds. Thus the time from the birth of a neutron to the instant at which it induces a fission, averaged over *all* neutrons, is of order $\tau = (0.011 \times 14 + 0.989 \times 10^{-3}) \simeq 0.15$ s which is significantly longer than $10^{-3}$ s. The reactor is operated so that only 98.9% of the fissions are produced by prompt (i.e. direct fission) neutrons and the remainder by delayed neutrons. When this condition is not observed the reactor becomes prompt-critical and cannot be controlled.

The power v. time curve for a nuclear reactor would typically look as in Fig. 5.11. At the start-up the neutron flux grows exponentially and after some overshoot remains constant. Under normal conditions when the reactor is being shut down the power and thus also the flux are reduced exponentially. It is important to realize that the reactor depends crucially on the cooling fluid to maintain steady operation because small variations in fuel temperature and density or in moderator density affect the reaction rate. Even if the chain reaction is *stopped*, for instance by the insertion of emergency control rods, the fuel elements are still extremely radioactive and generate enough heat to melt their cladding. To minimize the consequences of such possible accidents, reactors are surrounded by a structure designed to contain the radioactive materials that would be released from the core, and to a lesser extent from the primary coolant.

Commercial power reactors in the U.S. are mostly of the pressurized water or boiling water type shown in Fig. 5.10. In the pressurized water system the steam circuit is isolated from the core, while in the boiling water it is part of the primary coolant. Water reactors are designed to be thermally stable, that is if the temperature of the core increases the multiplication factor decreases, reducing the reactor power and thus also the temperature. When high neutron flux is desired, as in reactors used to produce weapons material, graphite rather than water is used for the moderator. Typically it takes a neutron approximately 114 collisions before it thermalizes in graphite as compared with only 18 collisions in water. However in water, neutrons are absorbed by the free protons in the reaction $n + p \rightarrow d + \gamma$ which has a large cross-section. To avoid this loss, heavy water reactors using $D_2O$ as a moderator have been built, and they can operate with natural uranium. A different approach is to use a liquid metal as the coolant since this permits operation at higher temperatures. In this case however the moderating effect of the water is absent and the fuel must be considerably enriched in fissile material.

The enrichment of uranium fuel can be achieved by diffusion of $UF_6$ gas through a long sequence of porous membranes. Alternative methods are centrifugal separation and a proposed technique based on selective laser ionization but they have not found much practical use. In general these methods are expensive, whereas fissile material can be produced in a reactor at much lower cost. We have already indicated the plutonium-producing reaction in Eq. (5.17). Another fissile material is $^{233}_{92}U$ which can be produced from thorium when it is bombarded by fast

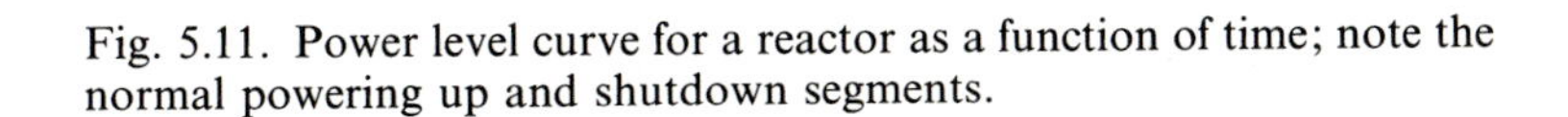
Fig. 5.11. Power level curve for a reactor as a function of time; note the normal powering up and shutdown segments.

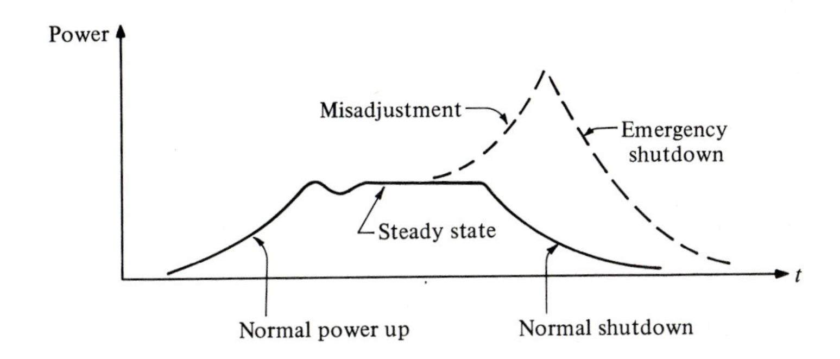

(1–2 MeV) neutrons

$$ {}^{1}_{0}n + {}^{232}_{90}\mathrm{Th} \rightarrow {}^{233}_{90}\mathrm{Th} \rightarrow {}^{233}_{91}\mathrm{Pa} + e^- + \bar{\nu}_e $$

$$ {}^{233}_{91}\mathrm{Pa} \rightarrow {}^{233}_{92}\mathrm{U} + e^- + \bar{\nu}_e \quad (5.20) $$

$^{233}$U has a lifetime $\tau \sim 160\,000$ years. Producing fuel in a reactor has the advantage of preserving energy resources and in particular of using the much more abundant elements $^{238}$U and $^{232}$Th as primary sources. Reactors used in this mode are named *breeders*. The obvious disadvantage is that the fissile material that is produced is of very high grade and thus suitable for use in weapons.

In the early days of nuclear power it was believed that fission reactors could satisfy the world energy demand well into the future. This hope has not materialized for two reasons. First, the available uranium reserves are limited, and secondly the construction of nuclear reactors and the necessary safety measures are expensive both in energy and capital. For instance the total reserves of uranium in the U.S. are estimated at $M_R \sim 2.5 \times 10^7$ tons, including highly uneconomical mining. From our example we estimated that 40 tons of 3% enriched uranium could yield about 1 GW-year of electric power. Since natural uranium contains only 0.7% of $^{235}$U, the reserves are equivalent to $W_T \sim 1.5 \times 10^5$ GW-years. Given that the U.S. energy consumption is of order $P \sim 2 \times 10^3$ GW, we see that the uranium supply would last for $\sim 75$ years. This estimate is by far too optimistic because only 10% of the uranium reserves can be easily mined; furthermore reactors are developed only for electric power generation. Recovery of unspent fuel could double the above figure and in particular a breeder program would alleviate the need for an extended supply of uranium.

We can compare the potential energy from uranium to the existing fossil fuel resources. In the U.S. the reserve estimates are

| | | |
|---|---|---|
| oil reserves | $\sim 500 \times 10^9$ barrels | $\rightarrow 3 \times 10^{21}$ J |
| coal reserves | $\sim 1500 \times 10^9$ tons | $\rightarrow 4 \times 10^{22}$ J |
| uranium reserves | $\sim 2.5 \times 10^7$ tons | $\rightarrow 5 \times 10^{21}$ J |

This energy content is to be compared to the present total yearly energy consumption in the U.S. which is $\sim 7 \times 10^{19}$ J. Note that coal could satisfy our energy needs for several centuries provided we can overcome the ecological consequences of its exploitation (greenhouse effect, acid rain) and use it in suitable form in transportation as an alternative to liquid fuels. There are also apparently vast reserves of natural gas on earth. The final determining factor in the choice of an energy source is the cost of production and one point of view holds that inevitably we will have to return to nuclear power in spite of the great hazards that it poses.

## 5.6 Radioactivity

By radioactivity, we mean the spontaneous emission of penetrating radiation by certain nuclei, a phenomenon first observed in 1896 by H. Becquerel. We discussed in Section 3 that the radiation may consist of $\gamma$-rays, that is high frequency em radiation, of electrons or positrons ($\beta$-rays), or of helium nuclei ($\alpha$-particles). Typically the emitted photons or particles have energies in the MeV range which allows them to penetrate through matter. However, as the particles traverse the material, they interact with individual atoms and ionize them, break up molecules and on occasion cause a nuclear interaction. These interactions result in changes in the structure of the material and are particularly detrimental to living systems.

For X-rays and $\gamma$-rays the primary type of interaction is ionization, that is the creation of electron–ion pairs; the energy necessary to ionize the atoms is supplied by the radiation. The *Roentgen* is defined as the amount of radiation that produces in $1\,\text{cm}^3$ of air an ionization equal to $1\,\text{esu} = 3.3 \times 10^{-10}$ coulombs, namely $2 \times 10^9$ electron–ion pairs. Since the density of air is $\sim 1.29 \times 10^{-3}\,\text{g/cm}^3$, 1 Roentgen of radiation produces $\sim 1.6 \times 10^{12}$ electron–ion pairs/g of air. The ionization potential of air can be taken as $E_\text{I} \sim 30\,\text{eV}$ so that 1 R corresponds to a deposition of $48 \times 10^{12}\,\text{eV/g}$ in air, or

$$1\,\text{R} \rightarrow 78\,\text{ergs/g} \quad \text{in air}$$

Because we are interested in the effects of radiation in living systems, the standard unit is the *rad* which is defined as the amount of X-ray radiation that deposits 100 ergs/g in tissue

$$1\,\text{rad} \rightarrow 100\,\text{ergs/g} \quad \text{in tissue}$$

Finally, since different types of radiation have different biological efficiency, a new unit is introduced; roentgen equivalent mammal, or *rem*. One rem of radiation causes the same biological effect when absorbed by a mammal as would 1 *rad* of X-rays. The radiation dose in rem equals the radiation dose in rad multiplied by the relative biological effectiveness (RBE). Charged particles and fast neutrons have RBE $\sim 10$ because they are more effective than X-rays in causing biological damage.

While the roentgen is a unit of radiation, we need also a measure for the intensity of a radioactive source. One Curie (Ci) is defined as a source that undergoes $3.7 \times 10^{10}$ disintegrations per second. This is the activity of 1 g of radium (including the activity of the surviving radium decay products). There is no exact conversion between the curie and the roentgen because the relation depends on the type and energy of the radiation involved, as well as on the geometry. However, an approximate connection is that 1 Ci of $^{60}$Co will produce at a distance of 1 m a radiation of 1.3 rem/hr.

$$1\ \text{Ci} = 3.7 \times 10^{10}\ \text{disintegrations/s}$$

$$1\ \text{Ci of } ^{60}\text{Co at } 1\ \text{m} = 1.3\ \text{rem/hr} \tag{5.21}$$

The above units of radiation have been in use since the discovery of radiation phenomena and are of historic origin. Recently, MKS units have been adopted to replace the previous units as follows: one *becquerel* (Bq) measures the intensity of a radioactive source and corresponds to one disintegration per second; therefore

$$1\ \text{Ci} = 3.7 \times 10^{10}\ \text{Bq}$$

The radiation dose is measured in *gray* (Gy) which is equal to 100 rad. It corresponds to the deposition of 1 J of energy per kg of tissue; therefore

$$1\ \text{Gy} = 100\ \text{rad}$$

Finally the *sievert* (Sv) replaces the rem, where

$$1\ \text{Sv} = 100\ \text{rem}$$

Radiation is attenuated when passing through matter; for X-rays or $\gamma$-rays the intensity after traversing a distance $l$ is given by

$$I = I_0 \mathrm{e}^{-l/\lambda}$$

Here $\lambda$ is called the radiation length; for MeV $\gamma$-rays passing through concrete $\lambda \sim 10$ cm, whereas for lead $\lambda = 0.5$ cm. Charged particles have a definite range to which they can penetrate; typically they lose 2 MeV per (g cm$^{-2}$) of material traversed. Thus a proton of 100 MeV kinetic energy will not penetrate beyond approximately 50 g/cm$^2$ which corresponds to 5.5 cm of copper.

As noted previously the lifetime for radioactive decay varies by many orders of magnitude for different nuclei. If the lifetime is $\tau$, then given a sample of $N_0$ nuclei, there will be only $N(t)$ surviving after a time $t$.

$$N(t) = N_0 \mathrm{e}^{-t/\tau} \tag{5.21$'$}$$

Thus the intensity of the source is

$$\frac{\mathrm{d}N}{\mathrm{d}t} = \frac{1}{\tau} N_0 \mathrm{e}^{-t/\tau} = \frac{N(t)}{\tau} \tag{5.21$''$}$$

Sources with short lifetimes have the advantage of decaying rapidly, even though the intensity is high. Sources with long lifetime on the other hand are less intense, but can last for an extremely long time.

The danger that radioactive substances pose to living organisms is not only from direct exposure but also from the possibility of ingestion of long-lived radioactive materials. These substances can then locate in various parts of the body and subject it to continuous radiation. The short-term effects of radiation on humans are fairly well documented: for whole body exposure received in a few hours the following effects occur

| | |
|---|---|
| 10 rem | detectable blood changes |
| 200 rem | injury and some disability |
| 400 rem | 50% deaths in 30 days |
| 600 rem | 100% deaths in 30 days |

The present safety standards require that occasional exposure of the public not exceed 0.5 rem/year. For radiation workers the limit is 3 rem in any consecutive 13 weeks and less than 5 rem/year. Cosmic rays and natural radioactivity contribute about 100 mrem/year while man-made sources contribute another 70 mrem/yr. The maximum permissible quantities of ingested radioisotopes depend on the lifetime of the isotope, but are typically of the order of 10–100 $\mu$Ci.

We can apply these considerations to the operation of a nuclear reactor with the parameters used in the example of the previous section. We had found that the fission rate was $R_F = 3 \times 10^{19}$/s; thus the radioactive intensity of the core is about $I = 10^9$ Ci. *Without* shielding, the flux at a distance of 100 m would be $F \simeq 10^5$ rem/hour (see Eq. (5.21)). Since 600 rem is a lethal dose, exposure to the reactor flux for $\Delta t = 6 \times 10^{-3}$ hr $\simeq$ 20 s would bring certain death. This example slightly exaggerates the problem because in practice most of the radioactive materials are contained inside the fuel elements and it is $\gamma$-rays and neutrons that are the primary components of the radiation; it shows however the importance of shielding and the potential for exposure to those working near the core.

When a nucleus fissions the fragments are neutron rich and therefore radioactive. They return to the region of stable nuclei primarily by $\beta^-$ decay but also by neutron emission. Among the many isotopes produced during fission are radioactive gases such as xenon-135 ($^{135}_{54}$Xe), which has a 9.2 hour lifetime, and long-lived elements such as cesium-137 ($^{137}_{55}$Cs), or strontium-90 ($^{90}_{38}$Sr), the latter having a lifetime of 28 years. If these elements escape from the reactor they will be distributed not only over the immediate area, but can also be transported over long distances in the form of radioactive fallout. For instance during the Chernobyl accident in 1986 the release of radioactive material amounted to $10^7$ Ci on the first day and persisted at the level of $\sim 2 \times 10^6$ Ci/day for several days; this amounted to about 3% of the total activity of the core. Since the maximum burden for ingestion by humans is of order 10–100 $\mu$Ci, this amount of activity posed a considerable threat to human life in the area near the reactor. Similar considerations apply to the disposal of the used fuel elements and other nuclear wastes because of the danger that the long-lived isotopes will, over many years, find their way back into the ecosystem.

## 5.7 Controlled fusion

If one takes a long-term view of the future of our technological civilization – on the scale of millennia – one concludes that fusion must be made to work. The fuel is limitless, there is no $CO_2$ produced and there are no fusion products to be disposed of. Of course no energy creation process is totally risk-free. A fusion reactor would be radioactive and there would still be a waste disposal problem, but the reactor cannot 'run away' as in the case of fission. Thus it is important to persist in attacking the challenging problems that have stood in the way of controlled fusion for so many years.

The energy of the sun is produced from the fusion of its hydrogen into helium. It is therefore instructive to review the processes that take place in the sun and lead to a continuous, as contrasted to an explosive, fusion reaction. There is convincing evidence to indicate that the first step in the fusion cycle is the production of deuterons through

$$p + p \rightarrow d + e^{+} + \nu_{e} \qquad Q = 0.93 \text{ MeV} \tag{5.22}$$

Once deuterons are produced there are various paths that can lead to the tightly bound ${}^{4}_{2}\text{He}$ nucleus. For instance

$$p + d \rightarrow {}^{3}_{2}\text{He} + \gamma \qquad Q = 5.49 \text{ MeV} \tag{5.23}$$

$${}^{3}_{2}\text{He} + {}^{3}_{2}\text{He} \rightarrow {}^{4}_{2}\text{He} + p + p \qquad Q = 12.86 \text{ MeV} \tag{5.23$'$}$$

In this case, six protons have fused into one ${}^{4}_{2}\text{He}$ nucleus with 2 protons and 2 positrons free in the final state. The total energy release is 25.7 MeV. Another possible path is

$$d + d \rightarrow {}^{3}_{2}\text{He} + n \qquad Q = 3.3 \text{ MeV} \tag{5.24}$$

followed by the reaction of Eq. (5.23′). The $(d + d)$ reaction leads also to the formation of tritons (${}^{3}_{1}\text{H}$), for which we will use the symbol $t$

$$d + d \rightarrow t + p \qquad Q = 4.0 \text{ MeV} \tag{5.24$'$}$$

This is then followed by the reaction

$$d + t \rightarrow {}^{4}_{2}\text{He} + n \qquad Q = 17.6 \text{ MeV} \tag{5.25}$$

All of the above reactions, except for the first one, involve only the nuclear force and have a reasonably large cross-section. The first reaction however (Eq. (5.22)) involves a positron and thus the weak force, therefore it is not very probable. It is the probability for the *weak* reaction that sets the rate at which the sun burns its fuel. In stars larger than our sun another cycle involving heavier nuclei, the carbon cycle, is the primary mechanism for the fusion of protons to ${}^{4}_{2}\text{He}$.

For any of the above reactions to take place the charged particles must have sufficient energy to overcome the repulsive Coulomb force in order to reach within the range of the nuclear force. As an example we calculate the electrostatic potential energy between a deuteron and a triton at a distance $r = 10$ F from one another; setting $Q_1 = Q_2 = 1$ we obtain

$$V = \frac{e^2}{4\pi\varepsilon_0}\frac{Q_1 Q_2}{r} \simeq 150 \text{ keV}$$

Each particle must have kinetic energy (K.E.) $\sim 75$ keV. This energy will be due to the thermal motion of the particles. Thus by setting $kT_f = (\text{K.E.})_f$, we find a temperature

$$T_f \sim 10^9 \text{ K} \tag{5.26}$$

This is much higher than the temperature of the interior of the sun, which has the value

$$T_0(\text{interior}) \sim 1.6 \times 10^7 \text{ K} \tag{5.26'}$$

The apparent contradiction between Eqs. (5.26) is resolved if we recall that the Maxwellian thermal distribution has a long tail, as shown in Fig. 5.12. Thus, even when the mean value $\langle kT \rangle = 0.75$ keV there are sufficient particles with kinetic energy $\sim 10^3 \langle kT \rangle$ to sustain the reaction of Eq. (5.25). At these temperatures the atoms are completely ionized and the sun's interior is a plasma held together by the gravitational force.

The rate at which a reaction proceeds is usually given by the product of the cross-section and of the incident flux. In the case of fusion we do not have a well-defined beam and it is preferable to give the rate of reactions per unit volume; we introduce the symbol $R$ for the number of fusions per cm$^3$ per s. If the density (cm$^{-3}$) of particles of species 1 is $n_1$ and that of species 2 is $n_2$ then we write

$$R = n_1 n_2 \langle \sigma v \rangle_{12} \tag{5.27}$$

where $\langle \sigma v \rangle_{12}$ is the *reactivity* and is expressed in (cm$^3$/s). It is the average value of the cross-section multiplied by the relative velocity between the

Fig. 5.12. Maxwell distribution of the velocity $v$ of particles in thermal equilibrium at temperature $T$.

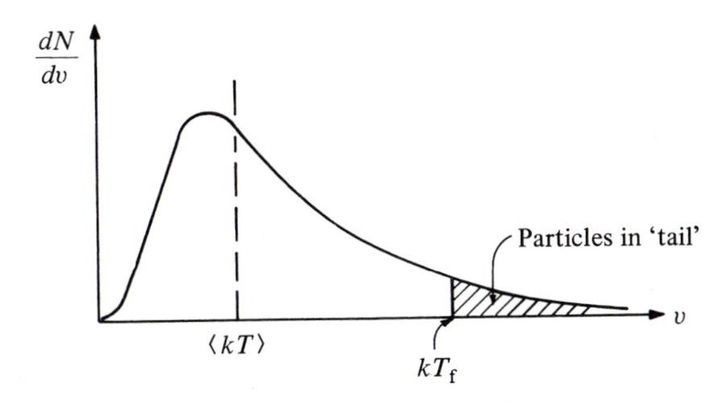

particles of the two species. Note that the thermal velocity is not unique and that the cross-section depends exponentially on the velocity; thus we must use an average value such as $\langle\sigma v\rangle_{12}$. The reactivity for the *d–d*, *d–t* and *d*–$^3$He reactions (Eqs. (5.24, 25)) is shown in Fig. 5.13 as a function of the kinetic energy of the particles.

From the data in Fig. 5.13 we see that the (*d–t*) reaction is the most favored one at the lower temperatures. As a numerical example let us assume the peak reactivity, $\langle\sigma v\rangle_{12} = 10^{-15}$ cm$^3$/s and a density of particles, $n_1 = n_2 = n = 10^{14}$/cm$^3$. This would result in a fusion rate

$$R = n_1 n_2 \langle\sigma v\rangle_{12} = 10^{13}/\text{cm}^3\text{-s}$$

Thus, it would take 10 seconds for all the particles to fuse. Therefore the plasma must be *contained* for that length of time in order to fuse all the fuel. In general, if the containment time is $\tau$, we want the fusion rate density $R$ to be such that $R\tau \geqslant n$, or

$$n\tau \geqslant \frac{1}{\langle\sigma v\rangle_{12}} \tag{5.28}$$

The requirements of Eq. (5.28) are to some extent contradictory and point out the key problem in achieving controlled fusion. One needs high

Fig. 5.13. The reactivity (cross-section times velocity) for various fusion reactions as a function of particle energy.

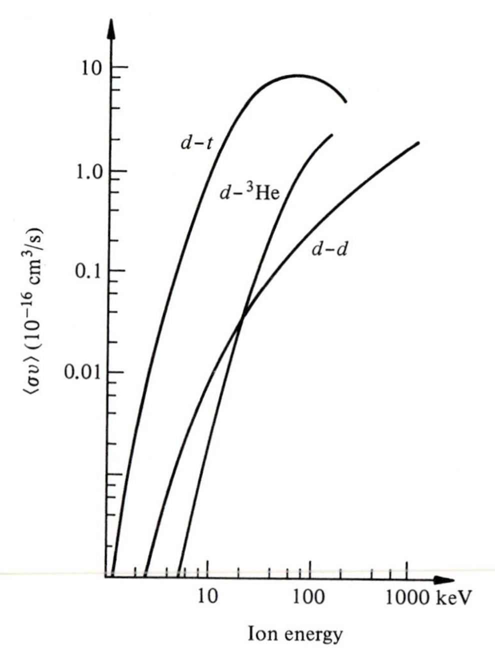

density, but as soon as the fuel ignites the available energy disperses it, lowering the density in a very short time. In fact, the energy released by the fusion reaction is proportional to $n^2\tau$

$$E_F = an^2\tau$$

Whereas the energy required to heat the fuel is proportional to $n$

$$E_H = bn$$

with $a, b$ coefficients that can be calculated from the basic properties of the reaction. If the ratio

$$Q = \frac{E_F}{E_H} = \frac{a}{b} n\tau \tag{5.29}$$

exceeds unity there is energy gain; equivalently

$$n\tau > \frac{b}{a} \tag{5.30}$$

This relation is called the Lawson criterion and for the $d$–$t$ reaction $b/a \sim 10^{14}\ \text{cm}^{-3}$-s. The Lawson criterion is less stringent than the inequality of Eq. (5.28) where it was assumed that all the fuel would fuse. Nevertheless, we must heat the plasma to temperatures of order $T \sim 7 \times 10^8$ K and contain it for about one second. The typical pressure in the plasma is of order $\sim 10^{-5}$ atmospheres, which corresponds to the density of $10^{14}$ atoms/cm$^3$.

Various methods have been proposed for confining the hot plasma based on the use of magnetic fields in various configurations. For instance charged particles will spiral around the magnetic field lines as shown in Fig. 5.14($a$). If the magnetic field is configured as a torus, then the particles will be trapped as indicated in ($b$) of the figure. While this idea is simple in principle, its practical realization involves large, complex and highly sophisticated apparatus. The heating of the plasma requires special

Fig. 5.14. Confinement of charged particles by a magnetic field: ($a$) linear field, ($b$) toroidal field.

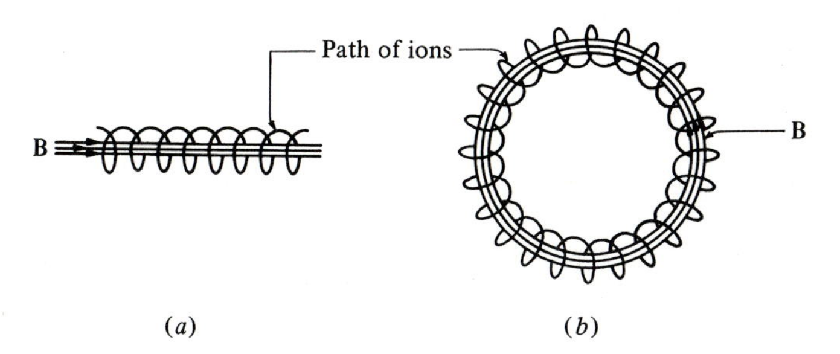

techniques and obviously the particles must not be allowed to touch the vacuum chamber wall.

Considerable effort is being devoted toward achieving controlled fusion, but the gain factor (see Eq. (5.29)) is still below unity. When $Q > 1$ is reached the released energy will have to be converted to useful electrical energy. This can be achieved by stopping the fusion products in a vessel surrounding the fusion chamber and using the heat to drive a thermal engine. Since the fusion products include neutrons (see Eqs. (5.22–25)), a possible use of a fusion reactor could be to breed fuel for fission reactors.

An alternative to magnetic confinement of the plasma is *inertial* fusion, where high density (solid) targets are bombarded by intense laser beams or by charged particle beams. These techniques must also reach $Q \geqslant 1$ and in addition compensate for the efficiency with which the laser or particle beams are produced. The fuel that is envisaged is deuterium and tritium. Deuterium is abundant and can easily be extracted from sea water. Tritium is radioactive with a lifetime of 12.3 years. It is produced in nuclear reactors where it is bred from the reaction $n + {}^6\mathrm{Li} \rightarrow {}^4\mathrm{He} + t$. The first task is to demonstrate in the laboratory that controlled fusion is feasible, beyond that considerable engineering effort will be required to produce operating fusion reactors.

In 1989 it was reported in the press that fusion had been achieved at room temperature, by a simple chemical reaction. These reports turned out to be completely unfounded but there exists a known reaction where fusion takes place without reaching the high temperature needed to overcome the Coulomb barrier. This process is mediated by $\mu$-mesons, particles which are like heavy electrons ($m_\mu \simeq 210 m_e$) but which are short-lived; the lifetime of the muon is $\tau_\mu = 2 \times 10^{-6}$ s. The process is known as muon catalysis and was discovered in 1956 by the Alvarez group at Berkeley.

Because the negative muon has the same properties as the electron, if it is stopped in a mixture of deuterium ($D_2$) and tritium ($T_2$) gas, it can substitute for one of the electrons in these molecules, forming a 'muonic atom' or 'muonic molecule'. The following chain occurs

$$\mu^- + \mathrm{D} \rightarrow \mathrm{d}\mu + \mathrm{e}^-$$
$$\mathrm{d}\mu + \mathrm{t} \rightarrow \mathrm{t}\mu + \mathrm{d}$$
$$\mathrm{t}\mu + \mathrm{D}_2 \rightarrow \mathrm{dt}\mu + \mathrm{D} + \mathrm{e}^-$$

However, because of the large mass of the muon, the 'Bohr orbit' in the $\mathrm{dt}\mu$ muonic molecular ion is 200 times smaller than for the normal electronic ion. As a result, the deuteron and triton have a high probability of being close enough to fuse according to the reaction.

$$\mathrm{d}t\mu \rightarrow (^5\mathrm{He})\mu \rightarrow {}^4\mathrm{He} + \mu^- + \mathrm{n} + 17.6\ \mathrm{MeV}$$

Another way of thinking of this process is to realize that the $t\mu$ atom is electrically neutral (acts like a neutron) and thus can approach the deuteron nucleus without seeing a Coulomb barrier.

Muon catalyzed fusion is efficient and the same muon can initiate as many as 150 fusions, before decaying. This corresponds to an energy release of $\sim 3$ GeV which is near the break-even point for the production of the muon (from $\pi^-$-decay) in energetic particle collisions. Energy production by muon catalyzed fusion is still at an early stage of exploration but could be promising.

## 5.8 Solar energy

In principle, solar energy is a non-polluting inexhaustible source of energy. Wind power, the growth of organic materials that can be used as fuels, as well as the fossil fuels available are derived indirectly from the energy delivered to the earth by the sun. The direct exploitation of solar energy is made difficult by its relatively low concentration, the day/night effect and the dependence on cloud cover coupled with the complexity of energy storage. At present, the conversion efficiency is relatively low so that solar power is significantly more expensive than power derived from fossil fuels. As a result the motivation for exploiting solar energy and for research on new conversion techniques is damped.

As indicated in Eq. (5.8) the solar constant has the value

$$S = 1.36\ \mathrm{kW/m^2} \tag{5.8}$$

To estimate the average flux incident per unit surface of land area we must account for the inclination of the sun's rays with respect to the vertical, and consider the night/day effect, the atmospheric absorption and the cloud cover at any particular location. As a result the average flux incident per square meter of land area in the U.S. is

$$S = 0.20\ \mathrm{kW/m^2} \tag{5.31}$$

From Table 5.1 the total energy consumption in the U.S. in 1970 was approximately $7 \times 10^{19}$ J, corresponding to a power rating

$$P_{\mathrm{vs}} \simeq 2 \times 10^9\ \mathrm{kW}$$

If all of this power was to be obtained from the sun, even with 100% efficiency, the collector would have to cover an area of $10^{10}$ m$^2$, that is, a region 60 by 60 miles squared.

The above estimate does not account for conversion efficiency, which presently averages ~10%. One could place the collector in a favorable location, such as a desert, but then the power would have to be transported to the urban centers. The investment in a plant of that size as well as its maintenance could prove prohibitively costly. Exotic schemes such as placing the collectors in space in a geostationary orbit, face similar problems of very large capital investment. On the other hand, use of solar energy to reduce in part the dependence on fossil and nuclear fuels is practical, and would add significantly to the conservation of the non-renewable sources of energy. It can be argued that short of the realization of abundant fusion energy, the present growth of global energy consumption cannot be sustained into the long-term future.

The most efficient exploitation of solar energy is by direct conversion to electrical power. This can be achieved by using 'photovoltaic cells'; silicon cells are extensively used on spacecraft and in other specialized applications. Another approach involves the focussing of sunlight onto small areas so that temperatures sufficiently high to drive thermal engines can be reached. Yet the simplest use of solar energy is in heating buildings and water for household use and in other applications of 'low quality' heat. We recall that these needs are almost 25% of the total energy consumption; of course where heating is most needed, the insolation is the lowest.

*The basic collector.* The central idea in a collector of solar energy is the same as in the heating of the earth by the sun. The visible light enters the collector and is absorbed, but the infrared emitted by the collecting surface is prevented from being radiated away. A sketch of a simple collector is shown in Fig. 5.15. Glass surfaces are adequate to trap the IR while admitting the visible; the absorber can be a coil through which water is circulated, and the whole assembly is insulated to reduce leakage by

Fig. 5.15. Simple solar collector for heating water in household applications.

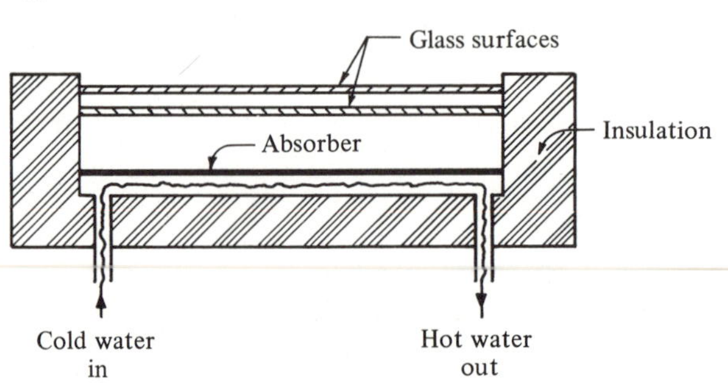

conduction and convection. Selective coating of the glass surfaces can improve the performance of the collector by reflecting wavelengths longer than $\lambda > 2\ \mu$m.

Simple collectors can achieve fairly high temperature differentials. When no energy is withdrawn the collector reaches its maximum temperature difference, $\Delta T_{max}$; this is shown below for various levels of insolation. The last column is calculated for a collector area of 50 m$^2$ and a 50% efficiency; in that case the temperature rise of the circulating water reaches half of $\Delta T_{max}$.

| Insolation | $\Delta T_{max}$ | Useful energy (50 m$^2$) |
|---|---|---|
| 0.32 kW/m$^2$ | 48°C | 8 kW |
| 0.64 | 85°C | 16 |
| 1.00 | 120°C | 25 |

The average household power needs are estimated at 6 kW whereas typical commercial collectors units have 5 m$^2$ areas. Thus for single dwellings it would be possible to find sufficient area to install adequate solar heating. In contrast, this could not work in densely populated cities.

*The basic concentrator.* Sunlight is highly collimated and therefore it can be precisely focussed. Everyone has toyed with lenses or mirrors to show that sunlight can ignite a fire at the focus. The sun subtends at the earth an angle

$$\theta_\odot = 9 \times 10^{-3} \text{ rad}$$

so that the image produced by a mirror of focal length $f$ has area $a = \pi(f\theta_\odot/2)^2$. If further the mirror has a radial aperture $R = f$, the resulting concentration of the solar flux is

$$C = \frac{A}{a} = \frac{\pi R^2}{\pi(R\theta_\odot/2)^2} = \frac{4}{\theta_\odot^2} \sim 5 \times 10^4 \tag{5.32}$$

Thus, very high energy densities can be reached.

The technical difficulty with spherical concentrators is that the large area mirror must track the sun. Alternate schemes such as cylindrical troughs, or plane mirrors have been used in different installations. The most ambitious such plant was built by the French in the Pyrenees in 1970 and generated 1 MW of power. At that altitude (2000 m) an insolation of 1 kW/m$^2$ can be expected with sunshine for half of the days of the year. The collector was fixed as shown in Fig. 5.16 and consisted of 10 000 mirrors, each of area $0.4 \times 0.4$ m$^2$. A movable array of plane mirrors on the side of the mountain tracked the sun to illuminate the collector mirrors; it comprised some 12 000 mirrors of $0.5 \times 0.5$ m$^2$ area. Temperatures as high as $T = 3500$°C were easily obtained at the focus. However the support of the mirrors and other engineering problems made the operation of the plant unprofitable.

An alternative to exact focussing of the sunlight is the use of fixed collectors as shown in Fig. 5.17. These are named 'Winston cones' and light incident on the aperture of the cone is guided to the collecting area, irrespective of its angle of entrance.

*Photovoltaic cells.* Solar energy is highly ordered, as compared to random thermal energy and for this reason it is possible to convert it directly to electrical energy. Visible photons can easily produce electron–hole pairs in a semiconductor where the energy gap is of order $E_g \sim 1\,\text{eV}$ (see Section 1.1). By using *p*-type silicon with an *n*-type layer on its surface we can construct a junction such that electrons liberated in the silicon will move toward the surface layer. This is indicated in Fig. 5.18 which gives the energy diagram for the junction (see also Section 1.4) with no bias applied. Electrons generated in the *p*-type material near the junction will cross over the junction into the *n*-type layer, producing a voltage difference between the two surfaces of the cell.

Fig. 5.16. High concentration solar collector installation in the French Pyrenees; note the dimensions of the mirror system.

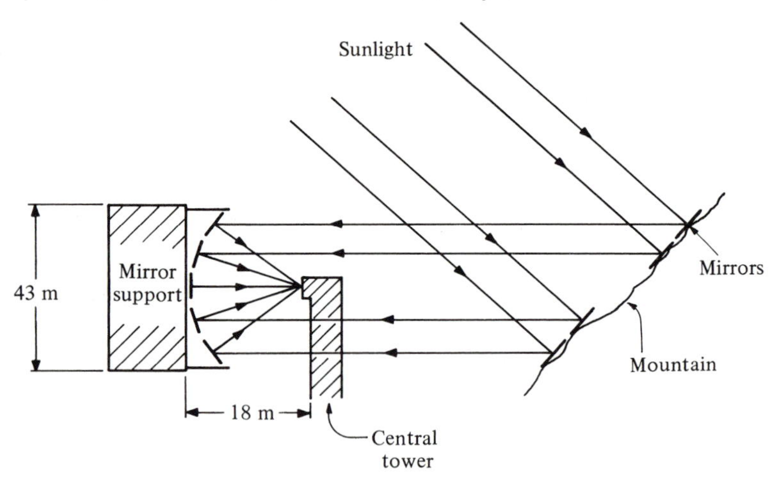


Fig. 5.17. A reflecting 'Winston' cone for trapping and collecting solar radiation.

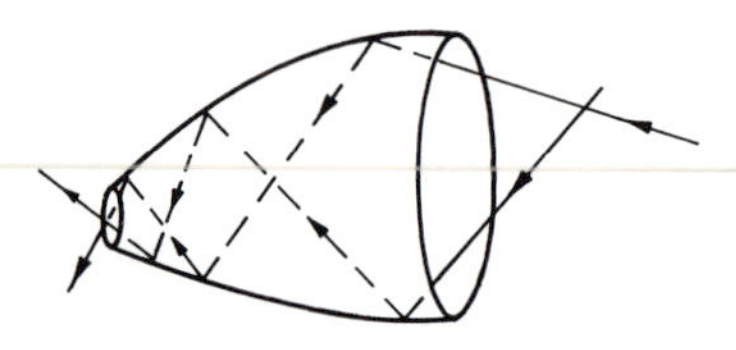

The $I$–$V$ characteristic for a typical silicon cell of $10 \times 10\,\text{cm}^2$ area is shown in Fig. 5.19. It can support a voltage of 2.5 to 3 V and at an insolation of $\sim 1\,\text{kW/m}^2$ it can deliver about 1.5 W; this corresponds to an efficiency of $\varepsilon \sim 14\%$.

To reduce the cost, solar cells are manufactured from amorphous silicon, but even so they are still priced at ~$20/cell. To be economically competitive the price would have to be further reduced by a factor of 100. For instance, a 3.5 kW system built in the Papago Indian Reservation in Arizona cost $330 000 including the storage batteries. If the capital is to be amortized at 6% per year, the cost of the electrical energy comes to 70 cents per kWh; this is to be compared to commercial rates of ~10 cents per kWh. The lifetime of the cells is not known and furthermore present manufacturing techniques are still very energy intensive.

Fig. 5.18. Energy band diagram for a photovoltaic cell; the cell is a diode made of $p$-type and $n$-type semiconductor with an energy gap $E_g \sim 1\,\text{eV}$.

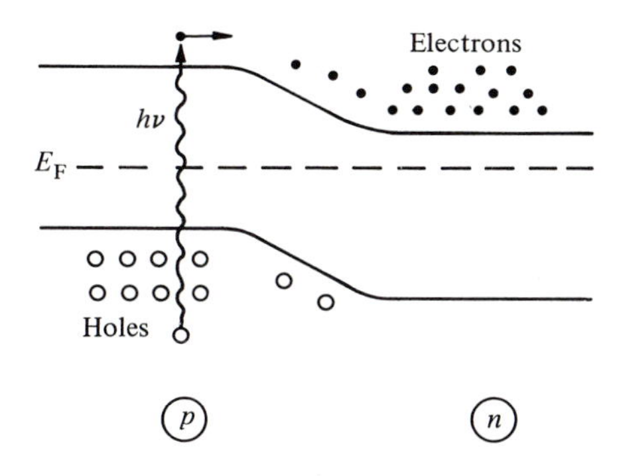


Fig. 5.19. $I$–$V$ characteristic for a typical photovoltaic cell made of silicon.

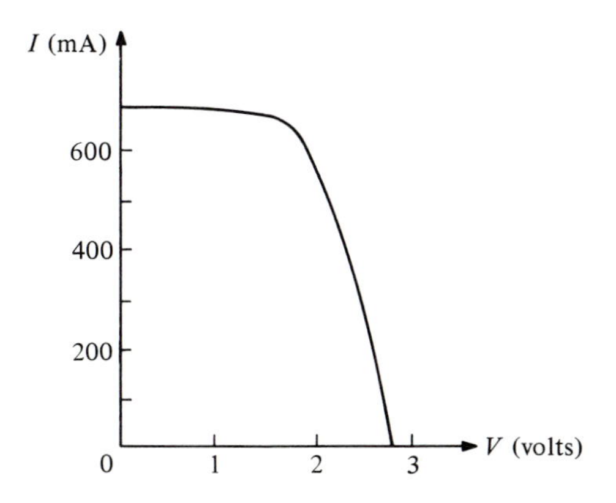

*Generating stations in space.* It has been proposed to place the converters in geostationary orbit and to transport the generated power to earth. This is an elegant solution to the insolation and area coverage problems, and the energy would be transported by a high power microwave link. The capital investment has been estimated at $2000/kW, approximately the same as for nuclear reactors; for the solar plant the largest part of the cost is in placing the generator in orbit. Notwithstanding the above arguments, the system does not seem practical. If we assume a 10 GW plant we will need at least $10^4$ tons in orbit; at the present rate this corresponds to 300 trips of the shuttle! Furthermore, an efficient 10 GW continuous microwave link is a very difficult undertaking.

In conclusion, solar energy will certainly be used in the future to cover a part of the energy needs of our civilization. Yet it does not seem probable that it can fully replace our immediate dependence on fossil fuels. In particular for the long term future a civilization based on energy consumption will have to rely on thermonuclear fusion, rather than depend only on solar energy.

## Exercises

### *Exercise 5.1*

A nuclear reactor uses 1 ton of $^{235}U$ per year.

(a) Assume 40% thermal efficiency conversion, and calculate the *power* in watts produced by the reactor.

(b) If the *entire* U.S. demand in energy were to be satisfied by such reactors, how much uranium would have to be mined per year? What fraction of the total U.S. reserves in uranium does this represent?

### *Exercise 5.2*

The height of the orbit of geostationary satellites is 36 000 km.

(a) If such a satellite is to transmit at 1 W calculate the size of the solar panels needed. Assume reasonable numbers for solar power conversion efficiency.

(b) Calculate the size of the antenna of the satellite if it transmits at $f = 4$ GHz and the antenna gain is 500.

### *Exercise 5.3*

Good grassland will support one 1000 lb cow per acre; the cow in turn produces 4000 lb of milk per year. Milk is 88% water and 4% fat (9 kcal/g) and 8% non-fat solids (4 kcal/g). Calculate the ratio of the energy content (calories) of the milk to that of the solar energy incident on the field in one year. This gives a measure of the energy conversion efficiency in agriculture.

***Exercise 5.4***

(a) Assume that the proton is unstable and decays with lifetime $\tau$. Calculate the number of protons in a human being and note that the radiation level is less than 10 mrem/yr which corresponds to less than 1 $\mu$Ci of radioactivity in the body. From these two facts find *a limit* for the lifetime $\tau$ of the proton. In addition:
(b) Compare with the age of the universe.
(c) Do you know what the experimental limit is on $\tau$(?)

# 6

# NUCLEAR WEAPONS

## 6.1 Fission and fusion explosives

The realities of World War II led to the rapid development of nuclear technology in the U.S. culminating in the construction of the first nuclear weapons. The first nuclear explosion took place in the Alamogordo Desert in New Mexico on July 16, 1945 in a test code-named Trinity. Soon thereafter, on August 6, the first nuclear bomb was dropped over Hiroshima in Japan and totally destroyed the city, causing over 150 000 casualties. A second bomb was dropped on Nagasaki on August 9, and on August 14 Japan surrendered unconditionally thus ending W.W.II. Today the nuclear capability of the two superpowers (the U.S. and the U.S.S.R.) is more than a hundred thousand times that of W.W.II and many other nations have acquired nuclear weapons. Yet it is a sign of hope that no nuclear weapons have been used in hostilities since those first two explosions in 1945.

In the previous chapter and in particular in Sections 5.3 and 5.4 we discussed extensively the fission of heavy nuclei when bombarded by neutrons. In the fission of $^{235}U$, on the average 2.4 neutrons are released and if these neutrons could be used to initiate further fission, a chain reaction can develop. This process is exploited in nuclear reactors where, however, the reaction rate is kept constant. As already indicated by Eq. (5.18) the probability $P_F$ that a neutron will cause fission after traversing a length $\Delta l$ of the fuel is

$$P_F = 1 - e^{-n_F \sigma_F \Delta l} \simeq n_F \sigma_F \Delta l \qquad (6.1)$$

where $n_F$ is the number density of fuel nuclei and $\sigma_F$ the fission cross-section. This probability competes with the probability that a neutron will escape from the fuel volume or that it will be absorbed in a non-fissioning reaction. Thus we can introduce an efficiency factor

$$\eta = \frac{P_F}{P_T} = \frac{P_F}{P_F + P_{NF}} \tag{6.1'}$$

where $P_{NF}$ is the probability for escape and non-fissioning absorption. If the average number of neutrons produced in every fission is $m$, then the multiplication factor (see Section 5.3) is given by

$$K = \eta m = \frac{P_F}{P_F + P_{NF}} m \tag{6.2}$$

Recall that $P_{NF}$ depends on the geometry of the fuel volume but also on the quantity and type of the non-fissile material present.

In a weapon we want $K$ as large as possible, that is we must maximize $\eta$; to achieve this we wish to increase $P_F$ and decrease $P_{NF}$. Another necessary condition is that the reaction proceed quickly (in $10^{-6}$–$10^{-3}$ s) in order to create an explosion. This imposes the condition that we use *fast* neutrons in the chain reaction since the moderation process takes $10^{-3}$ s for each generation of neutrons. However, the fission cross-section for fast neutrons is smaller than that for thermal neutrons, typically $\sigma_F(E \sim 1\,\text{MeV}) \simeq 5 \times 10^{-24}\,\text{cm}^2$. To compensate for the small cross-section the density of fissile material should be high (see Eq. (6.1)), which is achieved by using highly enriched fuel. Furthermore, the purer the fuel the smaller the probability of non-fission captures, reducing $P_{NF}$. Typically, if a value of $\eta = 0.5$ can be achieved (for fast neutrons), $K = 1.2$ leading to rapid neutron multiplication.

The efficiency factor $\eta$ depends on the dimensions of the fuel volume through both $P_F$ and $P_{NF}$. Thus for every type of fuel assembly when a certain mass is exceeded it will spontaneously undergo a fission chain reaction. This mass is called the *critical mass* and we gave a rough estimate for it in Section 5.3. The fuel used in fission weapons is either $^{235}$U or $^{239}$P and we give below estimates for the critical mass according to D. Schroeer (*Science Technology and the Nuclear Arms Race*, J. Wiley, New York, 1984). The efficiency $\eta$ can be increased by surrounding the fuel volume with materials such as beryllium, which reflect escaping

Table 6.1. *Critical mass for different fissile materials*

| | |
|---|---|
| Uranium enriched in $^{235}$U by 20% | 160 kg |
| Uranium enriched in $^{235}$U by 50% | 68 kg |
| Uranium enriched in $^{235}$U by 100% | 47 kg |
| Plutonium-239 (pure) | 19 kg |

neutrons back into the volume. Consequently, when reflectors are used criticality can be reached for smaller masses of fuel.

An explosion is the result of energy release in a very short time interval. One of the most powerful chemical explosives is TNT (trinitrotoluene, $C_6H_2(CH_3)(NO_2)_3$); the explosion of 1 kg of TNT releases approximately 1000 kilocalories of energy. This corresponds to $W \sim 4 \times 10^6$ J which is the energy consumed by a 40 W light bulb burning for one day. Thus it is not a large amount of energy, but its sudden release produces dire results. The yield of nuclear weapons is measured in terms of the TNT tonnage that would create similar effects. Here 1 ton $= 10^3$ kg of TNT and we will use the notation kt and Mt as units of explosive yield.

The energy released in the fission of the $^{235}$U nucleus is $\sim 200$ MeV as compared with a few eV released in chemical reactions. Thus, for a given mass of explosive, nuclear materials release $10^7$ times more energy than conventional weapons. Furthermore the energy is released in a very short time as we show below. We designate by $\tau$ the time between the creation (birth) of a neutron and its subsequent absorption (leading to fission). For fast neutrons the time between 'generations' is of order $\tau \sim 10^{-8}$ s (as compared to $10^{-3}$ s for the moderated neutrons in a reactor). If the chain reaction starts at $t = 0$, then the number of nuclei that have fissioned by the time $t$ is

$$N(t) = N_0(K)^{t/\tau} \tag{6.3}$$

where $K$ is the multiplication factor.* Thus if $N_0 = 1$, the number of generations needed to completely fission $N$ nuclei is

$$\frac{t}{\tau} = \log(N)/\log(K) \tag{6.3'}$$

For 10 kg of fissile material, $N \sim 2.5 \times 10^{25}$ and if we take $K = 2$ the number of generations is $(t/\tau) \sim 85$. Thus the total time for the explosion is of order of $t \sim 10^{-6}$ s. Even if we had chosen $K = 1.1$ the explosion time would have been only six times longer. Most of the energy is of course released by the last few generations of neutrons as the entire fuel mass fissions. Thus it is important to contain the fuel and prevent its dispersion as the explosion proceeds. This is achieved in part because of the inertia of the fuel and because of the very short explosion time. The addition of 'tamper materials' is designed to increase the inertia and thus the confinement of the fuel in order to achieve a more efficient explosion.

The weapon used against Hiroshima was constructed from 70% enriched $^{235}$U and, apparently, was assembled in two subcritical masses

* The multiplication factor should be defined with reference to the intergeneration time $\tau$.

as shown conceptually in Fig. 6.1. The central cylindrical plug was fired by chemical explosives into a hollowed out sphere of $^{235}$U. When the two parts matched, critical mass was achieved and the assembly exploded. The yield of the weapon was $\sim$12 kt of TNT.

Enrichment of natural uranium to a high concentration of $^{235}$U, such as needed for weapons grade material, is costly and labor intensive. On the other hand plutonium-239 is also fissile and can be produced with comparative ease in high flux reactors. It appears that most present day nuclear weapons use plutonium as the fissile material; the device exploded in the Trinity test and the bomb used against Nagasaki were $^{239}$Pu weapons. Because the plutonium is rather pure and because of its large fission cross-section the realization of the critical mass has to occur more rapidly than is possible by the 'gun-firing' technique illustrated in Fig. 6.1. Therefore chemical explosives are used to compress the plutonium core to almost twice its natural density in a very short time. The arrangement is shown in Fig. 6.2 where all segments must be detonated simultaneously and the explosives are so shaped as to create an inward going shock-wave.

When the spherical core is compressed its volume changes from $V$ to $V'$, its density from $\rho$ to $\rho'$ and the radius from $r$ to $r'$; for instance if

$$V' = V/2, \qquad \text{then} \qquad r' = r/2^{1/3}, \qquad \rho' = 2\rho \tag{6.4}$$

From Eq. (6.1) we know that the critical length $l_c$ is proportional to $(1/\rho)$ and that the core will go critical if $r > l_c$. Before the implosion, $r/l_c = \alpha < 1$ so that the weapon is subcritical. The implosion compresses the core to a radius $r' = r/2^{1/3}$ and reduces the critical distance to $l'_c = l_c/2$ so that the ratio $r'/l'_c$ becomes

$$\frac{r'}{l'_c} = \frac{r}{(2)^{1/3}(l_c/2)} \simeq 1.6\frac{r}{l_c} = 1.6\alpha \tag{6.4'}$$

Thus if $\alpha$ is chosen such that $0.65 < \alpha < 1.0$ the compressed core is

Fig. 6.1. Possible design of a nuclear fission device based on enriched $^{235}$U. The chemical explosive drives the two subcritical masses together.

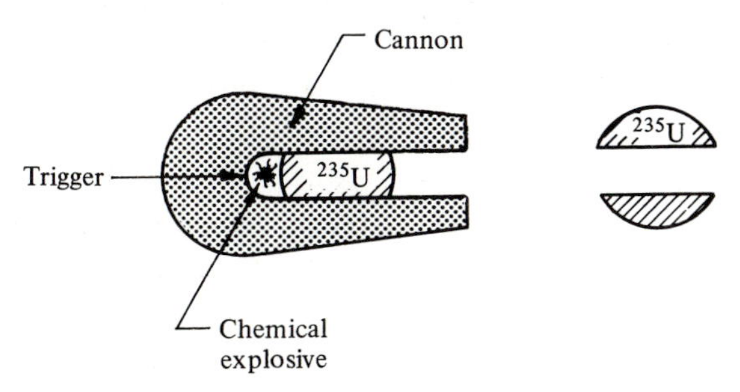

super-critical and will explode. Note also that the compression technique reduces the amount of material needed to reach criticality by the square of the compression ratio.

The Nagasaki bomb had an explosive yield of 22 kt, namely an energy release of $W \sim 10^{14}$ J. If we assume that the core had a mass of 10 kg of $^{239}$Pu the expected energy release from the fissioning of all the core material is

$$W = \frac{6 \times 10^{23}}{239} \left[ \frac{\text{nuclei}}{\text{g}} \right] \times [10^4 \text{ g}] \times 200 \text{ MeV} \sim 8 \times 10^{14} \text{ J}$$

This would indicate that only 12% of the core fissioned before being dispersed by the power of the explosion. Modern weapons are built with higher efficiency.

The existence of a critical mass places an upper limit on the size and yield of fission nuclear explosives. There is no such limitation on the fusion of light nuclei, but in this case the nuclei must have very high relative

Fig. 6.2. Possible design for the $^{239}$Pu implosion bomb; the carefully shaped chemical explosives compress the plutonium so as to reach critical density. (After D. Schroeer, *Science Technology and the Nuclear Arms Race*, J. Wiley, New York, 1984.)

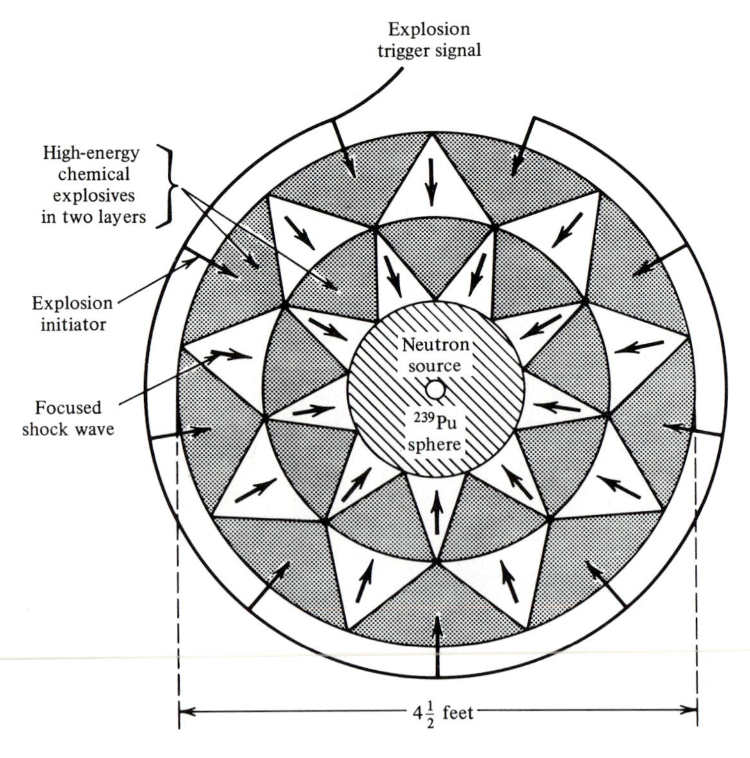

energies to overcome the repulsive Coulomb barrier; that is, the fuel must be heated to high temperature. As was discussed in Section 5.7, the most favorable reaction is the fusion of a deuteron and a triton into helium-4 according to the reaction

$$ {}^2_1d + {}^3_1t \rightarrow {}^4_2\text{He} + {}^1_0n + 17.6\ (\text{MeV}) \tag{6.5} $$

Since both deuterium and tritium are gaseous at normal temperatures it was first believed that it would not be possible to reach the necessary density for a self-sustaining fusion reaction. However, the explosion of a fission weapon by the U.S.S.R. in 1949 convinced the U.S. government to initiate a program for the construction of a fusion weapon. The first test of a thermonuclear device took place at the Eniwetok Atoll in the Pacific in late 1952; it involved cryogenic $d, t$ and had a yield of 10 Mt.

What has made fusion explosives practical is the use of lithium deuteride, LiD, as a fuel, a technique which, apparently, was first introduced by the Soviets. When bombarded by fast neutrons, lithium produces tritons according to the reaction

$$ {}^1_0n + {}^6_3\text{Li} \rightarrow {}^4_2\text{He} + {}^3_1t + 4.8\ (\text{MeV}) \tag{6.6} $$

and in an environment of high temperature the tritons fuse with the deuterons as shown in Eq. (6.5). The necessary neutron flux and the high

Fig. 6.3. Possible design for a thermonuclear weapon triggered by a fission explosive. (After H. Morland, *The Secret that Exploded*, Random House, New York, 1981.)

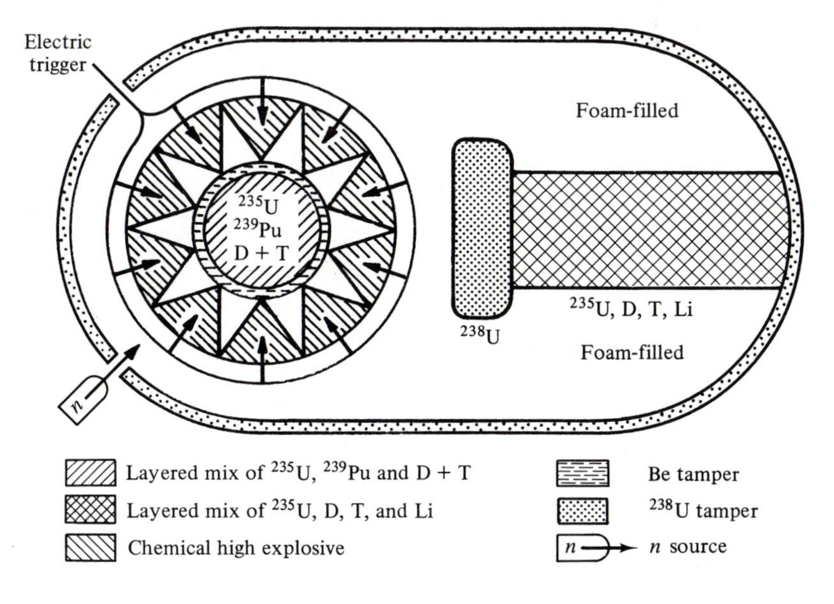

temperature $T \sim 10^7$ K are achieved by exploding a fission device in a LiD environment.

The LiD fuel is solid, and since we are seeking an explosion, the necessary confinement time is small. Thus the Lawson criterion of Eq. (5.30) can be easily satisfied. The density of LiD is of order 1 g/cm$^3$, or $n \simeq 8 \times 10^{22}$ nuclei/cm$^3$. For a confinement time $\tau \simeq 10^{-6}$ s we find $n\tau \simeq 10^{17}$ cm$^{-3}$ s, well in excess of the requirement $n\tau > 10^{14}$ cm$^{-3}$ s.

Fusion weapons, also called thermonuclear weapons or hydrogen bombs, always contain a fission device. One possible construction is shown in Fig. 6.3. The fuel consists of LiD, tritium gas (which must be replenished) as well as fissile material such as $^{239}$Pu or $^{235}$U. A plutonium fission explosive is used as a trigger. The X-rays from the fission trigger transfer the energy to the fuel igniting the fusion reaction. The whole assembly is encased in $^{238}$U which under the intense neutron flux fissions, contributing perhaps as much as 50% of the total yield. The containment of the fuel and the energy transfer between the components of the weapon are surely important details of the design. For strategic warheads the typical yield of fusion weapons is in the 1 Mt range.

## 6.2 The effects of nuclear weapons

The explosion of a nuclear weapon releases a large amount of energy in a very brief time interval. Following the explosion the energy is distributed roughly as follows

- 50% in the pressure blast
- 35% in the heat radiation
- 5% in prompt radiation
- 10% in delayed radiation

The pressure and heat are immediate effects that destroy structures, ignite fires, and, of course kill humans. The resulting prompt and delayed radiation is also lethal and can poison the environment for a very long time.

The cost in human life depends on the area that has been targeted. As an indication of the enormity of destructive power we consider the casualties from the two bombs used in W.W.II.

| | Yield | Killed | Injured |
|---|---|---|---|
| Hiroshima | 12 kt | 70 000 | 80 000 |
| Nagasaki | 22 kt | 40 000 | 20 000 |

Many present day weapons have a 1 Mt yield and it is estimated that in a full scale war between the U.S. and the U.S.S.R. 5000 Mt of explosives will be used. To emphasize the annihilation potential of such an exchange we can make a linear extrapolation from W.W.II casualties to find that

$$(210\,000)\frac{5\times 10^6\ \text{kt}}{34\ \text{kt}} \sim 30 \text{ billion people}$$

would be killed. Of course the linear extrapolation is inappropriate and the total population of the two nations is only $\sim 0.5 \times 10^9$ people. Nevertheless enough weapons have been built to eliminate the entire population of the earth several times over.

Nuclear explosives reach a much higher temperature than chemical explosives and for this reason the dominant form of energy release is in the form of radiation. From Eq. (5.6) we know that the total radiated energy per unit area per second from a black body at temperature $T$ is $S = \sigma T^4$ with $\sigma$ Stefan's constant. The energy density in the source is

$$u = \frac{4}{c} S = 7.6 \times 10^{-16} T^4 \ \text{J/m}^3\text{-(K)}^4 \tag{6.7}$$

If we assume a temperature in the explosive of order $T \simeq 4 \times 10^7$ K, we obtain $u \simeq 10^{15}$ J/m$^3$, which for a volume of 1 m$^3$ is a significant part of the total energy produced in the explosion. In contrast, in a chemical weapon $T \cong 5 \times 10^3$ K and the radiation is insignificant in comparison to the kinetic energy of the explosion products.

About half of the radiation energy is immediately transferred to the surrounding air (which is opaque to X-rays) and gives rise to the pressure blast. An overpressure of 5 psi destroys frame houses, whereas 50 psi is lethal to humans.* The em radiation is rapidly absorbed giving rise to thermal radiation where deposition of 8 cal/cm$^2$ is sufficient to cause 2nd degree burns. For a weapon with a 1 Mt yield the following effects are expected within a radius $R$ from the center of the explosion

$R \lessapprox$ 1500 m   50 psi overpressure
$R \lessapprox$ 5000 m   5 psi overpressure
$R \lessapprox$ 13 000 m   8 cal/cm$^2$

Because of the intense heat at the point of explosion a column of hot air rises (since it is less dense than cold air) from the surface, and this creates the characteristic 'mushroom cloud' of nuclear explosions. There is a very strong draft inside the column and correspondingly strong surface winds as shown in Fig. 6.4(*a*). In (*b*) of the figure is shown the actual cloud from the 10 Mt explosion of the 'Mike' test at the Eniwetok Atoll in 1952. For a 1 Mt explosion the top of the cloud reaches 60 000 ft and therefore radioactive debris and ashes are carried into the upper atmosphere from where they can be transported over long distances.

* 14.7 psi (pounds per square inch) = 1 atmosphere $\simeq 10^5$ N/m$^2$.

Furthermore, the intense surface winds cause fires to spread rapidly and to retain their intensity.

In the explosion process, the fuel becomes completely ionized and a large number of free electrons move with high velocity. This gives rise to an intense pulse of HF and VHF em radiation (EMP). This pulse is known to disrupt communications over long distances and to damage microcircuits.

Fig. 6.4. Thermonuclear explosion: (*a*) the cloud from a 1 Mt explosion, one minute after detonation. (After Glasstone and Dolan, *The Effects of Nuclear Weapons*, U.S. Government Printing Office, 1977.) (*b*) U.S. test shot of October 1952 at the Eniwetok Atoll; it demonstrated the feasibility of fusion weapons.

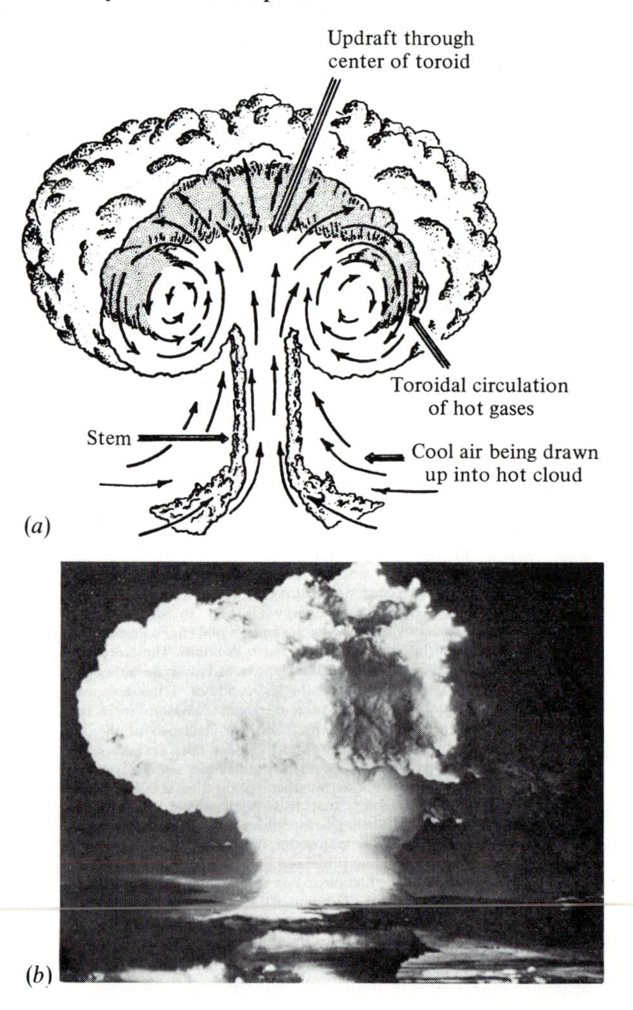

In addition to the intense heat radiation, nuclear radiation in the form of X-rays and neutral and charged particles is also present. We discussed in Section 5.6 that $\sim 500$ rem will cause death in 50% of the exposed population. For a 1 Mt explosion the integrated dose due to prompt radiation will exceed 500 rem within a radius $R \lesssim 2500$ m. More significantly, the fallout from the explosion will result in a 500 rem exposure in the first day due to latent radioactivity within a radius $R \lesssim 25\,000$ m. While a shelter with 1 m thick concrete walls could provide adequate protection against prompt radiation, survivors will have to deal with the problem of ground and food contamination.

The great danger from fallout is due to the long lived isotopes. Some of these are indicated in Table 6.2; the lifetime, the average time for retention in the body, and the maximum safe concentration in humans are indicated for each of the isotopes. In a fusion weapon the fallout comes from the fission trigger as well as from the other materials in the device that have become radioactive under the intense neutron flux. We can estimate the fallout from a 1 Mt weapon by assuming that a 10 kt trigger is used which implies $\sim 10^{25}$ fissions. Approximately 6% of these fissions lead to $^{90}$Sr, so that the amount of $^{90}$Sr produced is

$$6 \times 10^{23} \ (^{90}\text{Sr nuclei}) \qquad \text{or} \qquad \sim 1 \text{ kg}$$

The lifetime of $^{90}$Sr is $\tau = 28$ years $\simeq 9 \times 10^8$ s, and hence the activity is

$$\frac{dN}{dt} = \frac{N}{\tau} = \frac{6 \times 10^{23}}{9 \times 10^8} \simeq 6.7 \times 10^{14} \simeq 1.8 \times 10^4 \text{ Ci}$$

while this is less activity than released in the Chernobyl accident (see Section 5.6) it is a large amount as compared to the maximum safe concentration for humans.

Isotopes with shorter lifetimes result in much higher activity. If we take the mean lifetime as $\tau = 1$ day and consider that $10^{26}$ radioactive nuclei were produced, then the radioactivity on the first day after a 1 Mt explosion would be 30 000 MCi. By comparison, in the 1986 accident at the Chernobyl reactor in the U.S.S.R. the overall fallout reached 50 MCi

Table 6.2. *Long lived isotopes from weapons fallout*

| | $\tau$ | Retention | Limit of safe concentration |
|---|---|---|---|
| $^{90}$Sr | 28 years | 36 years | 20 $\mu$Ci |
| $^{137}$Cs | 30 years | 70 days | 30 $\mu$Ci |
| $^{14}$C | 5600 years | 10 days | 400 $\mu$Ci |
| $^{239}$Pu | 24 400 years | 180 years | 0.4 $\mu$Ci |

which was about 3% of the total activity of the core; this shows the magnitude of the problem that would be raised by even one 1 Mt explosion. That fallout can be transported over long distances was demonstrated dramatically in the 1954 U.S. test at Bikini. A Japanese fishing vessel located at a distance of over 100 miles from the explosion became coated with radioactive ash. Unaware of the radiation the crew made no attempt to decontaminate their ship. When they returned to port two weeks later they had received an integrated dose of 200 rem and were seriously sick from its effects. Ultimately in 1963, the U.S. and U.S.S.R. and many other (but not all) nations signed the 'Atmospheric Test Ban Treaty' which forbids nuclear tests in the atmosphere.

In the case of a major nuclear exchange, in addition to the destruction of the targeted areas one expects effects of global nature that will permanently change our ecosystem. The most serious effects come from the large amount of soot generated by the fires raging over cities and the wooded countryside. The soot will be transported to the higher atmosphere and will block the sunlight. We can estimate the change in the earth's surface temperature by a simple model along the lines indicated in Fig. 5.4(*b*). In addition to the greenhouse layer we will assume a layer of soot as shown in Fig. 6.5 which completely absorbs the visible and retransmits the energy in the infrared. The energy balance is then as indicated in the figure and the earth receives (and radiates) only $\alpha_e = (0.5\alpha_s)$ where $\alpha_s$ is the incident solar radiation. This is to be compared to $\alpha_e = 1.4\alpha_s$ in the model of Fig. 5.4 which resulted in a surface temperature $T = 303$ K. Thus in the presence of the soot layer

$$\frac{T'}{T} = \left(\frac{0.5}{1.4}\right)^{1/4} = 0.78$$

and therefore $T' = 236\text{ K} = -37°\text{C}$.

Fig. 6.5. Model for estimating the change in the earth's temperature due to a heavy layer of soot in the upper atmosphere, such as would be created by intense fires over large areas.

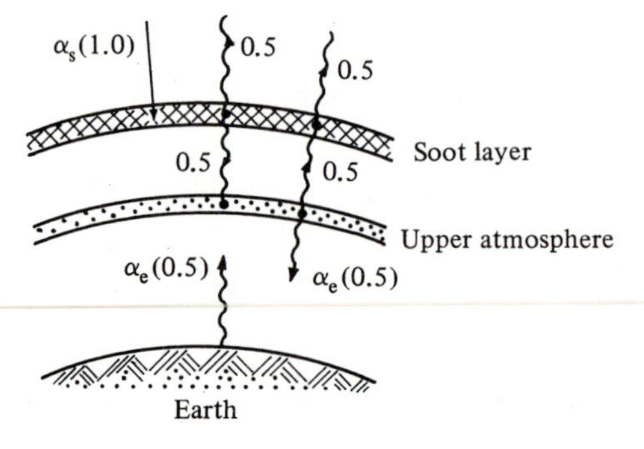

Like all climate calculations our model is subject to uncertainties especially since some of the input parameters are not known. It is however agreed among the experts that a global exchange will bring some form of 'nuclear winter' where sunlight will be blocked for a long time, affecting plant and animal life, and that temperatures will be severely depressed, affecting food production and human life. Furthermore the ozone layer will be partially destroyed so that when the soot is dispersed, UV radiation on the earth's surface will be intensified. These effects will be more serious in the Northern hemisphere where the conflict is expected to take place, but they will be felt over the entire globe as well. In Table 6.3 we summarize the predictions for the global consequences of a 10 000 Mt nuclear exchange as given by Ehrlich *et al.* (*Science*, Vol. 22, 1293, 1983). In the table NML = northern middle latitudes, and NH = northern hemisphere.

Table 6.3. *Global consequences of a 10 000 Mt exchange*

| | Effect* | Duration | Area affected |
|---|---|---|---|
| Sunlight intensity | ×0.01 | 1.5 months | NML |
| | ×0.05 | 3 months | NML |
| | ×0.25 | 5 months | NH |
| | ×0.50 | 8 months | NH |
| Surface temperature | −40°C | 4 months | NML |
| | −20°C | 9 months | NH |
| | − 3°C | 1 year | NH |
| UV radiation | ×4 | 1 year | NH |
| | ×3 | 3 years | NH |
| Radioactive fallout exposure | ⩾500 R | 1 hour–1 month | 30% of NML |
| | ⩾100 R | 1 day–1 month | 50% of NML |
| | ⩾ 10 R | ⩾1 month | 50% of NH |
| Fallout burden | $^{106}$Ru $10^4$ MCi | 1 year | NH |
| | $^{137}$Cs 650 MCi | 30 years | NH |
| | $^{90}$Sr 400 MCi | 30 years | NH |

* Multiply present value by indicated number

Estimates of the overall casualties, direct and indirect, from an all out 10 000 Mt nuclear exchange are given in Table 6.4 where it is assumed

Table 6.4. *Direct casualties from a 10 000 Mt exchange*

| | |
|---|---|
| Direct deaths from blast | 750 M |
| Deaths from blast and heat | 1100 M |
| Injuries needing treatment | 1000 M |
| | ~3000 M |

that the major cities of the two opposing nations and of their allies have been targeted. Thus about 50% of the globe's population, much more than the population of the nations at war, would be *wiped out*. Life for the survivors, as seen from Table 6.3, may be even more difficult or impossible.

## 6.3 Delivery systems and nuclear arsenals

The U.S. and the U.S.S.R. have built a vast nuclear arsenal. In addition Britain, France, China and possibly other nations have nuclear capability. Nuclear capability implies not only the possession of weapons but also the means to deliver them in enemy territory. The principal delivery vehicles are long-range rockets, referred to as *ballistic missiles*. The U.S. defense policy is based on three delivery systems – the so-called 'triad' of strategic weapons.

(1) Land based intercontinental ballistic missiles (ICBM)
(2) Long range bombers equipped with gravity bombs and cruise missiles
(3) Submarine launched ballistic missiles (SLBM).

Furthermore a variety of 'tactical nuclear weapons' are deployed in areas of potential conflict. The Soviets have similar capability.

Both nations have proclaimed that the nuclear arsenal is to be used only for defense, in case of an enemy attack. Thus the nuclear capability is such that even after a first strike by the adversary there will be enough weapons left to obliterate the attacking nation. The policy is referred to variedly as 'mutual assured destruction', 'deterrence', or 'massive retaliation'; it is vividly illustrated in the 1983 cartoon by Willis reproduced in Fig. 6.6. Unfortunately the policy of nuclear deterrence has led to a continuing escalation in armaments, as one or the other nation perceived that it was being overtaken by its enemy, so that its retaliatory power would be threatened or become ineffective. As a result, while it is estimated that only 400 Mt are sufficient to obliterate either country, present arsenals are ten times larger. This is referred to as 'overkill' capacity.

Technological improvements in the past two decades have made delivery systems highly efficient and extremely precise. One missile can now carry several warheads that are independently targeted, thus multiplying the effective destructive power; the acronym MIRV (for multiple independent reentry vehicles) is used for such payloads. The accuracy with which a warhead can be delivered is such that the circle of probable error (CEP) has radius $R < 300$ m. Thus cities and industrial complexes can be targeted with confidence. But even hardened missile sites can now be subject to direct hits. Table 6.5 lists the U.S. and U.S.S.R. arsenals as they were estimated in 1982.*

For these immense arsenals to be effective they must be accompanied by an extensive information, communication and command system. These include large radar and reconnaissance and navigation satellites and a complex computer system. To protect against a 'preemptive strike' by the enemy, land based missiles are placed in hardened silos and a mobile system has been proposed. There is continued debate on the value and effectiveness of a defensive system, as we will discuss in Section 6.5. In answer to these questions both nations have opted to increase their nuclear capacity. The growth in the number of warheads in the arsenals of the two superpowers is shown in Fig. 6.7 as a function of time. The sharp upturn around 1970 is attributed to the introduction of the MIRV concept.

The forerunner of modern long range missiles was the German V2 rocket used against London near the end of W.W.II. The exploration of space and also the present military missiles are based on many of the principles used in the V2. A ballistic missile, like an artillery shell, is given an initial velocity and then allowed to return (fall) to earth under the influence of gravity. However, while a shell is propelled in the gun barrel, a rocket sustains a continued thrust due to the exhaust of its fuel at very high relative velocity. Fig. 6.8($a$) shows the direct and indirect trajectories that can be followed by a ballistic missile to reach a target at range $x$.

Fig. 6.6. An accurate illustration of the principle of mutual deterrence (from *The Dallas Times Herald*, c. 1983).

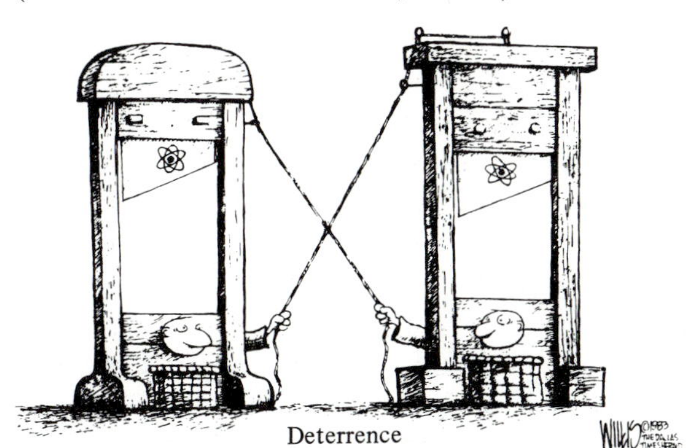


* B. G. Levi, The nuclear arsenals of the U.S. and U.S.S.R., *Physics Today*, March 1983.

Table 6.5. *Strategic nuclear arsenals*

| Delivery vehicle | Number of vehicles | Warheads per vehicle | Yield per warhead (Mt) | warheads ($N$) | Equivalent megatons ($NY^{2/3}$) | Range (km) | CEP (m) |
|---|---|---|---|---|---|---|---|
| ***U.S.*** | | | | | | | |
| *ICBMs* | | | | | | | |
| Titan II | 52 | 1 | 9.0 | 52 | 225 | 15 000 | 1300 |
| Minuteman II | 450 | 1 | 1.2 | 450 | 508 | 11 300 | 370 |
| Minuteman III | 250 | 3 | 0.17 | 750 | 230 | 13 000 | 280 |
| Minuteman III (improved) | 300 | 3 | 0.335 | 900 | 440 | n.a. | 220 |
| *SLBMs* | | | | | | | |
| Poseidon C-3 | 304 | 10 | 0.05 | 3040 | 413 | 4 600 | 450 |
| Trident D-4 | 216 | 8 | 0.1 | 1728 | 372 | 7 400 | 450 |
| *Aircraft* | | | | | | | Max. speed (Mach) |
| B-52D | 75 | | | 300 | | 9 900 | 0.95 |
| Gravity bombs | | 4 | ~1 | | | | |
| B-52G | 151 | | | 1208 | | 12 000 | 0.95 |
| Gravity bombs | | 4 | ~1 | | | | |
| SRAMs* | | 4 | 0.2 | | | | |
| B-52H | 90 | | | 720 | | 16 000 | 0.95 |
| Gravity bombs | | 4 | ~1 | | | | |
| SRAMs* | | 4 | 0.2 | | | | |
| Total gravity | | | | | 1114 | | |
| Total SRAMs | | | | | 330 | | |
| FB-111A | 60 | 2 | ~1 | 120 | 120 | 4 700 | 2.5 |

| | | | | | | | |
|---|---|---|---|---|---|---|---|
| ***U.S.S.R.*** | | | | | | | |
| *ICBMs* | | | | | | | CEP (m) |
| SS-11 Model 1 | 570 | 1 | 1.00 | 570 | 570 | 10 500 | 1400 |
| SS-11 Model 2 | Some | 3 | 0.1–0.3 | | | 8 800 | 1100 |
| SS-13 Model 1 | 60 | 1 | 0.75 | 60 | 50 | 10 000 | 2000 |
| SS-17 Model 1 | 150 | 4 | 0.75 | ~600 | ~495 | 10 000 | 450 |
| SS-17 Model 2 | Few | 1 | 6.00 | | | 11 000 | 450 |
| SS-18 Model 1 | | 1 | 20.00 | ~2500 | ~2300 | 12 000 | 450 |
| SS-18 Model 2 | 250 | 8 | 0.90 | | | 11 000 | 450 |
| SS-18 Model 3 | 58 | 1 | 20.00 | | | 10 500 | 350 |
| SS-18 Model 4 | | 10 | 0.50 | | | 9 000 | 300 |
| SS-18 Model 5 | | 10 | 0.75 | | | 9 000 | 250 |
| SS-19 Model 2 | Few | 1 | 5.00 | | | 10 000 | 300 |
| SS-19 Model 3 | 310 | 6 | 0.55 | ~1500 | ~1200 | 10 000 | 300 |
| *SLBMs* | | | | | | | |
| SS-N-5 | 57 | 1 | 1.00 | 57 | 57 | 1 400 | 2800 |
| SS-N-6 Model 1 | | 1 | 1.00 | | | 2 400 | 900 |
| SS-N-6 Model 2 | 400 | 1 | 1.00 | ~400 | ~400 | 3 000 | 900 |
| SS-N-6 Model 3 | | 2 | 0.2 | | | 3 000 | 1400 |
| SS-N-8 Model 1 | | 1 | 1.0 | | | 7 800 | 1300 |
| SS-N-8 Model 2 | 292 | 1 | 0.8 | ~300 | ~250 | 9 100 | 900 |
| SS-N-8 Model 3 | | 3 | 0.2 | | | n.a. | 450 |
| SS-NX-17 | 12 | 1 | 1.0 | 12 | 12 | 3 900 | 1500 |
| SS-N-18 Model 1 | | 3 | 0.45 | | | 7 400 | 1400 |
| SS-N-18 Model 2 | 208 | 1 | 0.45 | ~1040 | ~430 | 8 300 | 600 |
| SS-N-18 Model 3 | | 7 | 0.2 | | | 6 500 | 600 |
| *Aircraft* | | | | | | | Max. speed (Mach) |
| TU-95 (Bear) | 105 | 2 | 1 | 210 | 210 | 12 800 | 0.78 |
| Mya-4 (Bison) | 45 | 2 | 1 | 90 | 90 | 11 200 | 0.87 |

Data from *The Military Balance 1981–2*, The International Institute for Strategic Studies, London (1982).

* Short range air missile.

Assuming that the earth is flat and neglecting air resistance, the parameters of the trajectory as a function of the elevation angle $\theta$ are

$$\left.\begin{aligned} &\text{Range} && x = \frac{1}{g} v_0^2 \sin(2\theta) \\ &\text{Maximum height} && y = \frac{1}{2g} v_0^2 \sin^2 \theta \\ &\text{Time of flight} && \Delta t = \frac{x}{v_0} \cos \theta \end{aligned}\right\} \tag{6.8}$$

where $v_0$ is the initial velocity, and $g$ the acceleration of gravity on the earth's surface, $g = 9.81\,\text{m/s}^2$; maximum range occurs for $\theta = 45°$.

Fig. 6.7. Evolution of the number of strategic nuclear warheads of the U.S. and the U.S.S.R.; the dotted curves are projections. (After P. Craig and J. Jungerman, *Nuclear Arms Race*, McGraw-Hill, New York, 1986.)

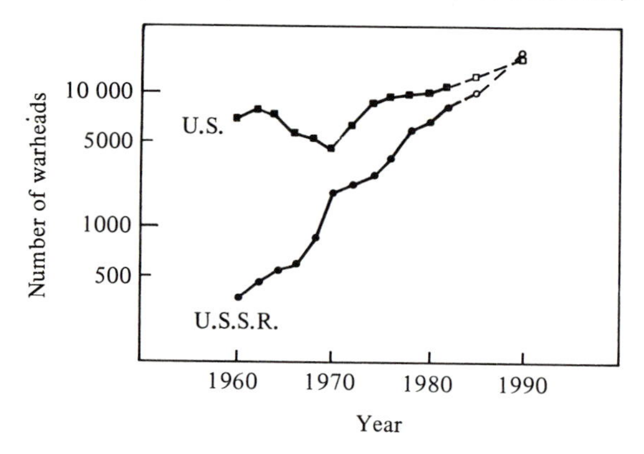


Fig. 6.8. (*a*) Ballistic trajectory in a uniform gravitational field. (*b*) Typical trajectories of ballistic missiles as a function of the desired range; note that the main part of the trajectory is outside the atmosphere.

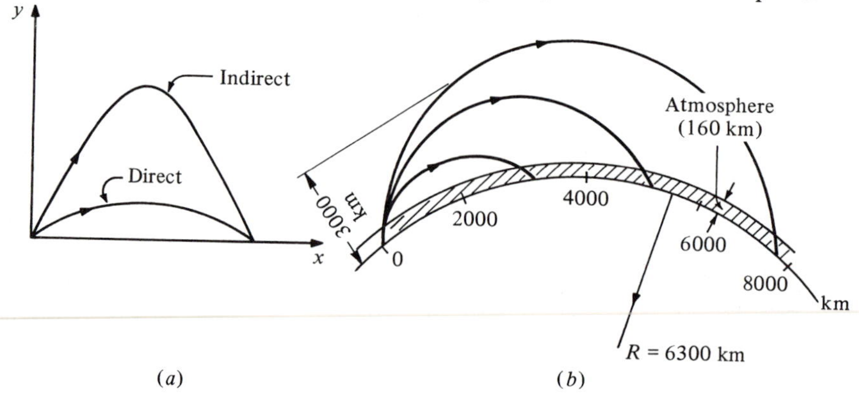

Trajectories typical of ballistic missiles are sketched in Fig. 6.8(*b*). We note below the range, initial velocity and flight time for these trajectories

(1) $R = 3000$ km $\quad v_0 = 5.5$ km/s $\quad \Delta t = 21$ min
(2) $R = 5000$ km $\quad v_0 = 6.4$ km/s $\quad \Delta t = 32$ min
(3) $R = 8000$ km $\quad v_0 = 7.4$ km/s $\quad \Delta t = 50$ min

For reference, we recall that the initial velocity required for achieving geostationary orbit is $v_0 = 10.7$ km/s and the escape velocity is $v_s = 11.2$ km/s.

We will discuss rocket propulsion in the following chapter in some detail. Suffice it to say here that in order to impart high velocity to a payload it is essential to use several *stages* so as not to waste thrust in accelerating parts of the spent rocket. We can derive the equation governing rocket propulsion, as shown pictorially in Fig. 6.9. The instantaneous mass and velocity of the rocket (as seen in the earth's frame) are $M$ and $V$. The velocity of the exhaust gases *relative* to the rocket is $v_e$ and $dm_e/dt$ is the mass of gas exhausted per unit time; note that

$$-\frac{dM}{dt} = \frac{dm_e}{dt} \tag{6.9}$$

In the absence of gravity the rocket together with the exhaust gases are an *isolated system* and thus their total momentum is conserved, $P_t =$ constant, or $dP_t/dt = 0$. Therefore

$$\frac{dP_t}{dt} = \frac{d}{dt}(MV) + \left(\frac{dm_e}{dt}\right)(V - v_e) = 0 \tag{6.10}$$

We find

$$\left(\frac{dM}{dt}\right)V + M\frac{dV}{dt} + \left(\frac{dm_e}{dt}\right)V - \frac{dm_e}{dt}v_e = 0$$

and because of Eq. (6.9) the first and third terms cancel so that

$$M\frac{dV}{dt} = v_e\frac{dm_e}{dt} = F_{\text{thrust}} \tag{6.11}$$

Fig. 6.9. Rocket propulsion: (*a*) as seen in the rocket's rest frame, (*b*) as seen in the earth's frame, (*c*) for vertical launch.

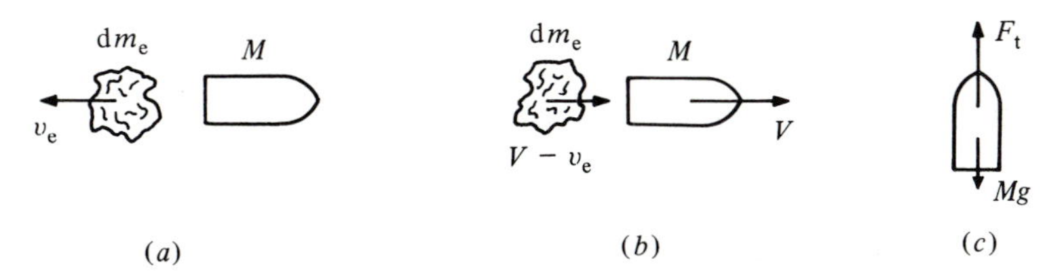

The term $v_e(\mathrm{d}m_e/\mathrm{d}t)$ has dimensions of force and is called the *thrust* of the rocket; it is *as if* a force of that magnitude were acting on the rocket imparting to it the acceleration $\mathrm{d}V/\mathrm{d}t$.

If the rocket is launched vertically against the force of gravity as shown in Fig. 6.9($c$), the equation of motion becomes

$$M\frac{\mathrm{d}V}{\mathrm{d}t}+\frac{\mathrm{d}M}{\mathrm{d}t}v_e=-Mg \tag{6.12}$$

which we can easily integrate. We multiply by $\mathrm{d}t$ and divide by $M$ so that

$$\int_0^{V(t)}\mathrm{d}V+v_e\int_{M_0}^{M(t)}\frac{\mathrm{d}M}{M}=-g\int_0^t\mathrm{d}t$$

Here $M_0$ is the initial mass of the rocket at time $t=0$ when $V=0$. The velocity attained at time $t$ is $V(t)$ and the mass of the rocket is $M(t)$, where

$$V(t)=v_e\ln\left(\frac{M_0}{M(t)}\right)-gt \tag{6.13}$$

It is clear that to achieve large final velocities $V_f$ we must have large exhaust velocity $v_e$, but also large values of $\ln(M_0/M_f)$, where the final mass $M_f$ includes the payload as well as the empty rocket structure. For given $M_0$ and payload, the best way to minimize $M_f$ is by using several stages.

As an example the Minuteman III missile has the following parameters

$M_0=24.5$ tons
$M_f=1.5$ tons
$v_e\sim 4$ km/s

and the burn time is $t=200$ s. The Minuteman is a three stage rocket but we will use the parameters of Eq. (6.13) to obtain an upper limit on the final velocity reached after the launch

$$V_f=\left(4\frac{\mathrm{km}}{\mathrm{s}}\right)\ln\left(\frac{24.5}{1.5}\right)-\left(0.01\frac{\mathrm{km}}{\mathrm{s}^2}\right)200=9.3\frac{\mathrm{km}}{\mathrm{s}}$$

In reality the achieved launch velocity is $V_f\sim 7.3$ km/s. We can also calculate the thrust developed by the rocket if we assume that the $(24.5-1.5)=23$ tons of fuel are exhausted uniformly in $t=200$ s.

$$F_{\mathrm{thrust}}=\frac{\mathrm{d}m_e}{\mathrm{d}t}v_e=\left(\frac{23\times 10^3}{200}\frac{\mathrm{kg}}{\mathrm{s}}\right)\left(4\times 10^3\frac{\mathrm{m}}{\mathrm{s}}\right)=4.6\times 10^5\ \mathrm{N}$$

## 6.4 Reconnaissance satellites

The collection of military intelligence about one's adversary is an important activity going on both in war and in peace time. Reconnaissance satellites, which by international agreement can overfly any part of the globe, have become the primary tool for such operations. Equally importantly, reconnaissance satellites play an essential role in the verification of the various arms limitation treaties that are in effect or that are being negotiated. Satellites are also used for navigation, for communications, and as part of an early warning system against an enemy attack.

To obtain the best resolution the satellite should be at low altitude when it overflies the territory that is to be photographed; this however implies increased drag and therefore shortened time in orbit. To minimize this effect reconnaissance satellites are launched into elliptical near-polar orbits with the perigee (lowest point to the earth) at ~160 km, which places them just outside the atmosphere. As the earth rotates, different parts of it pass under a polar orbit as can be seen in Fig. 6.10; thus in 24 hours the entire globe can be scanned by the satellite. The time for completing such low orbits is of order of 90 minutes.

We can easily calculate the period of a circular orbit at height $h$ above the earth's surface. The angular velocity of the satellite is designated by $\omega$, its mass by $m$ and the radius of the orbit by $r$; $M_\oplus$, $R_\oplus$ stand for the mass and radius of the earth and $G$ is Newton's constant. Then

$$m\omega^2 r = G\frac{M_\oplus m}{r^2} \tag{6.14}$$

or

$$\omega^2 = G\frac{M_\oplus}{r^3} = G\frac{M_\oplus}{R_\oplus^2}\frac{1}{R_\oplus}\left(\frac{R_\oplus}{r}\right)^3 = g\frac{1}{R_\oplus}\frac{1}{(1+h/R_\oplus)^3}$$

Fig. 6.10. A satellite in polar orbit will eventually cover all of the earth's surface.

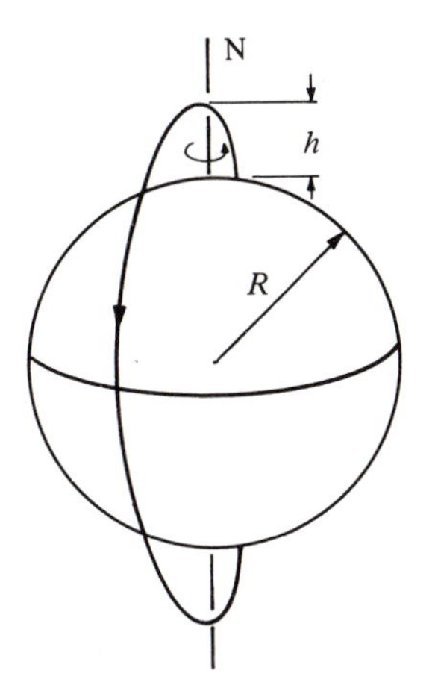

Here we introduced the acceleration of gravity at the earth's surface

$$g = G\frac{M_\oplus}{R_\oplus^2} = 9.81\,\text{m/s}^2 \tag{6.15}$$

and used $r = R_\oplus + h$. The radius of the earth is $R_\oplus = 6370$ km; thus we can expand in the small quantity $(h/R_\oplus)$ and we find for the period $T$,

$$T = \frac{2\pi}{\omega} \simeq 2\pi\left(\frac{g}{R_\oplus}\right)^{1/2}\left(1 + \frac{3}{2}\frac{h}{R_\oplus}\right) \tag{6.16}$$

Note that $T_0 = 2\pi(g/R_\oplus)^{1/2} = 84.4$ minutes is the period of oscillation of a body dropped through the center of the earth; it is often encountered in geodesic problems. If we assume $h \simeq 160$ km, then the term $(3/2)(h/R_\oplus)$ in Eq. (6.16) contributes only a 4% correction to $T_0$.

The resolution with which objects on the ground can be observed depends on the distance from the object, the resolution of the recording medium, be it film or an electronic digitizer, and on the quality of the optics. The relation between these variables can be found with the help of Fig. 6.11 where $s$ and $s'$ are the object and image size respectively; $F$ is the focal length of the lens and $A$ is the distance to the object. The resolution of the film is characterized by $R$ in lines/mm; thus the smallest distance that can be resolved in the image plane is $\delta = 1/R$ (mm). From the geometry of the figure $s' = s(F/A)$, and if we set $s' = \delta$, then $s$ is the ground resolution $G$, or

$$G = \delta\frac{A}{F} = \frac{A}{F}\frac{1}{R} \tag{6.17}$$

Fig. 6.11. An imaging lens and typical ray traces.

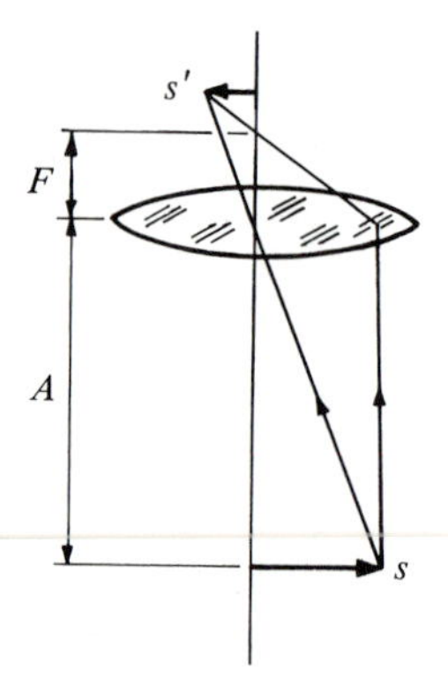

As an example let us take the focal length to be $F = 1$ m (slightly over 3 feet), the distance $A = 160$ km, and the resolution $R = 200$ lines/mm, that is a 5 $\mu$m resolution in the film plane. The ground resolution is then

$$G = \frac{160 \times 10^3}{1} \frac{1}{200 \times 10^3/(\text{m}^{-1})} = 0.8 \text{ m}$$

This is rather remarkable and we see that such images can reveal military targets in great detail. Much better resolution is to be expected from modern reconnaissance cameras with increased focal length.

In deriving Eq. (6.17) and in the examples we did not consider distortions introduced by the optics or the atmosphere which can be limiting factors. Furthermore, large focal length implies lenses of large diameter because of the limits imposed by diffraction. We recall from Eq. (4.20′) that the smallest angle that can be resolved using radiation at wavelength $\lambda$, with a lens of diameter $D$ is

$$\theta_{\min} = 1.2 \frac{\lambda}{D} \tag{6.18}$$

Thus, since $\theta = s'/F$, we find that

$$s' > F\theta_{\min} = 1.2\lambda(F/D)$$

If, as before we set $s' = \delta = 1/R$, then the ratio $(F/D)$, the so-called $f$-stop of the lens, must be smaller than

$$\frac{F}{D} < \frac{1}{1.2} \frac{\delta}{\lambda} \simeq 0.8 \frac{1}{\lambda R}$$

If we assume green light, $\lambda = 500$ nm, to achieve the resolution calculated in our previous example the lens diameter must exceed $D$, where

$$\frac{F}{D} < \frac{0.8}{(500 \times 10^{-9})(200 \times 10^3)} = 8$$

namely, $D > 12.5$ cm, since we had $F = 1$ m. A good quality lens of such diameter is only moderately difficult to manufacture. For longer focal lengths, however, the demands on the optics are severe.

The high resolution, combined with the large field of view results in a very rapid stream of data that must be recorded. Furthermore the recording time is limited because of the fast motion of the satellite. Typically, the velocity of the satellite is $v = 8.2$ km/s, and for a 3° field of view the area covered has a radius $l(\theta = 3°) = A \times \theta \simeq 160 \times 0.05 = 8$ km. Thus for complete coverage, images must be recorded at a rate of approximately one per second. For the focal length used in the example, the image radius is $l' = l(F/A) = F \times \theta = 5$ cm; given the film resolution $\delta = 5\ \mu$m, a $10 \times 10\ \text{cm}^2$ frame contains $(20\,000)^2 = 4 \times 10^8$ pixels. To preserve the ground resolution the frame must be 'shot' in a time shorter than $\tau = G/v = 10^{-4}$ s. For these reasons fast film is the preferred medium for recording the image; it has both the speed and the resolution. More recently, CCD cameras (see Section 2.7) are being developed and used as substitutes for film.

In the early reconnaissance satellites the film was dropped in canisters that deployed a parachute in the atmosphere and were collected by special airplanes. As larger payloads were placed in orbit, the film was developed in the satellite and either dropped to earth or scanned on board and transmitted by a communications link. Electronically recorded and transmitted images have less resolution than film but can be seen in real time and thus are useful for early warning systems. Note that the analysis of high resolution images is in itself a tedious and lengthy task especially when large areas are being investigated.

The atmosphere is transparent not only to visible light but also to certain bands in the infrared. This provides another window for reconnaissance photography and has the advantage that it is not obstructed by cloud cover, and can be used by day or by night. IR photography is particularly sensitive to sources of heat so that rocket and jet engine exhausts are easily located. Extensive radar installations are also part of the intelligence gathering network. While conventional radar is limited to line of sight targets, over the horizon (OTH) radar uses the reflection from the ionosphere (see Section 4.5) to increase its range. Finally, tracking missile launches and their trajectories, and listening in on radio transmissions are other important components in the effort to assess the adversary's intentions and his technical progress.

Two examples of the capabilities of aerial and ground based photography are shown in Fig. 6.12. Part (*a*) is a section from a photograph taken from the Apollo 15 spacecraft at a distance of approximately 100 km from the surface of the moon. The lunar landing module which has dimensions of order 5 m can be distinguished at the center of the photograph. Note that the resolution of the original film was much better than shown here and that the Apollo camera did not have as high a resolution as cameras carried by reconnaissance satellites. Part (*b*) is a photo of the multiple independently targeted reentry vehicles from an MX launch as they pass through the atmosphere and impact in the western Pacific near Kwajalein Atoll. The photo is taken from a Navy P-3 *Orion* aircraft at high altitude. Similar photos of Soviet missiles targeted in the Pacific are routinely obtained by U.S. reconnaissance and vice versa.

Fig. 6.12. (*a*) Section of a photograph taken from the Apollo 15 spacecraft at a distance of ~100 km; the lunar landing module which has dimensions of ~5 m can be easily distinguished. (Courtesy ITEK Corporation). (*b*) Picture of the multiple independently targeted reentry vehicles from an MX launch as they pass through the atmosphere and impact in the western Pacific near Kwajalein Atoll. The missile was launched from the Vandenberg Air Force Base in California. This photo is a time exposure (thus the wavy lines) taken by a Navy P-3 *Orion* aircraft from a high altitude. (Courtesy U.S. Department of the Air Force.)

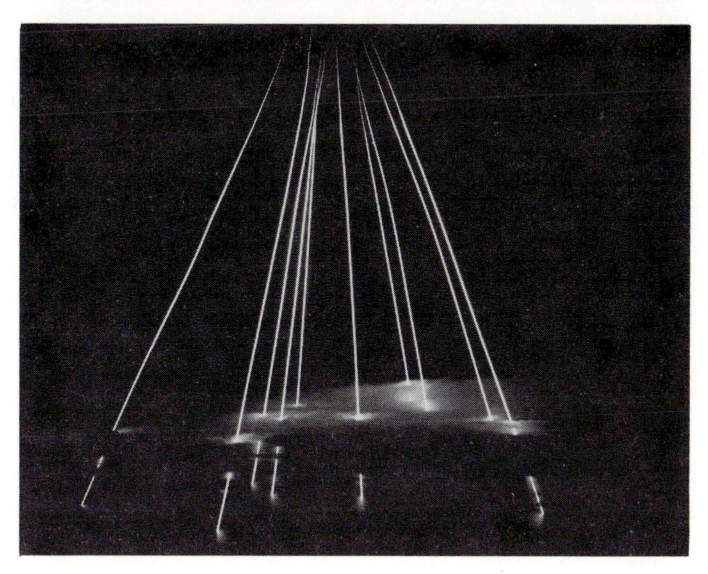

## 6.5 Proposed defense systems

The possibility of defending a nation against a nuclear attack has been often considered and much discussed. Systems designed to destroy incoming ballistic warheads were deployed in the U.S.S.R. in 1968 and by the U.S. in 1974 to protect specific geographic regions. These defense installations were not considered to be effective against a major nuclear attack, and in 1972 a treaty was signed, limiting the further deployment of such anti-ballistic missile system (ABM treaty). In 1983 the U.S. announced the development of a comprehensive defense against a Soviet nuclear attack; it was named the Strategic Defense Initiative (SDI), or more commonly 'Star Wars' because it envisioned the use of a large number of space-based weapons.

The pursuit of a defensive alternative has great moral and ideological appeal. However there is strong divergence of views as to whether such a system is at all technically feasible. There are obvious questions that can be debated: a system designed to preserve the retaliatory force has different characteristics from one that would protect the population. Even if the defense was 90% effective, it would hardly prevent an almost complete destruction of the target. How easily can the system be foiled by countermeasures or by building more attack capability? Finally, is the deployment of a defense by one of the superpowers 'destabilizing' in the sense that it would lead to further escalation in the arms race?

The typical trajectory of a ballistic missile launched from the Soviet Union and targeted against the U.S. mainland is shown in Fig. 6.13. Four phases are outlined: (*a*) The *boost* phase during which the fuel is burnt and the missile acquires its launch velocity; it lasts 200–400 s. During this time the missile is most vulnerable because of its low velocity, its large size, and the intense heat from the rocket engine exhaust which can be used to sense its position. (*b*) The *busing* phase during which the payload travels on its ascending trajectory but the warheads have not yet separated. (*c*) A *mid-course* phase in which warheads and decoys have been released and fall freely under the earth's gravitational attraction; this is the longest part of the flight and lasts approximately 40 minutes. Finally (*d*) is the *re-entry* phase when the warheads are entering the atmosphere above their respective targets.

Depending on the phase during which the missile is to be attacked, different problems arise. In the boost phase, because of the short time that is available, the weapon must be in place – over enemy territory – and one has to rely on directed energy weapons. Since the locations of enemy silos are known and because of the heat plume, targeting is relatively easy. An attack during the busing phase is more difficult in terms of tracking, aiming at, and hitting a small target in a short time. During mid-course there is sufficient time but the deployment of multiple warheads and decoys may require the targeting and destruction of some 10 000 objects. For the reentry phase, schemes such as exploding a nuclear weapon in the path of the incoming warheads have been considered; they have obvious drawbacks in terms of radiation and fallout over the area that is to be defended.

The weapons proposed for missile defense are either *directed energy* weapons such as intense laser beams or particle beams, or *kinetic energy* weapons. The former have the advantage that they propagate with the speed of light while kinetic energy weapons are much more effective in destrcying a missile. In a 1984 test, the U.S. demonstrated that a Minuteman missile launched from California could be intercepted during its reentry over Hawaii and destroyed in flight by the impact of a kinetic energy weapon. In general, the time available for ascertaining a threat and launching the defensive weapons – especially in the boost phase – is only a few minutes, posing serious command problems. Furthermore, since many of the proposed weapons cannot be tested under real conditions it will be difficult to know their effectiveness beforehand.

As an example of ballistic defense weapons we will consider lasers and estimate the required power levels. To destroy a missile a laser beam must deliver to it $10^4$ J/cm$^2$ in 1 second or less. Such a pulse would vaporize a 3 mm thick sheet of copper. Of course warheads or missiles can be hardened but we will assume that the 'lethal flux' is

$$F_l = 10^8 \text{ W/m}^2 \tag{6.19}$$

Fig. 6.13. Typical trajectory and the various phases of the flight of an ICBM.

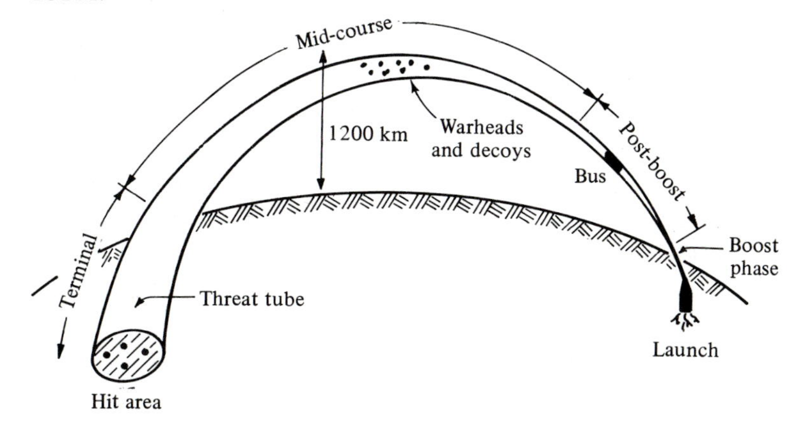

For any source of radiation the *brightness* $B$ is defined as

$$B = \frac{\text{energy/s}}{\text{steradian}} \tag{6.20}$$

and the total radiated power is $P = \int B \, d\Omega$. If the source is well collimated so that all the energy is contained in a small solid angle $\Delta\Omega$, we can write

$$P \simeq B\Delta\Omega \tag{6.20$'$}$$

and small $\Delta\Omega$ leads to high brightness. The flux $F$ at a distance $R$ from the source is given by

$$F = B/R^2 \tag{6.20$''$}$$

as follows from the definition of Eq. (6.20).

If the laser is placed in orbit at a height of $\sim 1000$ km over the enemy's launching sites the required brightness is

$$B = F_l R^2 = 10^8 \times (10^6)^2 = 10^{20} \text{ W/sr} \tag{6.21}$$

Alternately the laser can be earth based and the beam reflected from a geostationary mirror to a fighting mirror and from there onto the target as shown in Fig. 6.14. In this case the distance involved is $R \sim 2 \times 36\,000$ km and therefore the required brightness is of the order of $B \simeq 10^{24}$ W/sr; in addition losses in the atmosphere must be compensated for.

The power needed to reach the required brightness depends on how well the beam is collimated. The smallest possible solid angle is given by the diffraction limit and therefore determined by the diameter $D$ of the last focussing mirror

$$(\Delta\Omega)_{\min} = \pi(\theta_{\min})^2 = 1.44\pi\left(\frac{\lambda}{D}\right)^2 \tag{6.22}$$

Fig. 6.14. Proposed scheme for using a high power earth based laser, to attack enemy missiles far beyond the horizon.

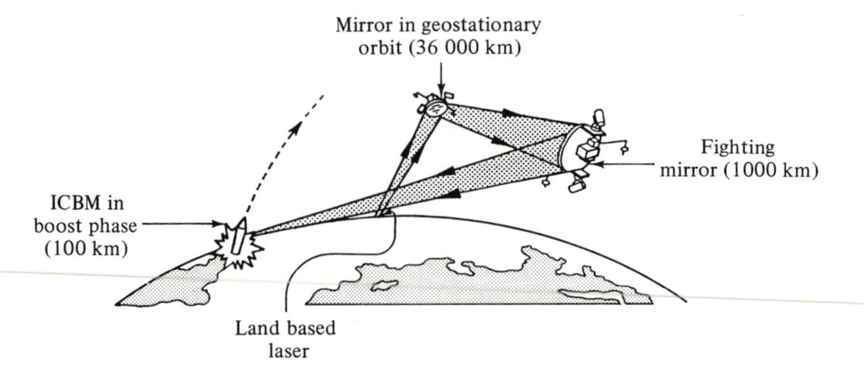

We assume that $D = 5$ m and that near-UV light, $\lambda = 300$ nm will be used. Then to achieve a brightness of $10^{20}$ W/sr

$$P_{\min} = B\Delta\Omega_{\min} = 10^{20} \times 4.5\left(\frac{3 \times 10^{-7}}{5}\right)^2 = 1.6 \times 10^6 \text{ W}$$

and this power level must be sustained for 1 second, leading to 1.6 MJ of optical energy. For earth based lasers, the required power is $10^4$ times higher, a level of performance that has not been achieved as yet.

Various high power lasers have been considered for the SDI program and are being further developed. The $CO_2$ laser is not a prime contender because of its long wavelength, $\lambda \simeq 10$ $\mu$m. In chemical lasers population inversion is established when the molecules are first formed in a reaction such as $F + H_2 \rightarrow HF + H$; lasing takes place among the vibrational levels of the HF molecule resulting in wavelengths $\lambda = 2.7$–$2.9$ $\mu$m and power levels of 1 MW have been achieved. In the excimer laser the KF (Krypton Fluoride) molecule is pumped by an electron beam and radiates in the visible and near UV; power levels of order 10 kW have been reached.

When SDI was first proposed there was great confidence that an X-ray laser could be constructed. The idea is sketched in Fig. 6.15 and consists of an assembly of thin rods which are pumped by the X-rays from a small nuclear explosive (30 kt). Because of the geometry, the X-rays emitted by stimulated emission are collimated along the direction of the rods and even though there is only a single pass there is substantial gain. The advantage of such a device – if it works – is that it does not require a bulky power source and can be easily placed in orbit. However, X-rays are absorbed by the atmosphere so that the enemy missile would have to be intercepted during the busing phase. Placing an X-ray laser, and therefore a nuclear weapon in space would be in violation of the U.S.–U.S.S.R. anti-ballistic missile treaty; an alternative is to launch the devices when needed.

A different very bright source of radiation is the recently developed *free electron laser* (FEL). In the FEL a relativistic electron beam of energy $E$

Fig. 6.15. Schematic drawing of a possible X-ray laser driven by a nuclear explosive.

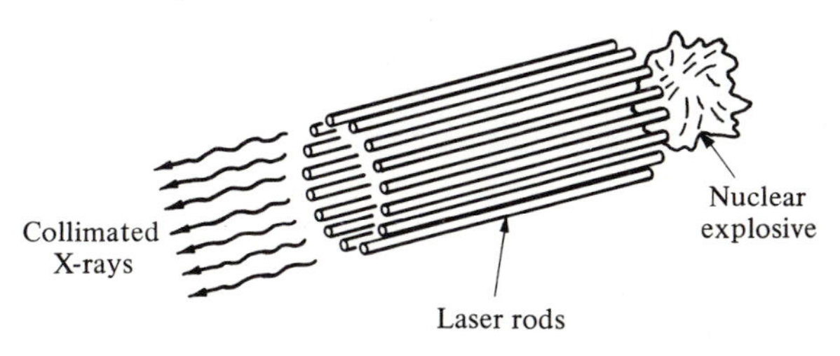

is passed through a 'wiggler', a region of an alternating transverse magnetic field. The trajectory of the electron is then curved as shown in Fig. 6.16(*a*) and therefore the electrons emit synchrotron radiation. By using an optical cavity the emission of radiation can be stimulated resulting in an intense coherent beam. The wavelength of the radiation is given by

$$\lambda \sim \frac{\lambda_w}{2\gamma^2} = \frac{\lambda_w}{2(E/mc^2)^2} \tag{6.23}$$

where $\lambda_w$ is the wavelength of the wiggler, and $\gamma = E/m$ is the ratio of energy to mass for the electrons. Thus the laser can be tuned by changing the electron energy. As an example we take $\lambda_w = 4$ cm and $E = 100$ MeV and this leads to $\lambda \simeq 500$ nm, which is in the visible. Results from the free electron laser operated at the French laboratory at Orsay are shown in Fig. 6.16(*b*). The principal peak at $\lambda = 648$ nm is accompanied by two sidebands. The lower trace shows the radiation spectrum without the optical cavity and is ~1000 times less intense.

We have considered only a few of the various proposals for antiballistic weapons, and tried to illustrate the problems of power requirements, timing, and of vulnerability of a space-based defense system. A highly detailed technical discussion can be found in the report of a study group formed by the American Physical Society, which was published in *Reviews*

Fig. 6.16. The free electron laser. (*a*) A beam of high energy electrons 'wiggles' as it passes through a spatially periodic alternating magnetic field; the emitted radiation is coherent and with the addition of mirrors can have a significant component of stimulated emission. (*b*) The emitted radiation as a function of wavelength; in this case the presence of the optical cavity enhanced the radiated power by a factor of $10^3$.

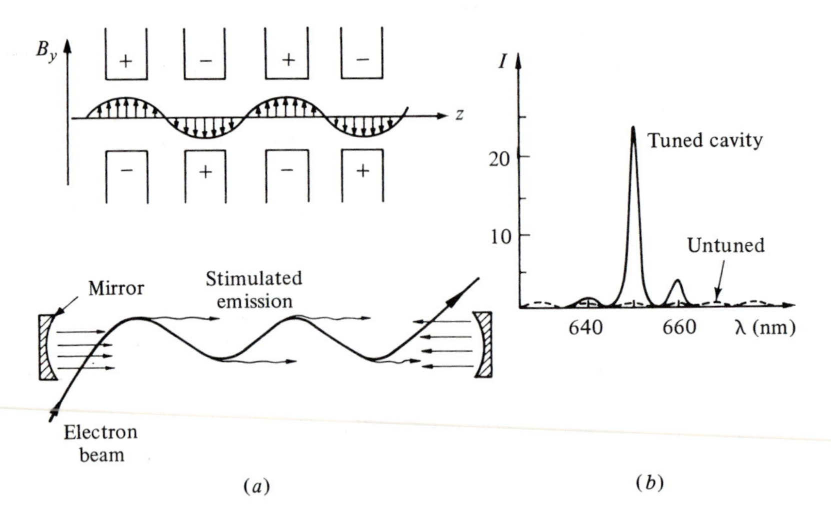

*of Modern Physics*, Vol. 59, July 1987. Furthermore, defense against submarine launched missiles, with their much shorter flight time, and against radar-evading low-flying planes and cruise missiles, introduces a new set of parameters and problems. How effective a defense system can be, and the political consequences of its deployment, are also important factors and are briefly touched upon in the following section.

## 6.6 Arms limitation treaties

Ever since the first nuclear explosions in 1945, efforts have been made to limit the spread of nuclear weapons and the capability to manufacture them. Nuclear weapons technology has been kept secret for military reasons, and while such secrecy cannot be effective over the long run it inhibits the spread of information. A more effective approach is through the adherence to multilateral agreements among nations and by the establishment of bilateral agreements between the U.S. and U.S.S.R. In spite of these efforts the stockpile of nuclear weapons has grown to the absurdly large quantities discussed in Section 6.3. The best hope for the avoidance of a nuclear catastrophe is a continuing dialogue between the superpowers leading eventually to a de-escalation and reduction of the inventories of nuclear weapons.

One of the reasons for the present predicament is the philosophy of escalation, coupled with the doctrine of massive retaliation. Furthermore, the proliferation of nuclear weapons among other nations increases the danger of a nuclear accident which could then precipitate a full-scale exchange. As has been discussed, such a conflict would create global effects that could permanently destroy life on earth as we know it. The agreements and treaties that are in effect have helped in checking the uncontrolled growth of nuclear materials, nuclear weapons and of their testing; but to assure a future free of nuclear war the existing agreements must be expanded and strengthened.

Below we list the most important treaties that are currently in effect and/or are being negotiated.

(1) The *limited test ban* treaty, signed in 1963. This was an important milestone in that it stopped all tests in the atmosphere and in the sea (with only minor exceptions). It curtailed the spread of radioactive fallout from nuclear testing and effectively limited the size of the weapons that are being tested.

(2) The *outer space* treaty (1967–72). This treaty prohibits the use

of outer space and of the seabeds for the deployment of nuclear weapons. It may be of major importance in preventing the placement of weapons in orbit around the earth as envisioned in certain strategic scenarios for the future.

(3) The *non-proliferation* agreement, negotiated between 1968 and 1970 but not signed by all nations. According to this treaty no new nations can join the 'nuclear club', and nuclear fuel technology is to be strictly controlled by those countries that do possess it. So far this treaty has been reasonably effective but it is evident that several nations have secretly acquired or are seeking a nuclear capability.

The bilateral agreements between the U.S. and the U.S.S.R. are grouped under the 'Strategic Arms Limitation Treaties' known as SALT I and SALT II. The latter has not as yet been ratified by the U.S. Senate but with some exceptions its provisions are respected by both countries. Included in SALT I were:

(1) The *antiballistic missile* (ABM) treaty of 1972. This prevents the construction of new ABM systems and forbids interference with reconnaissance satellites. This last step is essential in the efforts for verification of any arms limitation agreement.
(2) The agreement on limits for certain offensive weapons which resulted in a freeze of the number of deployed missiles (but not warheads) at their level of October 1972.

SALT II which was signed in 1979 has two important provisions:

(1) Restrictions on MIRVs (multiple independently targeted reentry vehicles) and on other new weapons.
(2) The *threshold* treaty which limits tests of nuclear weapons to a level below 150 kt. There have been protests from both sides that the treaty has been occasionally violated, but it remains true that overall the testing has been limited to relatively small devices.

Finally, in 1987 an agreement on intermediate nuclear forces (INF) was signed. This agreement covers weapons of the NATO allies deployed in Europe and similar Soviet weapons deployed against Europe. The INF treaty may well be a landmark agreement because it calls for the *dismantling* and removal of weapons already deployed and operational.

Future proposed agreements are:

(1) The *comprehensive test ban* treaty whereby *all* testing of nuclear weapons would be eliminated. This would be an important step but it is opposed by the military who are concerned about the preparedness of their forces and the reliability of the weapons stockpile. Yet, on occasion, one or the other country has declared a unilateral moratorium on testing.
(2) A treaty against testing of antisatellite weapons (ASAT).
(3) A true arms reduction treaty and even possibly a 'freeze' on all nuclear weapons construction.

The willingness of a nation to sign a treaty and to adhere to it depends in large measure on the nation's ability to make sure that its adversary is complying with the provisions of the treaty. Reconnaissance satellites play a key role in treaty verification and more recently on-site inspection by foreign teams has been included in the agreements. Political considerations, economic pressures, the concern about local conflicts and even nuclear posturing are all factors making the negotiation and ratification of arms limitation treaties difficult and protracted.

Arms limitation concerns are not restricted to nuclear arms and should eventually cover conventional weapons. Biological and chemical weapons can also produce disaster on a global scale and a convention was signed in 1977 to limit their production. Unfortunately, war has been part of man's history but it has now reached the point where it could completely annihilate our civilization. It is incumbent upon all of us to control the super-weapons that have been built and to prevent their use under any circumstances.

## Exercises

### *Exercise 6.1*

Consider a fission weapon containing 20 kg of $^{235}$U. *Assume* that the $^{235}_{92}$U nucleus fissions to the following channel in 10% of the cases

$$^1_0n + ^{235}_{92}\mathrm{U} \rightarrow ^{90}_{38}\mathrm{Sr} + ^{143}_{54}\mathrm{Xe} + 3\,^1_0n$$

(a) Calculate the amount of $^{90}$Sr released.
(b) Given that the lifetime of $^{90}$Sr is 28 years, find the total radioactivity released.
(c) Assuming that the $^{90}$Sr is dispersed over 2000 (km)$^2$ find the contamination in $\mu$Ci/m$^2$.
(d) Estimate the yield of the weapon in kt.

### *Exercise 6.2*

Consider the trajectory of a ballistic missile in its most simplified form (ignore the curvature of the earth and air resistance). Let the missile be launched for maximum range ($\theta = 45°$), and let the range of the missile be the shortest distance from Moscow to New York (approximately 9000 km).

(a) Calculate the required launch velocity.
(b) Calculate the time of flight.
(c) Assuming an acceleration of $5g$ calculate the burn (or boost) time.
(d) Calculate the maximum height of the trajectory.

### *Exercise 6.3*

Consider the illumination of an ICBM booster by blue light $\lambda = 0.3\ \mu$m. A mirror 4 m in diameter is stationed at a distance of 1500 km.

(a) What is the minimum size of the spot on the missile due to the diffraction limit of the mirror.
(b) If the energy deposition (surface density) required to damage any part of the missile is $10^4$ J/cm$^2$ find the energy required to destroy the missile.
(c) How does this compare with the power of commercially available lasers.

### *Exercise 6.4*

Consider a fire that would destroy a city of one million inhabitants.

(a) How much soot do you expect to be generated?
(b) Estimate the thickness of the soot layer that would reduce the sun's intensity by a factor of 10.
(c) Given the thickness and the total mass of soot calculated in (a), what is the maximal area that could be covered?
(d) What fraction of the earth's surface is that?

For details see the article by B. Levi and T. Rothman in *Physics Today* (September 1985) – 'Nuclear Winter' and references therein.

### *Exercise 6.5*

An empirical formula for the radius of destruction of a missile silo hardened to $H$ ($H$ expressed in thousands of psi) by a weapon of yield $Y$ ($Y$ in megatons) is

$$R_B = 460(Y/H)^{1/3} \text{ (meters)}$$

The probability of hitting the target within a radius $R$ is given by

$$P(R) = e^{-0.7(R/\mathrm{CEP})^2}$$

given that $H = 2000$ psi, $Y = 1$ Mt, CEP $= 360$ m

(a) Find the probability that the silo is destroyed by one such missile.

(b) Estimate the long lived radioactivity that is released.

***Exercise 6.6***

An X-ray laser is placed in polar orbit at an altitude $h = 1000$ km. To destroy a missile the laser must deliver $10^4$ J/cm$^2$ on target.

(a) Find the required brightness.

(b) If the beam divergence is 1000 times the diffraction limit and if $\lambda = 0.01$ nm and the device diameter is $D = 1$ m, find the necessary power in the laser.

(c) Estimate the spot diameter at the target.

# PART D

# SPACE TRAVEL

Transportation, of people and materials, is among the major factors that have made our civilization possible. The harnessing of animals and the use of ships were exploited early in the history of man. The sailing boat was an extraordinary invention because the sea offered reduced friction to the point where the wind would suffice to propel the ship. Railroads provided the freight capacity that made possible the industrial revolution, to be followed by the introduction of the automobile in the 20th century. The first airplane flight by the Wright brothers took place in 1903, and today transportation has brought within easy access all parts of the globe. This speed and ease in transportation has had and continues to have a profound effect in shaping the social and economic structure of the world community. In 1969 man landed on the moon, and unmanned spacecraft have reached to the edge of our planetary system. More ambitious missions into space can be foreseen as technology advances and the desire to carry them out persists.

Chapter 7 is devoted to a discussion of airplane and rocket flight and propulsion. In contrast to ships which are buoyant, airplanes are heavier than air and are supported by the dynamically produced lift. The reduced friction in air allows airplanes to reach high velocities, even in excess of the speed of sound. While airplanes must fly in the atmosphere, rockets are not subject to such a restriction. Rocket velocities are ten times higher than those of airplanes, to the point that rockets can exceed the escape velocity and leave the gravitational field of the earth.

To understand airplane flight a knowledge of elements of fluid dynamics is necessary, and these are developed in the text. Interestingly, rocket engines also involve fluid dynamics since the exhaust gases emerge at very high temperature and at supersonic speed. As a final application of these ideas the NASA shuttle is analysed in the concluding section of the chapter. For completeness, a brief review of the basic equations of fluid dynamics is given in Appendix 3. The speed of sound is discussed in Appendix 4.

Space travel proper is considered in Chapter 8. We focus on travel within our planetary system since such missions have been successfully completed. The solar system and orbit theory are reviewed briefly, to be followed by a discussion of transfer orbits and related maneuvers necessary for planetary flight. The Voyager 2 mission is considered in some detail and exemplifies the application of the principles discussed previously. Next we discuss ideas and proposals for future travel outside the solar system; we emphasize first principles such as energy and momentum conservation since they help us distinguish the physical from the fictional world. The final section is devoted to inertial guidance in view of the essential role it plays in any rocket or space mission.

# 7

# AIRPLANE AND ROCKET FLIGHT

## 7.1 Fluid flow and dynamic lift

The motion of a fluid is extremely complex because the individual molecules are subject to random thermal motion as well as to the collective motion of the fluid as a whole. Thus we consider a small element $d\tau$ of the fluid and follow its motion as a function of time. We will assume that the fluid is *incompressible*, so that the mass $dm = \rho\, d\tau$ contained in the volume $d\tau$ remains fixed and the density $\rho$ is constant throughout the fluid; we will also assume that the fluid is *non-viscous*, that is there are no internal frictional forces. These two assumptions are applicable to motion through air when the velocity $v$ is small as compared to the velocity of sound $v_s$, i.e. $v \ll v_s$. The velocity of sound is a measure of the random thermal velocity of the molecules; its value for air at s.t.p. is $v_s \simeq 330$ m/s. When necessary we will relax these assumptions.

The simplest form of flow occurs when the velocity at each point of the liquid remains constant in time. This is illustrated in Fig. 7.1($a$) where the element $d\tau$ follows the path from the point $P$ to $Q$ to $R$ and has the velocity $\mathbf{v}_P$, $\mathbf{v}_Q$, $\mathbf{v}_R$; at a later time another element of the fluid will be at $P$ but it will again follow the path to $Q$ to $R$ and have the same velocity. The path followed by a fluid element is called a *streamline* and the velocity is always tangential to the streamlines; such flow is called steady or *laminar* flow.

Streamlines cannot cross one another because at the crossing point the velocity would be undefined; thus the fluid contained within a flow tube, such as shown in Fig. 7.1($b$) remains always within the tube. If the area at the entrance of the tube is $A$, the mass of fluid entering the tube per unit time is

$$Q_1 = \rho_1 A_1 v_1$$

and correspondingly the mass leaving the tube at the exit is $Q_2 = \rho_2 A_2 v_2$. The conservation of mass, also referred to as the *continuity* condition demands that $Q_1 = Q_2$. For an incompressible fluid the density is constant so that $\rho_1 = \rho_2$ and the continuity equation takes the form

$$A_1 v_1 = A_2 v_2 \tag{7.1}$$

Namely, the area of the flow tube is inversely proportional to the velocity at that point. Conversely, the density of the streamlines is proportional to the velocity of the field.*

Streamlines are analogous to electric or magnetic field lines and thus should begin on a source and terminate on a sink. In practice we draw streamlines by extending them to infinity; they can also close onto themselves. Some examples of steady flow are shown in Fig. 7.2. In (*a*) the fluid is flowing through a tube that has a (smooth) constriction and at that point the velocity of the fluid is greatly increased. As we shall see later the pressure is reduced at the throat and this arrangement, called a *venturi tube* has many applications as for instance in carburetors for internal combustion engines. Fig. 7.2(*b*) shows the streamlines for the steady flow past a cylinder. The flow divides itself above and below the cylinder, while the velocity at the singular points $S_1$, $S_2$ is zero; these are called stagnation points. Note that the pattern of the streamlines is the same whether the liquid flows by the cylinder with velocity $v$, or whether the cylinder moves with velocity $v$ through a stationary liquid.

*Bernoulli's equation.* In the absence of friction, we can use energy conservation to derive an important relation between the forces acting on a fluid element and its velocity along a streamline. As before we will consider an element of fluid enclosed in a flow tube as shown in Fig. 7.3.

Fig. 7.1. (*a*) Fluid flow can be visualized by 'streamlines', continuous curves that are tangent to the local velocity vector. (*b*) Streamlines form the boundary of 'flow tubes'.

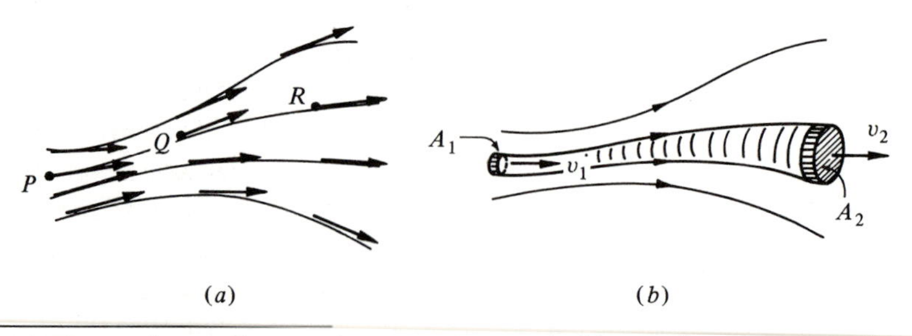


* Figures of streamlines are drawn, or photographed in two dimensions. In reality of course the pattern is three dimensional but often one dimension can be taken as infinitely long, or as having circular symmetry, etc.

At the entrance to the tube the area, pressure and velocity are $A_1$, $P_1$ and $v_1$, while at the exit they are $A_2$, $P_2$ and $v_2$. Furthermore we can let the entrance and exit areas be at different elevations, $h_1$ and $h_2$, with respect to the horizontal. The mass of fluid contained in an element crossing the entrance area for a time interval d$t$ is

$$dm = \rho_1 A_1 v_1 \, dt = \rho_2 A_2 v_2 \, dt$$

and must be equal at the entrance and exit regions for continuity reasons. We also assume that the fluid is incompressible, so that $\rho_1 = \rho_2 = \rho$. The work done on this element by the force due to the pressure is

$$\Delta W_P = W_1 - W_2 = F_1 \, ds_1 - F_2 \, ds_2$$
$$= (P_1 A_1)(v_1 \, dt) - (P_2 A_2)(v_2 \, dt_2)$$

where d$s_1$ and d$s_2$ are the thickness of the element along the streamline. The work done by the force of gravity is

$$\Delta W_g = dmgh_1 - dmgh_2 = (\rho A_1 v_1 \, dt)g(h_1 - h_2)$$

In the absence of friction the work must equal the change in the kinetic energy of the fluid element

$$\Delta(\text{K.E.}) = \tfrac{1}{2} \, dm(v_2^2 - v_1^2) = \tfrac{1}{2}(\rho A_1 v_1 \, dt)(v_2^2 - v_1^2)$$

Fig. 7.2. (*a*) Flow through a 'Venturi tube', i.e. a smooth constriction. (*b*) Streamlines for steady flow past a cylinder; note the two stagnation points.

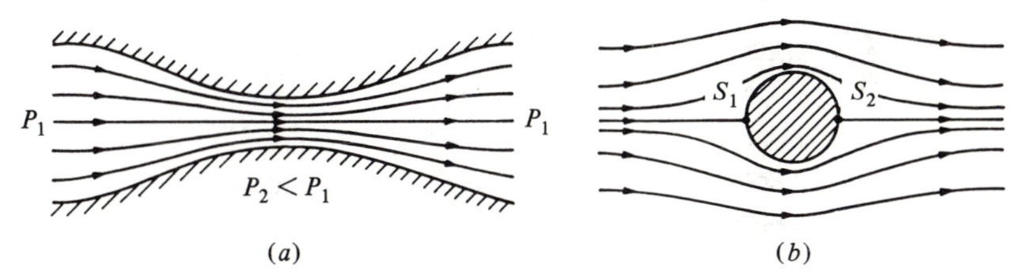


Fig. 7.3. Demonstration of Bernoulli's principle with the help of a flow tube of changing cross-section and of differing elevation at its two ends.

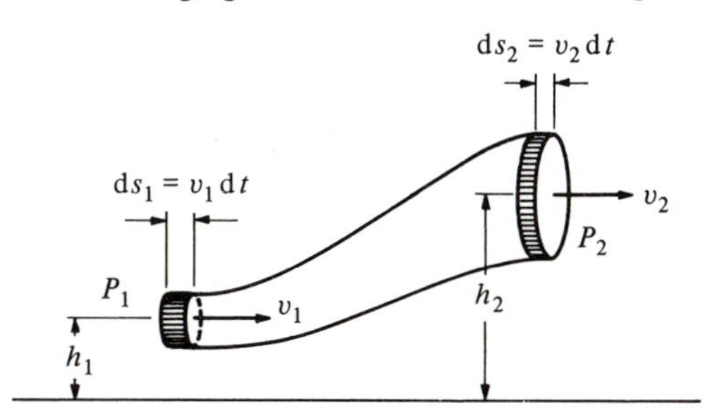

setting $\Delta(\text{K.E.}) = \Delta W_P + \Delta W_g$ we then find

$$P_1 + \rho g h_1 + \tfrac{1}{2}\rho v_1^2 = P_2 + \rho g h_2 + \tfrac{1}{2}\rho v_2^2 \tag{7.2}$$

which is known as Bernoulli's principle and is valid along a particular streamline.

Bernoulli's principle is often expressed by the statement

$$P + \rho g h + \tfrac{1}{2}\rho v^2 = \text{constant} \tag{7.2'}$$

which should be used carefully because the constant can have a different value for different streamlines. For horizontal flow or when the effects of gravity are negligible we write Eq. (7.2′) in the simpler form

$$P + \tfrac{1}{2}\rho v^2 = \text{constant} \tag{7.3}$$

From Eq. (7.3) we see that when the velocity increases the pressure drops and vice versa; this gives rise to slight paradoxes which can be observed in everyday life, as for instance the low pressure in a tornado, etc. Finally by taking the derivative of Eq. (7.3) we can obtain a differential form of Bernoulli's equation

$$\mathrm{d}P + \rho v\, \mathrm{d}v = 0 \tag{7.3'}$$

Again this equation should be used with care because $\mathrm{d}P$ and $\mathrm{d}v$ are the *total* differentials of $P$ and $v$ as a function of space and time; for steady flow $\partial P/\partial t$ and $\partial v/\partial t$ are of course zero.

As an important application of Bernoulli's principle let us re-examine the steady flow past a cylinder shown in Fig. 7.2(*b*) and reproduced in Fig. 7.4(*a*). At the stagnation points the velocity is zero so that the pressure $P_1 = P_2 = P_0$ and the resulting forces are equal and opposite. At points 3 and 4, the velocity is high and the pressure is low $P_3 = P_4 = P_0 - \tfrac{1}{2}\rho v_3^2 =$

Fig. 7.4. Flow past a cylinder. (*a*) Cylinder at rest; the pressure exerted on the cylinder surface is balanced in all directions. (*b*) When the cylinder rotates the streamline density is affected and the pressure is not anymore balanced.

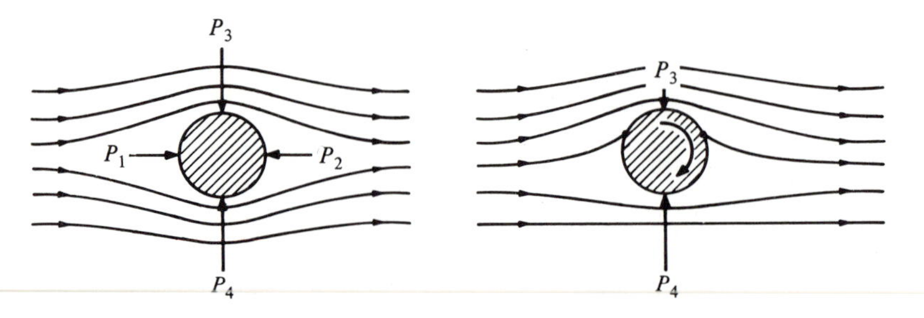

$P_0 - \frac{1}{2}\rho v_4^2$; however $P_3$ and $P_4$ give rise to equal and opposite forces so that the cylinder remains in equilibrium. Let us next assume that the cylinder rotates as indicated in Fig. 7.4(*b*). Then because of frictional forces, some of the fluid is carried along by the cylinder's motion and the streamlines are deformed as indicated. In this case $v_3 > v_4$ and therefore $P_3 < P_4$; as a result the cylinder is subject to a net force in the upward direction and will move accordingly.

The forces acting on a rotating cylinder or sphere that moves through a fluid or air are well known to baseball and tennis players where the effects of spin are quite pronounced. Lord Rayleigh appears to have been the first to appreciate that a tennis ball spinning in one direction (Fig. 7.5(*a*)) would reach much farther than when spinning in the opposite direction (Fig. 7.5(*b*)). In the first case the Bernoulli force reduces the effect of gravity, while in the second case it increases it. The same principle gives rise to the lift on an airplane wing, or *airfoil*.

An airfoil is designed so as to preserve the streamline flow around it, but in such a way that the velocity of the airstream above the foil is increased. This depends on the angle $\alpha$ of the airfoil with respect to the airstream. For instance for the position of Fig. 7.6(*a*) the flow is evenly

Fig. 7.5. A spinning tennis ball moving in the direction opposite to the streamlines: (*a*) if the ball spins in the sense indicated by the arrow it is subject to a 'lifting' force and can reach farther than a non-spinning ball, (*b*) if the spin is reversed the additional force drives the ball towards the ground and therefore its range is reduced.

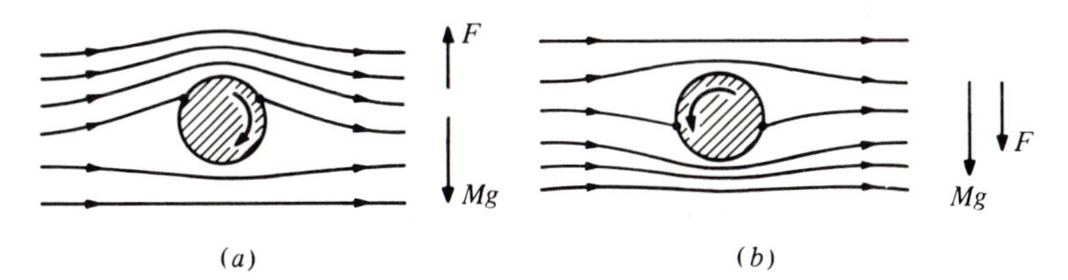


Fig. 7.6. An airfoil in a fluid stream: (*a*) the attack angle is zero and there is no lift, (*b*) for a finite attack angle $\alpha$ the airfoil experiences a lifting force.

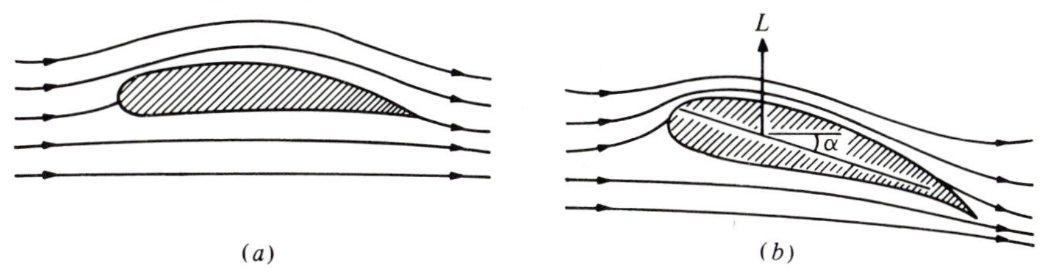

divided above and below the airfoil. When positioned as in (*b*) of the figure the flow pattern is changed and there is an upward lifting force acting on the airfoil. The angle between the airfoil axis and the velocity of the airstream is called the *angle of attack*.

*Circulation.* We are all familiar with eddies that are produced in the flow of water, in which case the streamlines are closed onto themselves as shown in Fig. 7.7(*a*). If the fluid elements are moving in a circular pattern there must be a force acting perpendicular to the direction of motion; namely a pressure gradient in the radial direction. We consider an element of radial thickness d$r$, width $w$ and of length $l$ (in the direction perpendicular to the plane of the paper) as shown in Fig. 7.7(*b*). The force acting on the fluid element is

$$F = A\,\mathrm{d}P = lw\,\mathrm{d}P$$

The mass of the element is $\mathrm{d}m = \rho\,\mathrm{d}V = \rho lw\,\mathrm{d}r$, and if the fluid element is moving in a circle of radius $r$, Newton's law demands that

$$\frac{v^2\,\mathrm{d}m}{r} = A\,\mathrm{d}P$$

or

$$\mathrm{d}P = \rho v^2 \frac{\mathrm{d}r}{r} \tag{7.4}$$

From Bernoulli's equation in its differential form (Eq. (7.2)) we can express $\mathrm{d}P = -\rho v\,\mathrm{d}v$ and therefore Eq. (7.4) is written as

$$-\rho v\,\mathrm{d}v = \rho v^2\left[\frac{\mathrm{d}r}{r}\right]$$

Fig. 7.7. Circulation around a vortex: (*a*) the streamlines are closed and their density increases as we approach the vortex, (*b*) an element of the fluid showing the forces acting on it.

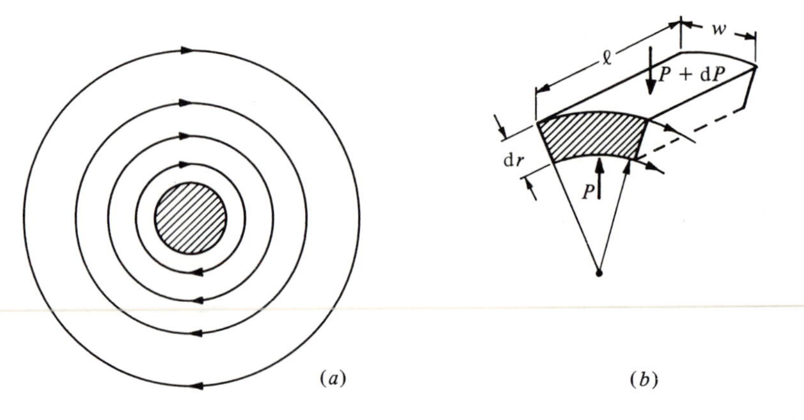

or

$$\frac{dv}{v} = -\frac{dr}{r} \tag{7.4'}$$

Eq. (7.4′) can be easily integrated to give

$$\ln v + \ln r = \text{constant}$$

or

$$v = \frac{\text{constant}}{r} \tag{7.5}$$

From Eq. (7.5) we see that the velocity at the center of the eddy becomes infinite, $v \to \infty$ as $r \to 0$; this singular point is called a *vortex*. A further consequence of Eq. (7.5) is that the line integral of the velocity taken along any streamline has always the same value. The line integral is defined through $\oint \mathbf{v} \cdot \mathbf{dl}$; if we take a circular streamline, $v$ is constant and $\mathbf{v}$ is always parallel to $\mathbf{dl}$. Thus $\oint \mathbf{v} \cdot \mathbf{dl} = v 2\pi r$ and since $vr = C$ (where $C$ is the constant appearing in Eq. (7.5)) we have the important result

$$\Gamma = \oint \mathbf{v} \cdot \mathbf{dl} = 2\pi C \tag{7.6}$$

While the result of Eq. (7.6) was derived for the special case of a circular path along a streamline it is generally true for any path as long as it *encloses* the vortex. If the vortex is not enclosed in the path, the line integral vanishes. Thus the strength of a vortex can be characterized by the value $\Gamma$ of the line integral of the fluid velocity as given by Eq. (7.6); $\Gamma$ is called the *circulation* of the vortex. The importance of this concept is that when referring to the flow around a rotating cylinder, or an airfoil we can represent it as the sum of a uniform flow and of a circulation $\Gamma$ surrounding them.

## 7.2 Airplane flight

We have seen in Fig. 7.6 that when an airfoil moves through air, a circulation develops around it and therefore it is subject to a lifting force. The lifting force per unit length of span, which we designate by $L'$ (see Fig. 7.8), is given by the Kutta–Jukowski law

$$L' = \rho v \Gamma \tag{7.7}$$

with $\rho$ the air density, $v$ the velocity of the airstream and $\Gamma$ the circulation. In turn the circulation can be expressed by

$$\Gamma = \pi \xi v c \alpha \tag{7.8}$$

where $c$ is the *chord* of the airfoil and $\alpha$ is the angle of attack defined in Fig. 7.6($b$); $\xi$ is an empirical factor of order unity and hereafter we will use $\xi = 1$.

The circulation is equivalent to a vortex as shown in the previous section and therefore we can replace the effect of the airfoil by imagining a *line of vortices* along its axis. However vortex lines, just as streamlines, must extend to infinity, and in the case of a *finite wing* only part of the vortex line moves along with the airplane. As a result the vortex line bends backwards as shown in Fig. 7.9($a$) and leaves a trail of vortices emanating from the wingtips. The air flow induced by the vortex is known as the *downwash* and it exerts a pressure on the wing as shown in ($b$) of the figure; it is strongest at the wing tip. The condensation trails that we see when airplanes fly at high altitude are due to the detached vortices, where because of the increased air velocity, the pressure and thus also the temperature, drop and condensation occurs. Fig. 7.10($a$) shows a vortex formed by flow around a sharp edge and in ($b$) of the figure the vortex trails of an airplane (emitting dust) can be clearly seen.

The presence of the downwash is unavoidable for any *finite* wing and has the important consequence that the airstream velocity is not parallel

Fig. 7.8. Flow and circulation around an airfoil.

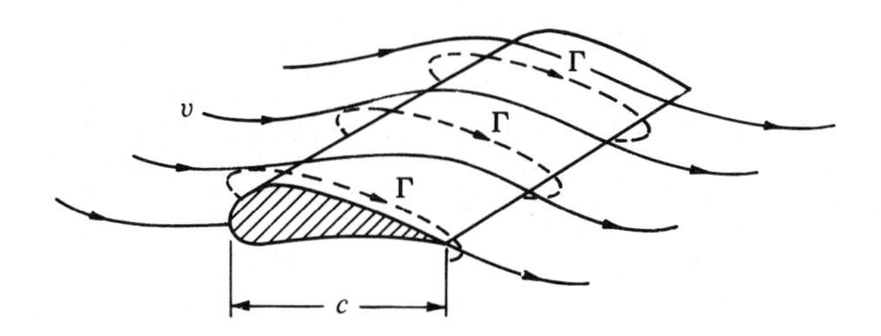


Fig. 7.9. ($a$) The line of vortices associated with an airfoil of finite length. ($b$) 'Downwash' forces acting on the airfoil as a result of vortex formation.

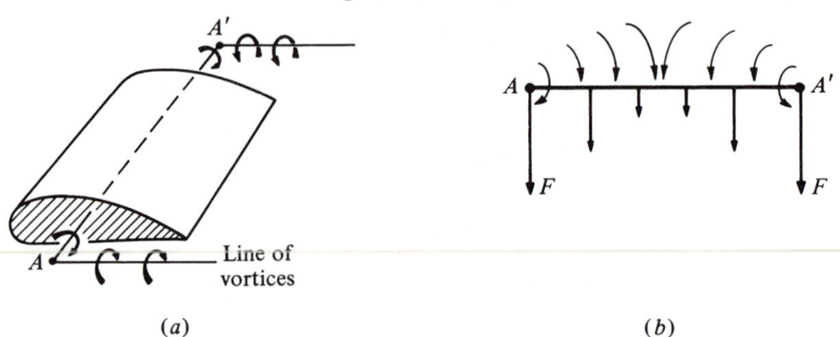

anymore to the motion of the airplane. This is illustrated in Fig. 7.11, where the effective velocity $\mathbf{v}'$ is the vector sum of the airplane velocity $\mathbf{v}$ and of the downwash $\mathbf{w}$. The aerodynamic force is always perpendicular to the airstream velocity, and therefore in the presence of the downwash, the force $\mathbf{R}$ is not normal but has a component $D$ which opposes the motion of the airplane. This is called the *induced drag*, and the airplane engine does work against the induced drag even in level flight. The induced drag per unit length of span, $D'$, is given by

$$\frac{D'}{L'} = \frac{w}{v} \quad \text{or} \quad D' = \frac{\rho v \Gamma w}{v} = \rho \Gamma w \tag{7.9}$$

The exact analytic calculation of lift, drag and other aerodynamic forces on any particular structure is obviously too complex for practical purposes. Instead very precise results can be obtained by measuring the same parameters on models in a wind tunnel and then scaling them to the full

Fig. 7.10. Demonstration of vortex flow: (*a*) due to streamline flow past a sharp object, (*b*) vortex trails emanating from the wing tips can be clearly seen in this picture because the plane was emitting dust. (From Th. von Karman, *Aerodynamics*, McGraw-Hill, 1954.)

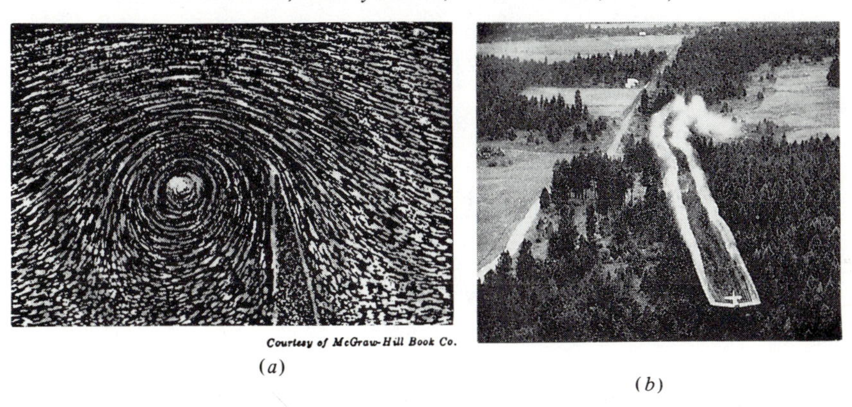


Fig. 7.11. Analysis of lift and drag forces acting on an airplane wing of finite length. The airplane velocity is $-\mathbf{v}$, but because of the downwash $\mathbf{w}$ the airstream velocity is $\mathbf{v}'$.

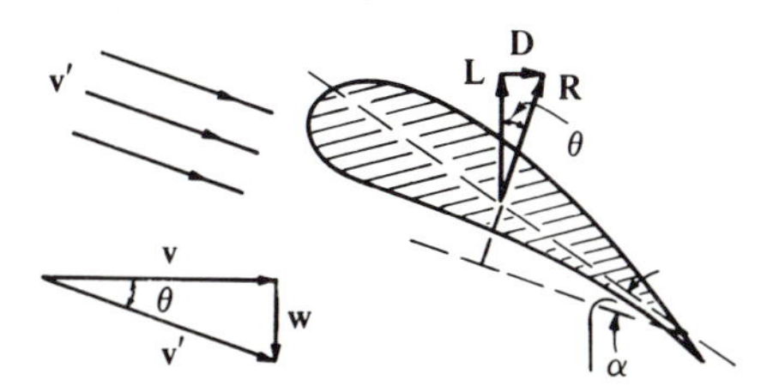

size of the airplane. It is therefore customary to express the forces in terms of *coefficients* defined as follows

$$C_L = \text{lift coefficient} = \frac{\text{Lift}}{\frac{1}{2}\rho v^2 S}$$
$$C_D = \text{drag coefficient} = \frac{\text{Drag}}{\frac{1}{2}\rho v^2 S} \tag{7.10}$$

Here $S$ is the area of the wing and $\frac{1}{2}\rho v^2$ is the 'dynamic pressure'. Finally a reasonable approximation for the downwash velocity is

$$\frac{w}{v} = \frac{2\alpha}{\text{Æ}} \simeq 2\alpha \frac{S}{b^2} \tag{7.10'}$$

where $\text{Æ} = b^2/S$ is the 'aspect ratio' of the wing with $b$ the length or 'span' of the wing. Thus we also have $\text{Æ} \sim b/c$ ($c$ is the chord), from where we see that, in general, $\text{Æ} > 1$.

From Eqs. (7.7) to (7.10) we can easily deduce in the same spirit of approximation that

$$C_L \simeq 2\pi\alpha \tag{7.11}$$

and that

$$C_D = 4\pi\alpha^2(S/b^2) \tag{7.11'}$$

Finally the necessary engine power to keep the plane in level flight is

$$P = \mathbf{D}\cdot\mathbf{v} = Lv\frac{C_L}{\pi\text{Æ}} \tag{7.12}$$

where $L$ is the lift force and $D$ the drag force.

## 7.3 The effects of viscosity

So far we have assumed that the fluid is incompressible and non-viscous. The effects of the compressibility of air become important when the velocity approaches the speed of sound and we will consider them in the following section. The viscosity of a fluid is due to internal friction and is always present; the fact that circulation is established around an airfoil is due to the frictional forces between the surface of the wing and the air molecules, so that a thin *boundary layer* is being carried along with the wing. At high velocities the flow ceases to be laminar, eddies develop and momentum is transferred between adjacent fluid elements; the flow has become *turbulent*.

As in our previous treatment we consider the fluid as consisting of small elements $d\tau$, carrying momentum $dp = \rho v \, d\tau$; the density and velocity are assumed to be continuous variables of position and time. We distinguish between 'body-forces' which act on the bulk of the material and 'surface-forces' which are specific to each element. Examples of body-forces are gravitation and, in the case of a plasma, electrical forces; we will not concern ourselves with such forces here. The surface-forces are the *normal stress*, that is the force resulting from the pressure in the fluid, and the tangential stress or *shear* which is due to the viscous forces in the liquid. In solids the shearing force depends on the corresponding normal force $F_N$, i.e. we write for the frictional force $F_f = \mu F_N$ where $\mu$ is the coefficient of friction. In liquids the shearing stress depends on the rate of change of the velocity through the fluid. This can be illustrated if we think of a container filled with a fluid at rest, up to a height $d$ as shown in Fig. 7.12. We now place a flat plate on the surface of the fluid and move it with constant velocity $v_0$ along the $x$-direction. As long as the flow remains laminar, the fluid layer next to the plate ($z = d$) has velocity $v = v_0$, whereas the layer at the bottom of the container ($z = 0$) must have velocity $v = 0$. If the area of the plate is $A$, it is found that the tangential force $F_t$ required to maintain the velocity $v_0$ is given by

$$F_t/A = \eta(v_0/d) \tag{7.13}$$

The proportionality factor $\eta$ is called the *coefficient of viscosity* of the fluid and can be defined more generally through

$$\eta = \frac{F_t/A}{dv_x/dz} \tag{7.13'}$$

The coefficient of viscosity has dimensions of kg/m-s as can be deduced from the defining equation and varies by orders of magnitude for different fluids. Some typical values are given in the Table 7.1. The viscosity of a fluid is due to the intermolecular forces and depends strongly on the temperature. Some materials such as tar or glass which at normal

Fig. 7.12. Demonstration of the effects of viscosity by moving a plate $A$ over a liquid of depth $d$. The velocity of the liquid as a function of depth is indicated by the arrows.

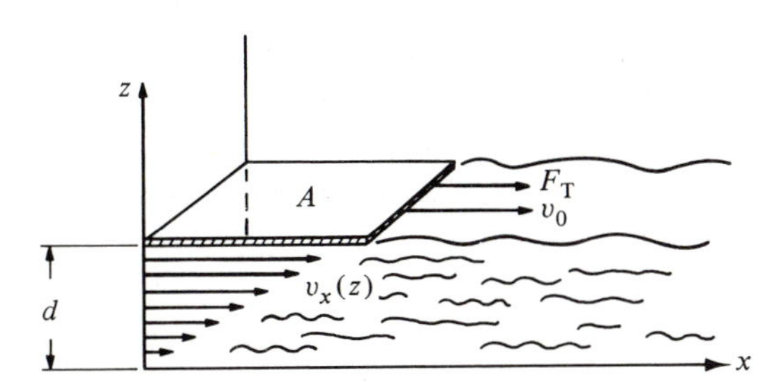

temperature appear to be solids can be thought of as extremely viscous fluids.

When we take frictional forces into account, Newton's equation for the fluid elements take the form

$$\mathbf{F}_\mathrm{P} + \mathbf{F}_\mathrm{f} = m\mathbf{a} \tag{7.14}$$

with $\mathbf{F}_\mathrm{f}$ the frictional force, $\mathbf{F}_\mathrm{P}$ the force due to pressure gradients and $\mathbf{a}$ the acceleration of the element. We can of course replace $m\mathbf{a}$ on the r.h.s. of Eq. (7.14) by introducing on the l.h.s. an *inertial* force $\mathbf{F}_\mathrm{I} = -m\mathbf{a}$; we then obtain an equilibrium condition between the three forces

$$\mathbf{F}_\mathrm{P} + \mathbf{F}_\mathrm{f} + \mathbf{F}_\mathrm{I} = 0 \tag{7.14'}$$

These forces are shown schematically in Fig. 7.13 and need not coincide with the velocity vector or be colinear. Eq. (7.14) makes it easy to visualize different regimes of fluid flow. For instance at low velocities, the acceleration is also low and therefore the inertial force is weak. Thus the pressure force must balance the frictional force. If on the other hand $\eta$ is low, then the pressure force balances the inertial force. In general, if the ratio of $\mathbf{F}_\mathrm{I}$ to $\mathbf{F}_\mathrm{f}$ is the same the flow pattern is identical, and geometrically similar objects have similar dynamic properties. The ratio of the inertial to the frictional force is called the *Reynolds number* $\mathscr{R}$.

To define the Reynolds number, which is dimensionless, we must introduce a scale factor or characteristic length $L$. For instance in flow through pipes, $L$ is the diameter of the pipe; for an airfoil, $L$ would be

Table 7.1. *Viscosity of selected fluids*

| | | |
|---|---|---|
| Lubricating oil | $\eta \simeq 1.0$ | kg m$^{-1}$ s$^{-1}$ |
| Water | $\eta = 0.9 \times 10^{-3}$ | |
| Air at s.t.p. | $\eta = 1.8 \times 10^{-5}$ | |

Fig. 7.13. The forces acting on a fluid element: $\mathbf{F}_\mathrm{p}$ is the static pressure force, $\mathbf{F}_\mathrm{I}$ the inertial force and $\mathbf{F}_\mathrm{f}$ the frictional force, and the three forces must balance.

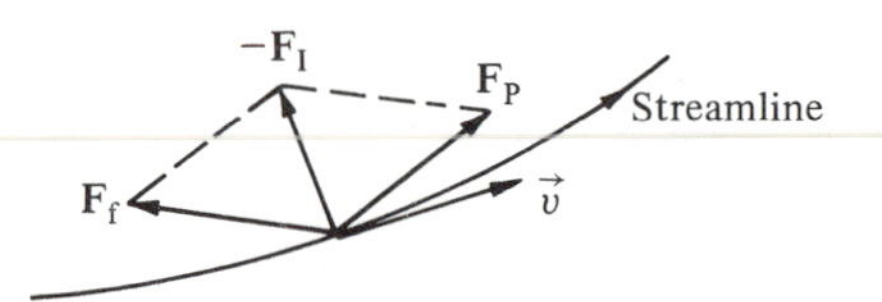

the chord. Now the frictional force (see Eq. (7.13)) is

$$F_f = \eta A \frac{v}{d} \sim \eta A \frac{v}{L} \tag{7.15}$$

while the inertial force is given by

$$F_I = m|\mathbf{a}| = (\rho A L)\left|\frac{d\mathbf{v}}{dt}\right| \tag{7.15'}$$

We assume that $\Delta t \sim L/v$ and that in that time $\Delta v \sim v$ leading to $a \sim v^2/L$ which is typical of circular motion. We then find

$$F_I \sim \rho A v^2 \tag{7.15''}$$

Therefore the Reynolds number is defined as

$$\mathcal{R} = \frac{F_I}{F_f} = \frac{\rho v L}{\eta} \tag{7.16}$$

From this result we see that:

$\mathcal{R}$ Small corresponds to viscous flow at low velocity. As we know, this is the regime of *laminar* flow.

$\mathcal{R}$ Large corresponds to low viscosity and high velocity. This is the regime of *turbulent* flow.

For flow in pipes the transition from laminar to turbulent flow occurs when $\mathcal{R} \sim 10^3$.

As an example of flow at low Reynolds number we consider a small sphere of radius $a$ which is dropped into a viscous fluid. Eventually the sphere reaches a terminal velocity $v_T$; at this point the inertial force is zero and the pressure gradient is small as compared to the force of gravity. Therefore the force of gravity $F_g = mg$ equals the frictional drag force $F_f$. From Eq. (7.13) we expect that $F_f$ will be proportional to $\eta$, to $v_T$ and to the dimensions of the sphere. In this case $F_f$ can be calculated and is given by Stokes' law

$$F_D = F_f = 6\pi a \eta v_T \tag{7.17}$$

Next, let us compare the flow at high Reynolds number typical of an airfoil and of a large fish. For the airfoil the following parameters are realistic:

| | |
|---|---|
| $v = 200$ km/hr | $\sim 55$ m/s |
| $L = 5$ m | (chord length) |
| $\rho = 1.29$ kg/m$^3$ | (density of air at s.t.p.) |
| $\eta = 2 \times 10^{-5}$ kg/s | (from Table 7.1) |

Then

$$\mathcal{R} = \frac{\rho v L}{\eta} = \frac{1.3 \times 55 \times 5}{2 \times 10^{-5}} \sim 1.8 \times 10^7 \quad \text{(airfoil)}$$

For the fish we can use

$$v = 20 \text{ knots} \quad \sim 10 \text{ m/s (fast fish)}$$
$$L = 1.5 \text{ m}$$
$$\rho = 10^3 \text{ kg/m}^3 \quad \text{(density of water)}$$
$$\eta = 10^{-3} \quad \text{(from Table 7.1)}$$

$$\mathscr{R} = \frac{\rho v L}{\eta} = \frac{10^3 \times 10 \times 1.5}{10^{-3}} \sim 1.5 \times 10^7 \quad \text{(fish)}$$

We see that the airfoil and the fish have similar Reynolds numbers and this explains why their cross-sectional shapes are so similar. In both cases 'streamlining' is important to avoid turbulence.

Some turbulence always develops behind an airfoil and as a result the pressure behind the airfoil is reduced. This produces a pressure drag in addition to the frictional drag. Fig. 7.14(*a*) shows the onset of turbulence behind an airfoil, whereas in (*b*) of the figure the attack angle has been increased to the point that the flow is no more laminar and therefore there is no lifting force; we say that the foil is in a stall. At large Reynolds number the drag is mainly due to the inertial force rather than to friction. Thus

$$F_D \sim \tfrac{1}{2}\rho v^2 A_\perp$$

where $A_\perp$ is the area perpendicular to the direction of flow. As a result, the power required to maintain constant velocity is

$$P = \mathbf{F}_D \cdot \mathbf{v} \sim \tfrac{1}{2}\rho v^3 A_\perp \tag{7.18}$$

as also found from Eq. (7.10), where $F_D = C_D \tfrac{1}{2}\rho v^2 S$. The dependence of the engine power on the cube of the velocity explains why increments in air speed were so dependent on more powerful propulsion systems rather than on refinements in aerodynamic design.

As we have already stated, the circulation that is established around an airfoil is due to the *boundary layer* that becomes attached to the wing. The thickness of the boundary layer $\delta$ depends on the viscosity of the fluid and for a typical linear dimension $L$, it holds that

$$\delta = L/\mathscr{R}^{1/2} \tag{7.19}$$

For instance for the airfoil analysed in the last example, $\mathscr{R} \simeq 10^7$ and the chord of the foil was $L = 5$ m. Then we find that

$$\delta = 5 \text{ m}/(10^7)^{1/2} \sim 1.6 \text{ mm}$$

Indeed the boundary layer is very thin and mechanical perturbations of the wing surface at that scale can cause the boundary layer to become *detached* with a consequent loss of lift. On the other hand, for extremely smooth surfaces the boundary layer becomes detached at lower Reynolds number than for a slightly irregular surface finish. This is exemplified by the 'dimpled' golf balls which, for the same stroke, have a range almost five times longer than that of a smooth ball.

## 7.4 Supersonic flight

Variations in the local density and pressure of a fluid can propagate through a medium and are detected by the ear giving rise to the sensation of sound. The velocity of sound is given in general by

$$v_s = (dP/d\rho)^{1/2} \tag{7.20}$$

where P is the pressure and $\rho$ the density. For nearly ideal gases Eq. (7.20) can be evaluated and because sound propagation is adiabatic

Fig. 7.14. Turbulence developing from the motion of an airfoil through a viscous fluid: (*a*) small angle of attack, (*b*) for a large angle of attack the turbulence results in loss of lift and thus to a stall. (From A. H. Shapiro, *Shape and Flow*, copyright 1961 by Educational Services Inc. used by permission of Doubleday, a division of Bantam, Doubleday, Dell Publishing Group Inc.)

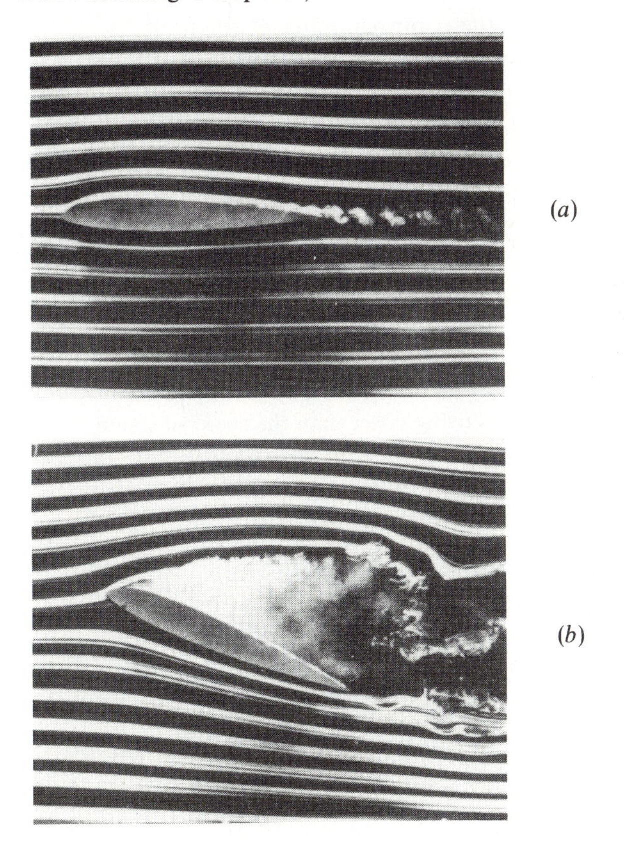

one finds that

$$v_s = (\gamma R T_0/M)^{1/2} \tag{7.20'}$$

where $\gamma = c_p/c_v$, $R = N_0 k$ is the universal gas constant, $M$ is the molecular weight of the gas and $T_0$ the ambient temperature. The derivation of Eqs. (7.20) is given in Appendix 4.

To evaluate $v_s$ for air at s.t.p. we use

$$R = 8.31 \text{ J/(g-mole)-K}$$
$$M = 28.8 \text{ g/mole} = 0.0288 \text{ kg/g-mole}$$
$$\gamma = 1.4$$
$$T = 4°\text{C} = 277 \text{ K}$$

and find

$$v_s = 335 \text{ m/s} \tag{7.21}$$

The speed of sound is a measure of the typical random velocity of the molecules in the gas. Thus, when an object moves through a gas at velocities approaching $v_s$ we cannot any more assume that the random velocity within each fluid element has averaged to zero; of course, we must also account for the compressibility of the fluid.

When the velocity of the body exceeds the speed of sound a *shock wave* is produced, namely a sharp discontinuity in pressure and density propagates through the medium as illustrated in Fig. 7.15(*a*). The creation of the shock wave can be understood with the help of the sketch in part (*b*) of the figure. The body moves uniformly along the $x$-axis starting from the point $x_A$ at time $t_A$; at the present time $t_D$ it is at the point $x_D$ and therefore its velocity $v = (x_D - x_B)/(t_D - t_A)$. The circles represent the locus at the time $t_D$ of the pressure waves produced when the body was passing

Fig. 7.15. (*a*) Pressure profile of a shock wave. (*b*) Generation of a shock wave by a body moving faster than the speed of sound.

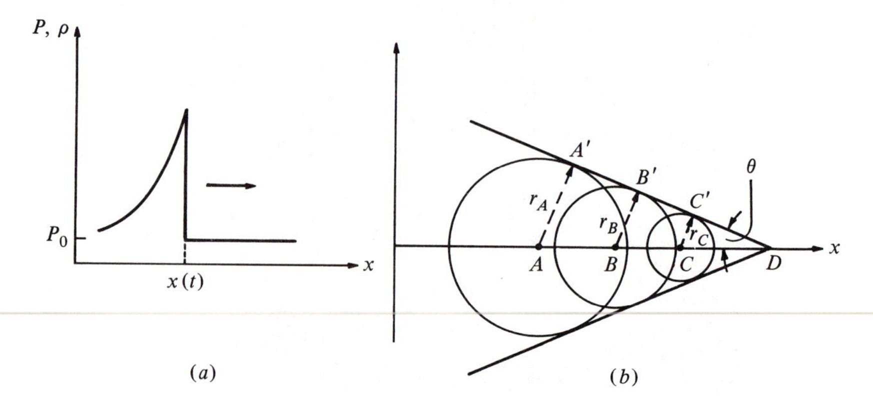

the points $x_A, x_B, \ldots$, etc. Since the pressure wave propagates with the speed of sound, the radii of these circles are given by $r_A = v_s(t_D - t_A)$, $r_B = v_s(t_D - t_B), \ldots$ etc. It is clear from the geometry that the pressure waves from all points of the trajectory arrive *simultaneously* on the surface of a cone with apex in the present position and of angle $\theta$, where

$$\sin\theta = \frac{AA'}{AC} = \frac{v_s(t_D - t_A)}{v(t_D - t_A)} = \frac{v_s}{v} \tag{7.22}$$

As long as $v > v_s$ the angle $\theta$ is real and a shock wave is produced. It is customary to call the ratio $v/v_s$ the *Mach number*, $\mathcal{M}$. We then find that

$$\cos\theta = (1 - \sin^2\theta)^{1/2} = \frac{1}{\mathcal{M}}(\mathcal{M}^2 - 1)^{1/2} \tag{7.22'}$$

The shock waves can be easily observed using an optical interference technique known as Schlieren photography. An example corresponding to $\mathcal{M} = 1.45$ is shown in Fig. 7.16; the angle of the shock wave approximately satisfies Eq. (7.22).

The creation of a shock wave requires energy and this must be provided by the airplane engine. Consequently the dıag is greatly increased as $v \to v_s$. Fig. 7.17($a$) shows the lift coefficient of a flat plate as a function of Mach number; it exhibits a resonant behavior tending to infinity at

Fig. 7.16. Shock wave produced by a wedge moving at Mach number 1.45; it is visualized by Schlieren photography. (From Th. von Karman, *Aerodynamics*, McGraw-Hill, 1954.)

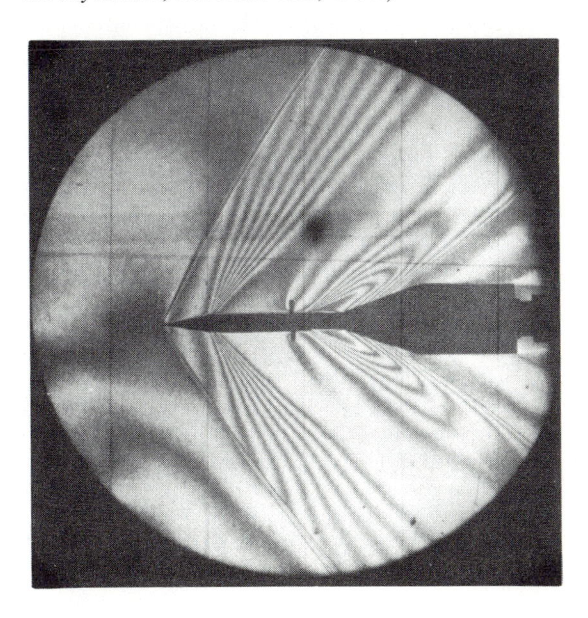

$v = v_s$. When the velocity $v \to v_s$ maximum stress is developed and this is why we speak of 'breaking through the sound barrier'. The lift and drag coefficients for a more realistic structure, the German V2 rocket used in W.W.II, are shown in (*b*) of the figure. Once $v > v_s$ the drag is mainly due to friction which at these velocities greatly exceeds the induced and pressure drag.

For supersonic flight the airfoil design differs from that which is optimum for subsonic velocities. Sharp edges are preferable to streamlined profiles and swept-back and delta wings offer better performance. At even higher velocities we speak of transonic flight where the main problem is how to cool the wing and body surface which are heated to very high temperature by air friction. The thrust required to reach supersonic velocities cannot be developed by propellers, and instead jet or rocket engines are necessary; we discuss this subject in the following section.

## 7.5 Propulsion dynamics

So far we have examined airplane flight by tacitly assuming that the plane was moving with a given velocity $v$ through the air. There are,

Fig. 7.17. Supersonic flight: (*a*) theoretical lift coefficient of a flat plate as a function of Mach number ($\alpha$ is the angle of attack), (*b*) lift and drag coefficients for the German V2 rocket used in W.W.II. (From G. P. Sutton, *Rocket Propulsion Elements*, J. Wiley, 1963.)

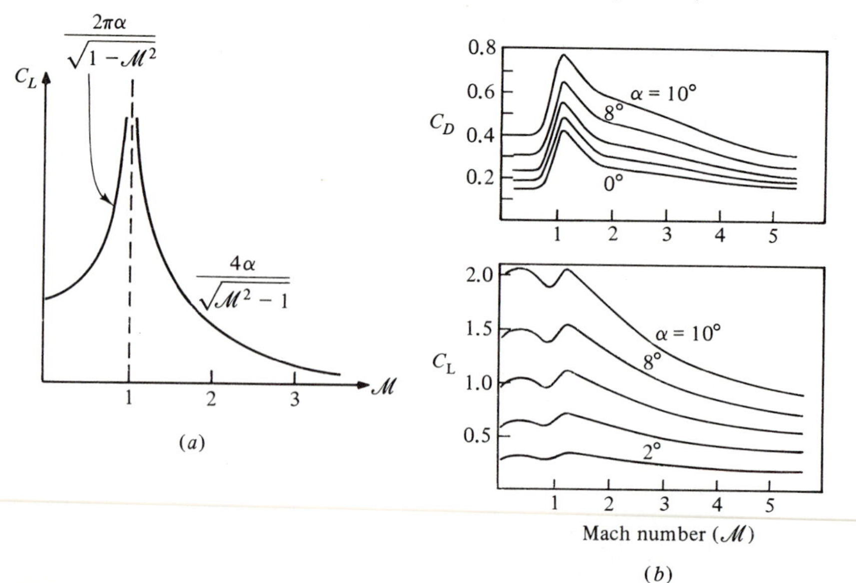

however, retarding forces acting on the airplane such as the induced and frictional drag and if the plane is climbing the force of gravity. Therefore a force must act on the plane not only to accelerate it but also to maintain its speed. In the first airplanes the *propulsive* force was obtained from the action of the propeller whereas at present jet engines are used for all but small airplanes.

In contrast, rockets are not accelerated continuously but reach their final velocity during a short time interval when they are launched. The thrust that propels a rocket is obtained from the exhaust at high velocity of a stream of gases; the gases are produced by the burning of the fuel in a high temperature chamber and are accelerated as they exit through a nozzle. Once launched, rockets continue on their flight without further major propulsive impulses except for course corrections and similar maneuvers.

One can think of the action of the propeller as that of a screw that is forced to advance through the medium because of its rotational motion. While this picture is suitable for describing the propulsion of a ship, it is not applicable in air where there is very large slippage. Instead, we consider the airstream produced by the propeller, which has acquired a momentum $\Delta p$; since momentum must be conserved an equal and opposite force $\Delta p/\Delta t$ acts on the airplane. In Fig. 7.18 we show the streamlines and flow tube ahead and beyond the propeller. The area swept out by the propeller is $S$ and its linear speed is $v$ (the linear speed is given by the pitch of the propeller multiplied by its angular velocity in rev/s); the velocity of the plane is $u$, so that the air velocity before the propeller is $u$ and after the propeller it is $u + v$. The mass of air moved by the propeller per unit time will be designated by $Q$ where

$$Q = \frac{\mathrm{d}m}{\mathrm{d}t} = \rho S v \tag{7.23}$$

Then the thrust acting on the plane is

$$F = \frac{\mathrm{d}p}{\mathrm{d}t} = v\frac{\mathrm{d}m}{\mathrm{d}t} = Qv \tag{7.23'}$$

Fig. 7.18. Streamlines and their modification by propeller action.

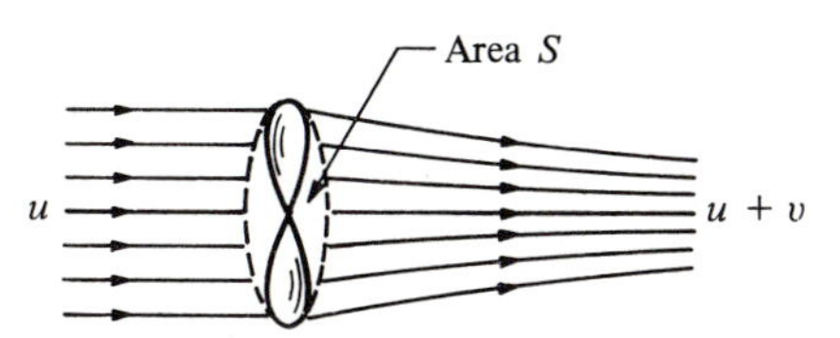

It is of interest to consider the *efficiency* of the propeller, namely the ratio of the power used to maintain the airplane in flight to the total engine power. The propulsive power is

$$P_p = Fu = Qvu \tag{7.24}$$

The total power expended is equal to the rate of increase of the kinetic energy of the airstream

$$P_t = \frac{dW_t}{dt} = \tfrac{1}{2}Q[(u+v)^2 - u^2] = Q\left[vu + \frac{v^2}{2}\right] \tag{7.24'}$$

Therefore the propulsion efficiency is

$$\eta = \frac{P_p}{P_t} = \frac{vu}{vu + v^2/2} = \frac{1}{1 + v/(2u)} \tag{7.24''}$$

Thus, to have high efficiency we want $(v/u)$ to be small; however for maximum thrust we need $v$ to be large. Consequently variable pitch propellers are used to provide large $v$ at take-off and when high thrust is required, but permitting $v$ to be reduced for level flight. We also see that the efficiency increases for high speed airplanes. The results of Eq. (7.24′) are of general validity and are also applicable to jet engines and to rockets.

In practical applications one is interested in the power required to maintain a given thrust. When $u = 0$ we can obtain a simple result by combining Eq. (7.24′) with Eqs. (7.23) and find

$$P_t = \tfrac{1}{2}F(F/\rho S)^{1/2} \tag{7.25}$$

Apart from the numerical factor of $\frac{1}{2}$ this result is valid for a hovering helicopter and gives the correct dependence of power on the thrust developed. As an example, a propeller with disk $S = 10\ \text{m}^2$ and delivering a thrust $F = 10^4$ N ($\sim$2200 lb) would require $P_t \sim 190$ horsepower.

Jet engines too rely on creating a high velocity airstream; this is achieved by heating the intake air in the combustion chamber and then exhausting it through the rear of the engine. A schematic of a *turbojet* engine is shown in Fig. 7.19. The airstream enters from the left and is compressed before being ignited; on their way to being exhausted the hot gases drive a turbine which provides the power for the compressor. In general, jet engines are simpler in construction than reciprocating (piston) engines, but require materials that can withstand high temperature and pressure. The first operational jet engine was used in a German fighter plane toward the end of W.W.II in 1945; it developed a thrust of $\sim$1000 lb. In contrast, typical jet engines today have thrust of 20 000 to 30 000 lb.

Rocket engines are in principle similar to jet engines in that the thrust is developed by the exhaust at very high velocity of a stream of gases.

Since rockets must be able to operate outside the atmosphere and because very large thrust is required, the gases that are heated and exhausted are carried by the rocket itself. The thrust is given by

$$F = \frac{dm}{dt} v_e \tag{7.26}$$

as derived in Eq. (6.11); $v_e$ is the exhaust velocity and typically $v_e \sim 2000$ m/s, namely a supersonic velocity, and $(dm/dt)$ is the mass of the gases exhausted per unit time. Eq. (7.16) is valid in the rocket's rest frame; recall that $v_e$ is the exhaust velocity with respect to the rocket.

To calculate the efficiency of a rocket engine we proceed as we did for the propeller (see Eqs. (7.24)). The power delivered to the rocket is

$$P_p = Fu = (dm/dt)v_e u \tag{7.27}$$

with $u$ the instantaneous velocity of the rocket in an *absolute* frame. The total power delivered by the engine is the rate of increase of the energy of the rocket and of the exhaust gases; note that in the absolute frame the velocity of the gases is $v' = u - v_e$. Thus

$$P_t = P_p + \frac{1}{2}\frac{dm}{dt}(v')^2 = \frac{dm}{dt} v_e u + \frac{1}{2}\frac{dm}{dt}(u - v_e)^2 \tag{7.27$'$}$$

And therefore the efficiency is

$$\eta_j = \frac{P_p}{P_t} = \frac{2uv_e}{u^2 + v_e^2} = \frac{2(u/v_e)}{1 + (u/v_e)^2} \tag{7.27$''$}$$

The efficiency is maximal and equals one when $u = v_e$, that is when the exhaust gases are at rest in the absolute frame after leaving the rocket.

It is customary to characterize rocket engines by their *specific impulse* $I_{sp}$, which is defined as the ratio of the thrust to the weight (as measured

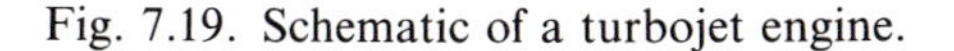

Fig. 7.19. Schematic of a turbojet engine.

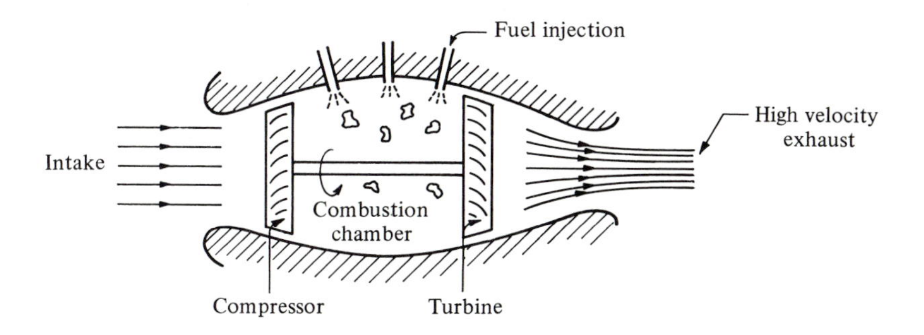

on the surface of the earth) of the fuel consumed per unit time*

$$I_{sp} = \frac{F}{dw/dt} = \frac{F}{g(dm/dt)} = \frac{v_e}{g} \tag{7.28}$$

Here $g$ is the acceleration of gravity at the *surface of the earth*. When the specific impulse is defined as in Eq. (7.28), it has dimensions of time and is given in seconds. For typical rocket engines $I_{sp} \sim 200$–$400$ seconds; this implies exhaust velocities of 2000–4000 m/s.

Rocket engines use solid or liquid propellants as fuel; for instance liquid oxygen and liquid hydrogen can be mixed in the reaction

$$2H + O \rightarrow H_2O$$

which releases 3 eV of energy. The energy released per unit mass of fuel is

$$Q_R = N_0 \times (3 \times 1.6 \times 10^{-19})\ \text{J/mole} = 3 \times 10^5\ \text{J/mole}$$
$$\sim 2 \times 10^7\ \text{J/kg}$$

Thus for a large rocket, carrying 100 tonnes of fuel the total energy release is of order $Q = Q_R M \eta_c$ where $M = 10^5$ kg is the mass of the fuel and the combustion efficiency $\eta_c \sim 0.4$–$0.7$; thus $Q \simeq 10^{12}$ J. Only a fraction of the energy $Q$ appears as kinetic energy of the payload; a large part is expended in the residual energy of the exhaust gases and in the kinetic energy of the rejected first stages of the rocket. We examine these considerations in the following sections.

## 7.6 Rocket engines

A simplified model of a rocket engine consists of a combustion chamber followed by a *nozzle* through which the gases exhaust; this is shown in the schematic of Fig. 7.20(*a*). Of course there must be provisions for supplying the fuel (usually under pressure) to the combustion chamber and controlling the flow of the two components; often part of the nozzle must be cooled in spite of the ability of the various components to withstand high temperatures. A more realistic view of a rocket engine is given in Fig. 7.21.

In the model of Fig. 7.20(*a*) we identify three regions: the combustion chamber (*c*), the throat (*t*) and the exhaust plane (*e*). The pressure, density and temperature of the gases will take different values at different positions,

* The specific impulse was first introduced as the ratio of the thrust, in pounds-force, divided by the mass of fuel, expressed in pounds-mass, per second. This was a convenient, even if not correct, definition; we will adopt the definition of Eq. (7.28) which shows that $I_{sp}$ is strictly equivalent to specifying the exhaust velocity.

but we can assume cylindrical symmetry about the axis. The force resulting from pressure always acts normal to the walls and we will designate the outside pressure by $P_o$. Differences between $P_o$ and the exhaust pressure $P_e$ modify Eq. (7.26) so that the thrust is given by

$$F = \frac{dm}{dt} v_e + (P_e - P_o)A_e \tag{7.29}$$

with $A_e$ the exhaust area. If $P_e > P_o$ we have increased thrust but we are not efficiently using the energy in the chamber. If $P_e < P_o$ a 'pressure drag' acts on the rocket, while for optimal design of the engine $P_e = P_o$. Since $P_o$ is a function of altitude it is difficult to maintain this condition throughout the flight.

We will now attempt to find the exhaust velocity $v_e$ in terms of the temperature $T_c$ in the combustion chamber. Once heated, the expansion of the exhaust gases is adiabatic and therefore

$$U + PV = \text{constant}$$

where $U$ is the internal energy of the gas, and $V$ the volume. The mechanical energy $PV$ is converted to the ordered kinetic energy of the gases $PV = \frac{1}{2}mv^2$ whereas $\Delta U = mc_p\Delta T$, with $c_p$ the specific heat under constant pressure. Thus

$$\tfrac{1}{2}(v_1^2 - v_2^2) + c_p(T_1 - T_2) = 0 \tag{7.30}$$

If we now compare the parameters in the chamber ($v_1 = 0$, $T_1 = T_c$) to those at the exhaust, we find

$$v_e^2 = 2c_p(T_c - T_e) \tag{7.30$'$}$$

We recall that $\gamma = c_p/c_v$, $c_p - c_v = R$ and that the velocity of sound is given by $v_s = (\gamma RT/M)^{1/2}$ with $M$ the molecular weight (see Eq. (7.20′)). We can

Fig. 7.20. Rocket engine: (*a*) schematic showing the combustion chamber and throat and exhaust channel, (*b*) flow through an idealized 'DeLaval' nozzle.

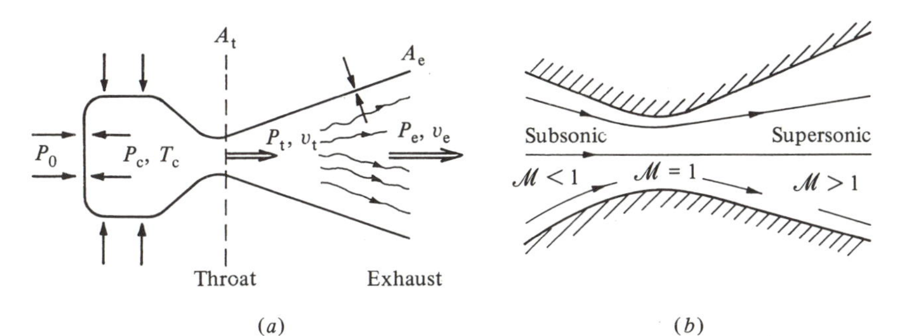

then write for Eq. (7.30′)

$$v_e = \left[\frac{2\gamma}{\gamma - 1}\frac{RT_c}{M}\left(1 - \frac{T_e}{T_c}\right)\right]^{1/2} = (v_s)_c\left(\frac{2}{\gamma - 1}\eta\right)^{1/2} \tag{7.31}$$

where $(v_s)_c$ is the velocity of sound in the combustion chamber and $\eta$ is an efficiency factor which in actual engines takes values 0.5–1.0. To evaluate $\eta$ we note that

$$\eta = \left[1 - \frac{T_e}{T_c}\right] = \left[1 - \left(\frac{P_e}{P_c}\right)^{(\gamma - 1)/\gamma}\right] \tag{7.31′}$$

where the last step follows for an adiabatic process in an ideal gas.

We know that for air, $\gamma \sim 1.4$, and for the oxygen–hydrogen mixture used in rocket engines $\gamma$ is even closer to unity, $\gamma \sim 1.25$. Thus even with $\eta = 0.5$, the exhaust velocity is $v_e \sim 2(v_s)_c$ namely supersonic. As the gases exit the combustion chamber they gain velocity, and the throat of the nozzle is defined as the plane where the Mach number $\mathscr{M} = 1$; in the exhaust region $\mathscr{M} > 1$. This is shown in Fig. 7.20(*b*) which is drawn for an idealized 'DeLaval' nozzle.

To obtain the thrust of the engine we must evaluate d$m$/d$t$, and we do so by considering the flow through the throat area $A_t$. Then

$$\frac{\mathrm{d}m}{\mathrm{d}t} = A_t \rho_t v_t \simeq A_t \frac{P_c}{(v_s)_c} \tag{7.32}$$

Fig. 7.21. (*a*) Schematic diagram of a liquid propellant rocket.

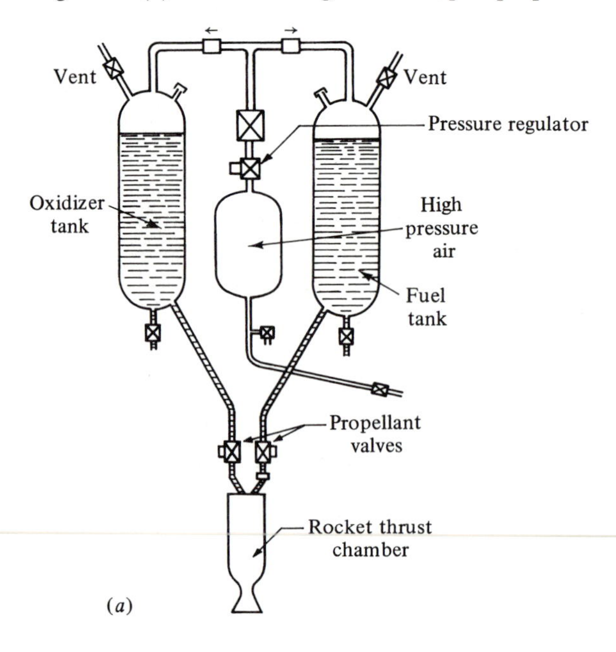

Since we also know that $v_e$ is of order $(v_s)_c$, we can write for the thrust (see Eq. (7.29))

$$F = v_e \frac{dm}{dt} \simeq C_{FX} A_t P_c \qquad (7.33)$$

where we ignored the pressure difference term. Here $C_{FX} \sim (1.5\text{–}2.0)$. Eq. (7.33) is useful for calculating the required throat area to achieve a given thrust.

Fig. 7.21. (*b*) F-1 liquid propellant turbopump-fed rocket engine manufactured by Rocketdyne; this engine develops a thrust of $\sim 7.5 \times 10^6$ N. (From G. P. Sutton, *Rocket Propulsion Elements*, J. Wiley, 1963.)

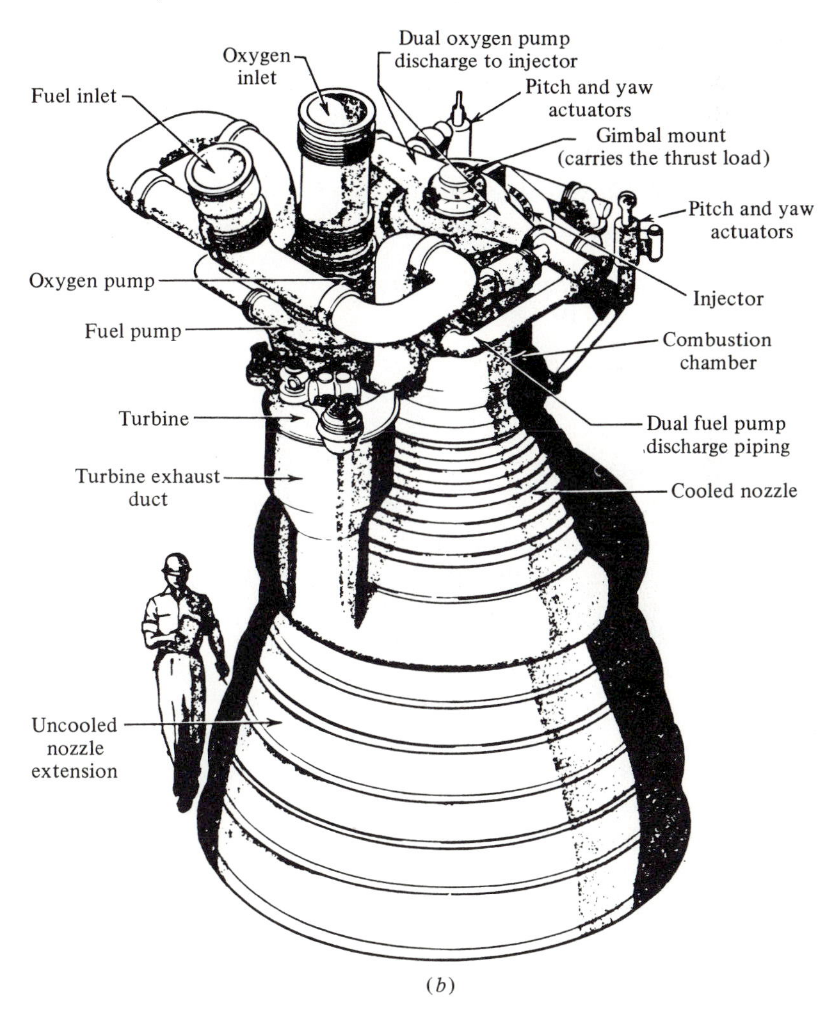


(*b*)

As an application of the relationships that we derived we consider a rocket engine operating at $T_c = 3000$ K and $P_c = 200$ atmospheres ($\sim 3000$ psi). If the throat area is $A_t = 0.2$ m$^2$ then the thrust is

$$F \simeq 1.5(0.2 \text{ m}^2) \times (2 \times 10^7 \, M/\text{m}^2) \sim 6 \times 10^6 \text{ N}$$

or equivalently a thrust of one million pounds. Next we calculate the exhaust velocity from Eqs. (7.31). Assuming that $P_e = 100$ psi, and that $\gamma = 1.25$, the efficiency factor $\eta$ is

$$\eta = 1 - (0.33)^{0.2} \simeq 0.5$$

The speed of sound in the combustion chamber can be estimated by assuming that the exhaust gases have the same properties as air, so that

$$(v_s)_c \sim v_s(T_c/T_{air})^{1/2} = v_s\sqrt{10} = 1000 \text{ m/s}$$

Thus from Eq. (7.31)

$$v_e \simeq 2(v_s)_c \sim 2 \times 10^3 \text{ m/s}$$

and the specific impulse for the engine is

$$I_{sp} = 200 \text{ seconds}$$

Finally by using the values for the thrust and the exhaust velocity, we can obtain the mass rate of propellant flow. We have

$$\frac{dm}{dt} = \frac{F}{v_e} = \frac{6 \times 10^6}{2 \times 10^3} = 3 \times 10^3 \text{ kg/s}$$

If the total mass of the fuel is 100 tonnes then the burn time would be $\Delta t \sim 33$ s.

## 7.7 Multistage rockets

We have already discussed the rocket equation in Section 6.3. For flight at an angle $\theta$ as shown in Fig. 7.22, Newton's equation takes the form

$$M\frac{du}{dt} = F - D - Mg\cos\theta \qquad (7.34)$$

where the thrust **F**, the drag force **D** and the velocity **u** are all along the flight direction at an angle $\theta$ to the vertical. By integrating Eq. (7.34) we find the velocity as a function of time, $u(t)$, and a second integral yields the displacement along the line of flight, $s(t)$.

We can apply Eq. (7.34) to the special case of a vertical launch, $\theta = 0$, and if we ignore the drag force we obtain

$$\frac{du}{dt} = \frac{F}{M} - g = I_{sp}g_0\frac{dm/dt}{M} - g = -I_{sp}g_0\frac{d}{dt}(\ln M) - g \qquad (7.35)$$

In the last step we used $dm/dt = -dM/dt$, where $M$ is the mass of the rocket. If $I_{sp}$ and $g$ are *constant*, then Eq. (7.35) can be directly integrated to yield

$$u_{bo} - u_0 = g_0 I_{sp} \ln\left[\frac{M_{full}}{M_{empty}}\right] - g t_p \tag{7.36}$$

which is the same expression as given by Eq. (6.13). Here $u_{bo}$ stands for the burn-out velocity and $t_p$ for the time of powered flight; we also used $g_0$ to indicate the acceleration of gravity at the surface of the earth, and $u_0$ is the initial velocity. Note that according to Eq. (7.36) the first term contributing to the velocity increment is independent of the specific value of $dm/dt$, but the second term is not, because $t_p$ depends on $dm/dt$. For instance, if we assume a *uniform burn*

$$t_p = \frac{M_f - M_e}{dm/dt} = \frac{M_f - M_e}{F} g_0 I_{sp} \tag{7.37}$$

The height at burn-out is obtained by integration of Eq. (7.36)

$$h_{bo} = g_0 I_{sp} t_p \left[1 - \frac{\ln(M_f/M_e)}{(M_f/M_e) - 1}\right] - \tfrac{1}{2} g t_p^2 + u_0 t_p + h_0 \tag{7.38}$$

The distance traveled after burn-out up to the highest point of the trajectory is called the *coasting height* $h_c$, and $h_m = h_c + h_{bo}$. If we set $u_0 = h_0 = 0$ but take into account the variation of $g$ with height above the earth's surface, we find

$$h_c = \frac{u_{bo}^2}{2g_0} \frac{(R_\oplus + h_{bo})^2}{[R_\oplus^2 - u_{bo}^2(R_\oplus + h_{bo})/2g_0]} \tag{7.39}$$

For flights near the earth's surface, $h_{bo} \ll R_\oplus$ and Eq. (7.39) reduces to the familiar expression $h_c = u_{bo}^2/2g_0$. Conversely, when $u_{bo}$ exceeds the escape velocity $u_{bo} > (2g_0 R_\oplus)^{1/2}$ the denominator becomes negative indicating that the rocket does not return to earth.

Fig. 7.22. Forces acting on a rocket flying at an angle $\theta$ with respect to the vertical.

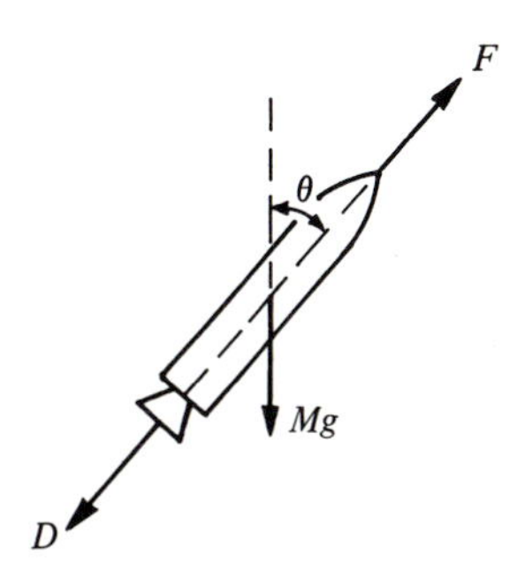

For engineering applications it is convenient and customary to introduce a notation in terms of dimensionless *fractions* (these are less than unity) and dimensionless *ratios* (which are greater than unity). If

$M_f$ is the full mass of the rocket

$M_s$ is the mass of the structure after the fuel has been consumed, and

$M_\ell$ is the mass of the payload,

we define the following

$s = M_s/M_f$ dead weight fraction

$l = M_\ell/M_f$ payload fraction

$r = F/g_0 M_f$ thrust to weight ratio

$R = M_f/(M_s + M_\ell) = 1/(s+l)$ mass ratio

In this notation we can rewrite Eq. (7.36) as

$$u_{bo} - u_0 = g_0 I_{sp}\left[\ln\left(\frac{1}{s+l}\right) - \frac{1-(s+l)}{r}\right] \tag{7.40}$$

where we assumed that $g = g_0$. The velocity increment $(u_{bo} - u_0)$ for a rocket with $I_{sp} = 300$ s is plotted in Fig. 7.23 as a function of the mass ratio $R$ and for different values of the thrust to weight ratio $r$. It is evident from the graph that increasing $r$ has only a small effect on the final velocity increment.

To optimize the velocity increment for fixed payload one would try to increase $I_{sp}$ or decrease $s$; however increasing $I_{sp}$ may involve a more sophisticated engine with a consequent increase in dead weight. The best way for increasing the final velocity is to use a multistage rocket, where

Fig. 7.23. Velocity increment given a rocket with $I_{sp} = 300$ s as a function of the mass ratio $R$; the curves are for different values of $r$, the thrust to weight ratio.

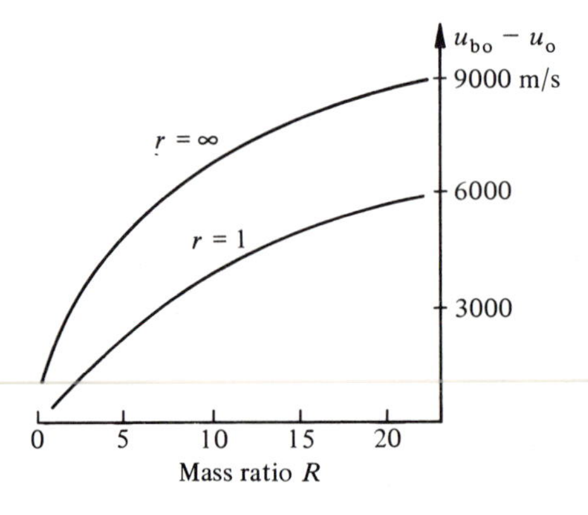

each stage contributes its own velocity increment to the payload. When a stage reaches burnout, it is detached and dropped off so that only the subsequent stages are accelerated further. After $n$ stages have been burned out, the velocity of the payload is

$$u = u_{bo}(n) + u_{bo}(n-1) + \cdots + u_{bo}(1) \tag{7.41}$$

and the overall payload ratio is given by

$$G = \frac{M_{full}}{M_\ell} = \frac{1}{l_n l_{n-1} \cdots l_1} \tag{7.41$'$}$$

We will illustrate these ideas by considering a two stage rocket. For simplicity we will restrict ourselves to level flight (or equivalently we can assume that $r \to \infty$); then the two consecutive velocity increments result in a payload velocity

$$u = I_1 \ln\left(\frac{1}{s_1 + l_1}\right) + I_2 \ln\left(\frac{1}{s_2 + l_2}\right) \tag{7.42}$$

We want to maximize $u$ while keeping the overall payload ratio fixed

$$G = \frac{1}{l_1 l_2} \tag{7.42$'$}$$

and under the assumption that the two stages are similar. The similarity can be expressed by requiring that the structure parameter

$$\delta = \frac{s}{1-l} \tag{7.42$''$}$$

be the same for both stages; $\delta$ is the ratio of the mass of the structure to the mass of the propellant plus structure. With these two conditions and without varying $I_1, I_2$ we find that maximum $u$ occurs when

$$I_1 = \frac{l_1(1-\delta_1)}{R_1} = I_2 \frac{l_2(1-\delta_2)}{R_2}$$

Thus, if the specific impulse is the same for both stages, the payload velocity is largest if the mass ratios are equal for both stages.

The effect of rocket staging is illustrated graphically in Fig. 7.24. The final velocity is given as a function of the overall payload ratio $G$ for a different number of stages $n = 1$–$4$ and for identical specific impulse for all stages. The curves have been obtained from Eq. (7.42) (and its extension for $n = 3, 4$) and show how important staging is when high payload velocity is required.

As an example consider a rocket fueled by liquid oxygen and kerosene with the following parameters

| | |
|---|---|
| Dead weight fraction | $s = 0.12$ |
| Payload fraction | $l = 0.08$ |

| | |
|---|---|
| Mass ratio | $R = 5.0$ |
| Specific impulse | $I_{sp} = 300$ s |
| Thrust to weight ratio | $r = 2.0$ |

Using Eq. (7.40) we then find for the final velocity

$$u_{bo} = 3600 \text{ m/s} \quad \text{for vertical launch}$$
$$= 4750 \text{ m/s} \quad \text{for horizontal launch}$$

In contrast, if we used two stages with the same parameters as given above we would have

$$\delta_1 = \delta_2 = \frac{s}{1-l} = \frac{0.12}{1-0.08} = 0.13$$

and

$$l_1 = l_2 = 0.08$$

so that $G = 156$. The final velocity for a horizontal launch would then be

$$u_{bo}^{(2)} = 2gI_1 \ln\left[\frac{1}{\delta(1-l)+l}\right] = 9500 \text{ m/s}$$

Fig. 7.24. Terminal velocity as a function of the number $n$ of stages used and of the mass to payload ratio $G$; the structure factor is fixed at $\delta = 0.10$. (From H. Seiffert, *Space Technology*, J. Wiley, 1959.)

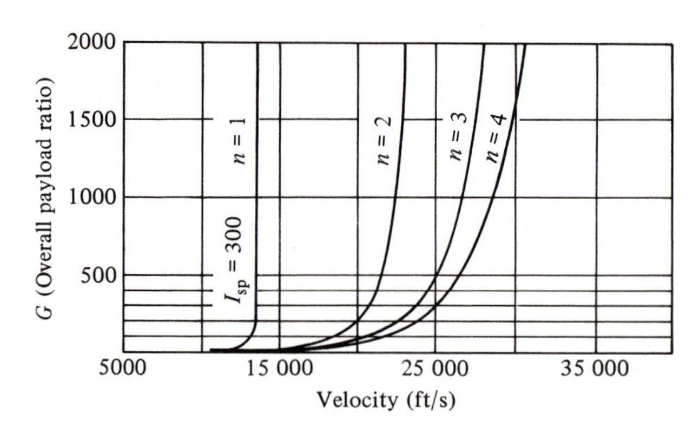


## 7.8 The NASA shuttle

The *space shuttle* was conceived by the 'National Aeronautics and Space Administration' (NASA) as a manned reusable vehicle that could reach low earth orbit and return from orbit by landing as a winged craft. The shuttle is launched by three rocket engines fueled by liquid propellants from an external tank and assisted by two solid propellant booster rockets. The boosters are separated and jettisoned after two minutes of flight; they are allowed to drop by parachute and are retrieved and reused. The external tank separates after eight minutes at which point the orbiter climbs into an earth orbit of altitude 200–300 km. On return to earth the orbiter enters the atmosphere at a large attack angle (34°) and is slowed down by atmospheric drag; it lands as a glider with a touchdown speed of approximately 150 miles/hr.

A schematic of the orbiter mounted on the main tank is shown in Fig. 7.25. The orbiter itself is 125 feet long and has a wing span of 78 feet and a weight of 270 000 lb including fuel and cargo. It is constructed similarly to commercial aircraft and is provided with a large cargo bay in which payloads up to 65 000 lb can be carried. Once launched the shuttle can maneuver by using its thrusters which are powered by hydrazine (MMH) and nitrogen tetroxide ($N_2O_4$); normally they can provide an on-orbit $\Delta V$ of $\sim$300 m/s. The first shuttle was launched in 1981 and many successful flights followed. On January 28, 1986 one of the shuttles exploded 74 seconds after lift-off, killing its crew. The tragedy slowed down the U.S. space program and raised questions about the use of manned flight as contrasted to simple rocket missions for the exploration of space. Shuttle flights were resumed in the U.S. in 1988.

As an application of the equations that we derived in the previous sections we can calculate the velocity of the orbiter from the known mass of fuel and the specific impulse of the engines. We first collect the pertinent data.

Fig. 7.25. Side view of the NASA shuttle mounted on its external fuel tank.

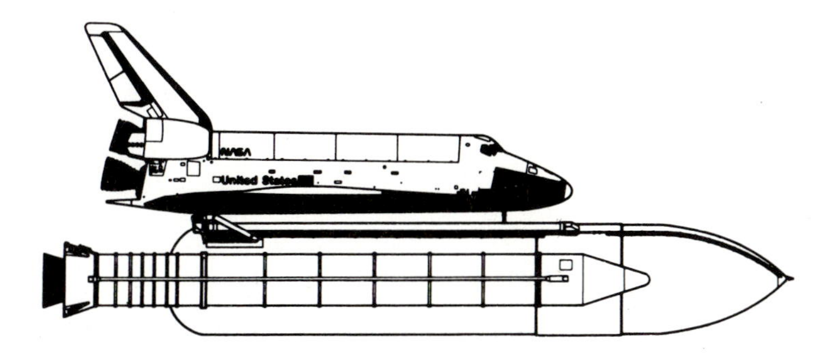

Boosters (two; parameters are for one booster)

| | |
|---|---|
| Mass | $M_b \sim 6 \times 10^5$ kg |
| Thrust | $F_b \sim 1.2 \times 10^7$ N |
| Burn time | $t_b \sim 120$ s |

Therefore, assuming a uniform burn, we estimate that $dm/dt = 5 \times 10^3$ kg/s and thus

$$I_{sp} = 240 \text{ s} \tag{7.43}$$

which is reasonable.

Main engines (three; parameters are for the total system)

| | |
|---|---|
| Liquid hydrogen (fuel) (383 000 gallons $\sim 10^8$ g-mole) | $M \sim 10^5$ kg |
| Liquid oxygen (oxidizer) (143 000 gallons $\sim 4 \times 10^7$ g-mole) | $M \sim 6 \times 10^5$ kg |
| External tank (empty) | $M \sim 3.5 \times 10^4$ kg |
| Thrust | $F_t \sim 5 \times 10^6$ N |
| Burn time | $t_t \sim 500$ s |

Assuming uniform burn we estimate $dm/dt = 1.4 \times 10^3$ kg/s and thus

$$I_{sp} = 360 \text{ s} \tag{7.44}$$

Mass of orbiter $M_o = 1.2 \times 10^5$ kg

The trajectory followed by the shuttle during launch is shown schematically in Fig. 7.26. Initially the trajectory is at 45° and at the time the boosters are burned out it is almost horizontal. Even though the main engines are ignited at launch, for simplicity of calculation we will assume that they are turned on after the boosters are jettisoned. For the booster phase we then have from Eq. (7.36)

$$u_{bo}^{(1)} = gI_{sp} \ln\left[\frac{M_f}{M_e}\right] - g \cos\theta \, t_p \tag{7.36}$$

where we use

$$M_f = 2M_b + M_t + M_o = 2.05 \times 10^6 \text{ kg}$$

$$M_e = \qquad M_t + M_o = 8.5 \times 10^5 \text{ kg}$$

and

$$I_{sp} = 240 \text{ s}, \qquad t_p = 120 \text{ s}, \qquad \theta = 60°$$

Thus we obtain

$$u_{bo}^{(1)} = 2100 - 600 = 1500 \text{ m/s} \tag{7.45}$$

The velocity increment obtained from the main engines can be calculated from the same equation where now we use

$$M_f = M_t + M_o = 8.5 \times 10^5$$
$$M_e = \qquad M_o = 1.2 \times 10^5$$

and

$$I_{sp} = 360 \text{ s}, \qquad t_p = 500 \text{ s}, \qquad \theta = 90^\circ$$

Thus we obtain

$$\Delta u = u_{bo}^{(2)} - u_0 = 6900 \text{ m/s} \tag{7.45'}$$

The velocity required for low earth orbit is

$$V_o = \sqrt{(gR_\oplus)} = 7900 \text{ m/s}$$

and we found $u_{bo}^{(2)} = u_{bo}^{(1)} + \Delta u = 8400$ m/s. Thus the orbiter can enter into a low orbit by maneuvering its thrusters after the main tank is jettisoned.

We can also check the value of the overall energy delivered by the main engines. In the reaction $2H + O \to H_2O$, the energy released is 3 eV and since $4 \times 10^7$ g-mole react, the total energy is

$$E = N_0 \times (4 \times 10^7) \times 3 \text{ eV} \sim 10^{13} \text{ J} \tag{7.46}$$

The kinetic energy of the orbiter is

$$\text{K.E.} = \tfrac{1}{2} M_o V_{\text{orbit}}^2 = 4 \times 10^{12} \text{ J} \tag{7.46'}$$

which indicates reasonable efficiency. Note that we ignored the energy expended by the boosters which is of the same order as in Eq. (7.46). We

Fig. 7.26. Trajectory and maneuvers of the shuttle during its launch; approximate separation times are indicated.

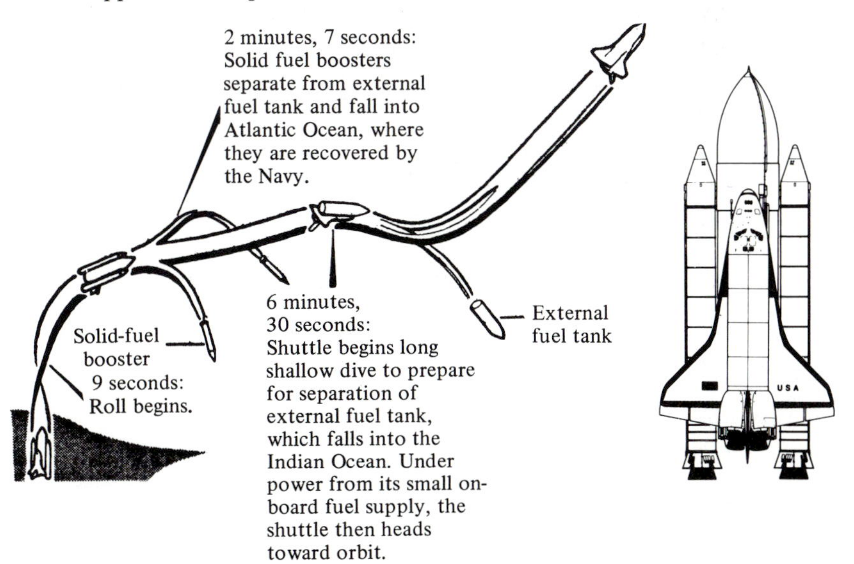

also ignored the gain in potential energy of the orbiter, but this is small: for instance for $h = 100$ km, $\Delta U = M_o gh = 1.2 \times 10^{11}$ J. The power developed by the main engines is

$$P = \frac{E}{t_p} = \frac{10^{13}}{500} = 2 \times 10^{10} \text{ W} \simeq 25 \text{ million horsepower} \tag{7.47}$$

This is to be compared to the power delivered to the craft, which is given by $P_o = F_t V$ where $F_t$ is the thrust. If we use an average velocity $V \sim 4$ km/s and $F_t = 5 \times 10^6$ N

$$P_o = F_t V = 2 \times 10^{10} \text{ W} \tag{7.47$'$}$$

in close agreement with the result of Eq. (7.47). In practice the propulsive power is not constant, but the average values we calculated here are typical of the flight.

Finally, we consider the propulsion parameters. Given that the combustion chamber pressure and temperature are

$$P_c = 3000 \text{ psi} = 220 \text{ atm} = 2.2 \times 10^7 \text{ N}$$

$$T_c = 2700°\text{C}, \qquad \rho_c = 16 \text{ kg/m}^3$$

we find from Eqs. (7.31) that the exhaust velocity is

$$v_e \sim 3.15 \times 10^3 \text{ m/s}$$

Since the main engine mass flow is $dm/dt = 7 \times 10^5/500 = 1.4 \times 10^3$ kg/s, the expected thrust is

$$F_t = v_e \frac{dm}{dt} = 3.15 \times 10^3 \times 1.4 \times 10^3 \sim 4.5 \times 10^6 \text{ N}$$

as already assumed.

In conclusion we see that the main parameters of the shuttle flights can be calculated from first principles. Yet the technical realization of these principles involves great ingenuity and effort and a malfunction of any one component can easily lead to disaster.

## Exercises

### *Exercise 7.1*

An airplane wing moves at a speed of 200 km/hr through air at standard density; the lift per meter of span is 3000 Newtons.

(a) Determine the circulation around the wing.
(b) Assume that the angle of attack was $\alpha \simeq 5°$ and find the chord of the wing.
(c) Calculate the Reynolds number for the air flow past the wing.

***Exercise 7.2***

A helicopter has a propeller with the following parameters: radius $R = 5$ m, pitch $d = 2$ m/turn, rotational speed 1200 rpm (turns/minute) and no slippage.

(a) Calculate the thrust of the propeller and therefore the maximum weight that can be lifted.
(b) Calculate the power delivered.
(c) Calculate the Reynolds number at the tip of the propeller, by using a reasonable estimate for the chord of the blade.

***Exercise 7.3***

Consider a rocket engine with the following parameters: $P_c = 300$ atmospheres, $T_c = 2200°C$, $dm/dt = 1$ kg/s, operating at sea level ($P_o = 1$ atm). Furthermore let $\gamma = 1.25$, $\bar{M} = 18$.

(a) Find the throat area from the simplified expression

$$\frac{dm}{dt} \simeq \frac{A_t P_c}{v_{sc}}$$

where $v_{sc}$ is the speed of sound in the combustion chamber.
(b) Use an *approximate* expression to calculate the thrust using the result of (a).
(c) If the nozzle is ideally designed ($P_e = P_o$), find the exhaust velocity $v_e$, and recalculate the thrust.

***Exercise 7.4***

From your own experience, or data, make a *log–log plot* of the maximum speed (in km/hr) versus the specific power (horsepower per ton of mass) for various modes of transportation. Note that on a log–log plot almost any relationship falls on a straight line.

***Exercise 7.5***

Evaluate the speed of sound in the earth's atmosphere: (a) as a function of temperature, (b) as a function of altitude above sea level.

***Exercise 7.6***

Determine the burnout velocity, burnout altitude and maximum altitude for a dragless projectile in vertical flight given the following parameters: $v_e = 7250$ ft/s; $M_p/M_o = 0.57$ (propellant mass/full mass); $t_p = 5.0$ s and $u_0 = h_0 = 0$.

# 8

# TO THE STARS

## 8.1 The solar system

We shall begin this chapter with a survey of our solar system of which the earth is one of its nine planets. The planets follow closed orbits around the sun because of the attractive gravitational force exerted between the sun and the planets. The planetary orbits are nearly circular and the influence of the other planets is much weaker than that of the sun. Furthermore, with the exception of Pluto, planetary orbits lie approximately in the same plane, the plane of the *ecliptic* which is defined by the earth's orbit. Planets spin about their own axis and are accompanied by satellites. In addition the solar system contains a large number of asteroids and an unknown number of comets.

The planets, their mean distance from the sun, their mass and their orbital period are listed in Table 8.1. Note that distances are given in astronomical units (AU), where one AU equals the length of the semi-major axis of the earth's orbit around the sun

$$1\,\mathrm{AU} = 1.495 \times 10^{11}\ \mathrm{m} \tag{8.1}$$

Masses are given in earth masses $M_\oplus$, where

$$M_\oplus = 5.978 \times 10^{24}\ \mathrm{kg} \tag{8.2}$$

The orbital period $\tau$ is related to the semi-major axis of the orbit $a$ through Kepler's third law

$$\tau = \frac{2\pi}{\omega} = 2\pi \frac{a^{3/2}}{(GM_\odot)^{1/2}} \tag{8.3}$$

Here $G$ is Newton's constant and $M_\odot$ the mass of the sun. The product $GM_\odot$ is determined to better accuracy than $G$ or $M_\odot$ individually, and we will use the symbol $K_{\odot}$ for it, so that

$$K_{\odot} = GM_{\odot} = 1.324 \times 10^{20}\ \text{m}^3/\text{s} \tag{8.4}$$

for reference we also give the accepted value of Newton's constant

$$G = 6.668 \times 10^{-11}\ \text{N-m}^2/\text{kg}^2 \tag{8.5}$$

and the main value of the earth's radius

$$R_{\oplus} = 6.378 \times 10^{6}\ \text{m} \tag{8.6}$$

From Eqs. (8.4, 8.5) we infer that the mass of the sun is

$$M_{\oplus} = 3.325 \times 10^{5}\ M_{\oplus} \tag{8.4$'$}$$

The orbits of the inner and of the outer planets are sketched approximately to scale in Figs. 8.1$(a, b)$. As seen from the north pole of the ecliptic all planets and most satellites move counterclockwise.

When a particle moves in a central field of force, the orbit lies in a plane and angular momentum about the center is conserved. Of course, the total energy is also conserved and for bound orbits the total energy is negative

$$E = T + U = T - |U| < 0 \tag{8.7}$$

It is understood that the potential energy is defined to be zero for a particle at infinite distance from the attracting center. If the force has a $1/r^2$ dependence, then the *average* value of the kinetic and potential energy are related by the virial theorem

$$\langle T \rangle = -\tfrac{1}{2}\langle U \rangle \tag{8.8}$$

Just as the planets are subject to the sun's attractive force, natural or artificial satellites are subject to the $1/r^2$ attractive force of the planet. For interplanetary probes one must consider the attraction of the sun and that of the planet from which and/or toward which the probe is launched.

Table 8.1. *The planets of the solar system*

| Planet | Symbol | Semi-major axis (AU) | Mass (in $M_{\oplus}$) | Orbital period (in earth years) |
|---|---|---|---|---|
| Mercury | ☿ | 0.387 | 0.054 | 0.241 |
| Venus | ♀ | 0.723 | 0.814 | 0.606 |
| Earth | ⊕ | 1.000 | 1.000 | 1.000 |
| Mars | ♂ | 1.524 | 0.108 | 1.88 |
| Jupiter | ♃ | 5.203 | 318.4 | 11.86 |
| Saturn | ♄ | 9.539 | 95.3 | 29.46 |
| Uranus | ⛢ | 19.18 | 14.6 | 84.0 |
| Neptune | ♆ | 30.06 | 17.3 | 164.8 |
| Pluto | ♇ | 39.52 | 0.83 | 247.7 |

Such probes follow, with respect to the planet, a hyperbolic trajectory. In general, motion in a $1/r^2$ central force field leads to trajectories which are *conic sections* and can be bound (circle, ellipse) or unbound (parabola, hyperbola). Because of the large mass of the sun as compared to that of the planets we can treat the sun as stationary and account for the effect of the other planets by perturbative techniques.

Conic sections are generated by the intersection of a cone with a plane surface as shown in Fig. 8.2. Let $\psi$ be the angle between the normal to the plane surface and the axis of the cone, whose apex half-angle is $\theta$. When $\psi = 0$ the intersection is a *circle*; when $\psi < \pi/2 - \theta$ it is an *ellipse*; when $\psi = \pi/2 - \theta$ it is a *parabola* and when $\psi > \pi/2 - \theta$ it is a *hyperbola*. This family of curves can be characterized by their semi-major axis $a$ and their eccentricity $e$. It is convenient to give the equation of the conics in polar coordinates $r$, $\phi$, and we will use the convention indicated in the sketches of Figs. 8.3.

*Ellipse*. The locus of points $P$, for which the sum of the distances from the two foci is constant

$$r + r' = 2a = \text{constant}$$

$$b = a(1 - e^2)^{1/2}$$

$$r = \frac{a(1 - e^2)}{1 + e \cos \phi} \tag{8.9}$$

$$e < 1$$

Fig. 8.1. The solar system: (*a*) the inner planets, (*b*) the outer planets.

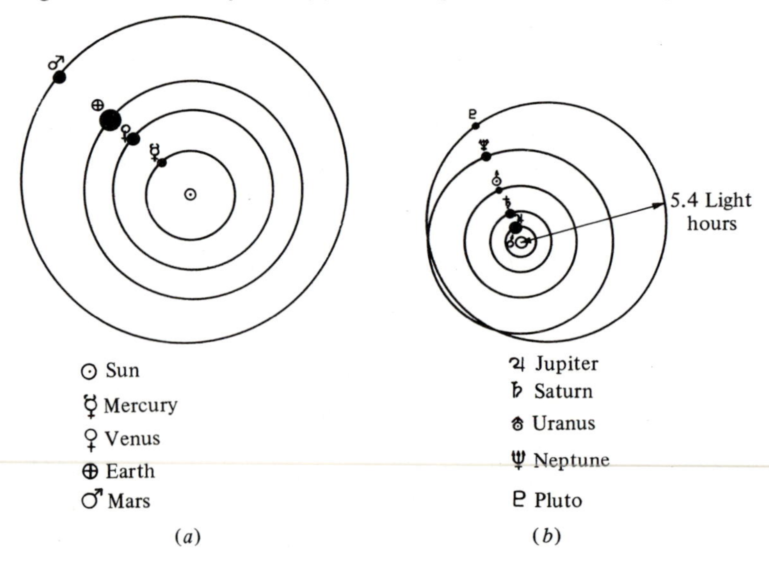

*Hyperbola.* The locus of points $P$ for which the differences of the distances from the two foci is constant. A hyperbola has two disconnected branches, and the asymptotes have a slope $\cos\phi = 1/e$

$$r' - r = 2a = \text{constant}$$

$$b = a(e^2 - 1)^{1/2}$$

$$r = \frac{a(e^2 - 1)}{1 - e\cos\phi} \tag{8.10}$$

$$e > 1$$

*Circle.* The locus of points $P$ for which the distance from the center is constant

$$r = a$$

Namely, the limiting case of an ellipse with $e = 0$.

*Parabola.* The locus of points $P$ which are at an equal distance from the directrix and the focus. It is the limiting case between an ellipse and a hyperbola, namely

$$e = 1$$

However, $a \to \infty$, while $a(e^2 - 1) = 2d$ remains finite

$$r = \frac{2d}{1 - \cos\phi} \tag{8.11}$$

For the Cartesian coordinates indicated in Fig. 8.3($c$) the equation of the parabola is

$$y^2 = 4d(x - d) \tag{8.11'}$$

from which it also follows that $x = r$.

According to the above discussion, the orbits of the planets must be ellipses; the eccentricity of the earth's orbit is very small, $e_{\oplus} = 0.0167$ giving rise to a difference between the aphelion and perihelion of $R_A - R_P = 0.033$ AU. Pluto and Mercury have the most eccentric orbits

Fig. 8.2. Conic sections: ($a$) ellipse, ($b$) parabola, ($c$) hyperbola.

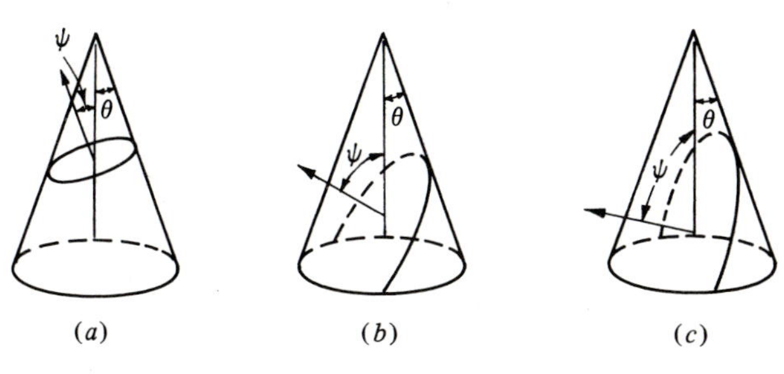

with $e \simeq 0.2$. Recurring comets also have elliptic orbits but with $e \to 1$. The axis of spin of most planets is normal to their orbit plane, with the notable exception of Uranus. For the earth the inclination of the spin axis with respect to the normal to the ecliptic is $\varepsilon = 23° \, 27'$ and this is the primary effect that gives rise to the seasons of the year.

While the earth has only one satellite, the moon, and Mars has two, the outer planets have a large number of satellites. The matter density of the inner planets is of the same order as that of the earth, $\rho \sim 5 \text{ g/cm}^3$

Fig. 8.3. Parameters characterizing conic sections: (*a*) ellipse, (*b*) hyperbola, (*c*) parabola.

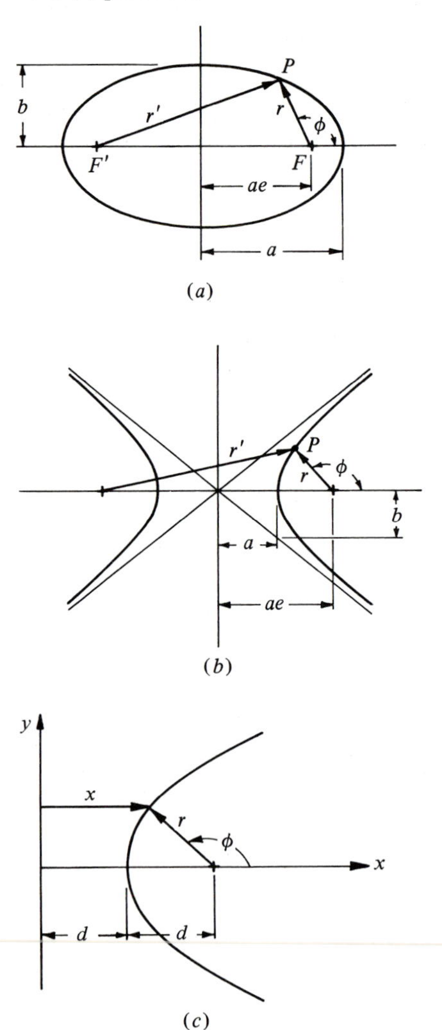

whereas for the outer planets as well as for the sun, the density is much lower, $\rho \sim 1\,\mathrm{g/cm^3}$. The surface temperature and the composition of the atmosphere of the planets depend crucially on their distance from the sun with Mars being the most 'hospitable' planet in terms of the conditions prevailing on earth.

## 8.2 Motion in a central field of force

Motion under the influence of a central force is a classical problem in mechanics, with which most readers will be familiar. Nevertheless we will give its formal solution and then derive the *velocity equation*, which is a first integral of the equations of motion. The formalism of the velocity equation is well adapted to the analysis of orbital maneuvers such as will be discussed in Sections 3 and 4. Since the force is central, the motion is confined to a plane and we will use polar coordinates $r$, $\phi$.

Gravitation is an inverse square law force and therefore the equation of motion is

$$\mathbf{F} = -G\frac{M_{\odot}m}{r^2}\hat{u}_{\mathrm{r}} = m\frac{\mathrm{d}^2\mathbf{r}}{\mathrm{d}t^2} \tag{8.12}$$

Here $\hat{u}_{\mathrm{r}}$ is the unit vector along the radial direction, and $\hat{u}_{\phi}$ the orthogonal unit vector. As can be seen from Fig. 8.4 the following relations hold

$$\frac{\mathrm{d}\hat{u}_{\mathrm{r}}}{\mathrm{d}t} = \frac{\mathrm{d}\phi}{\mathrm{d}t}\hat{u}_{\phi} \qquad \frac{\mathrm{d}\hat{u}_{\phi}}{\mathrm{d}t} = -\frac{\mathrm{d}\phi}{\mathrm{d}t}\hat{u}_{\mathrm{r}} \tag{8.13}$$

Since $\mathbf{r} = r\hat{u}_{\mathrm{r}}$, and using Eqs. (8.13) we find that

$$\frac{\mathrm{d}\mathbf{r}}{\mathrm{d}t} = \frac{\mathrm{d}r}{\mathrm{d}t}\hat{u}_{\mathrm{r}} + r\frac{\mathrm{d}\phi}{\mathrm{d}t}\hat{u}_{\phi} \tag{8.13$'$}$$

$$\frac{\mathrm{d}^2\mathbf{r}}{\mathrm{d}t^2} = \left[\frac{\mathrm{d}^2 r}{\mathrm{d}t^2} - r\left(\frac{\mathrm{d}\phi}{\mathrm{d}t}\right)^2\right]\hat{u}_{\mathrm{r}} + \left[r\frac{\mathrm{d}^2\phi}{\mathrm{d}t^2} + 2\frac{\mathrm{d}r}{\mathrm{d}t}\frac{\mathrm{d}\phi}{\mathrm{d}t}\right]\hat{u}_{\phi} \tag{8.13$''$}$$

Fig. 8.4. Polar coordinates for the description of arbitrary motion in a plane; $\hat{u}_{\mathrm{r}}$ and $\hat{u}_{\phi}$ are the unit vectors and are position dependent.

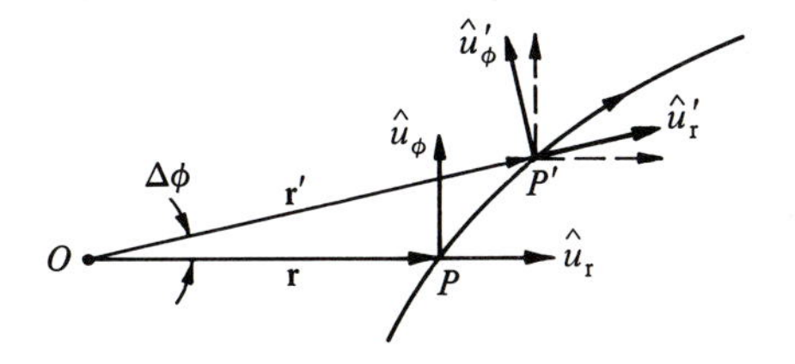

Therefore, the equation of motion (Eq. (8.12)) is equivalent to two scalar equations

$$\frac{d^2r}{dt^2} - r\left(\frac{d\phi}{dt}\right)^2 = -\frac{K}{r^2} \tag{8.14}$$

$$r\frac{d^2\phi}{dt^2} + 2\frac{dr}{dt}\frac{d\phi}{dt} = 0 \tag{8.14'}$$

where we used the notation $K = K_\odot = GM_\odot$. Hereafter we will use dots for total time derivatives, $\dot{r} = dr/dt$, $\dot{\phi} = d\phi/dt$, etc.

The angular momentum about the center of force is given by

$$l = |\mathbf{r} \times \mathbf{p}| = m|\mathbf{r} \times \dot{\mathbf{r}}| = mr^2\dot{\phi} \tag{8.15}$$

Taking the derivative of Eq. (8.15) we obtain

$$\frac{1}{m}\frac{dl}{dt} = \frac{1}{m}\dot{l} = 2r\dot{r}\dot{\phi} + r^2\ddot{\phi} = r(2\dot{r}\dot{\phi} + r\ddot{\phi})$$

which is equal to zero by virtue of Eq. (8.14′). Thus

$$l = \text{constant} \tag{8.16}$$

as was to be expected since a central force exerts no torque on the particle. Eq. (8.16) is a first integral of the equation of motion, and since the equation is of second order there must also exist a second integral. This is the total energy of the particle

$$E = T + U \tag{8.17}$$

To express the total energy in suitable form we return to Eq. (8.14) which we write in terms of the angular momentum, as

$$\ddot{r} = -\frac{K}{r^2} + \frac{l^2}{m^2r^3} \tag{8.18}$$

Multiplying both sides by $m\dot{r}$, Eq. (8.18) can be cast as a total derivative

$$\frac{d}{dt}\left[\tfrac{1}{2}m\dot{r}^2 - \frac{Km}{r} + \frac{l^2}{2mr^2}\right] = 0$$

The integral is therefore a conserved quantity which we identify with the total energy

$$E = \tfrac{1}{2}m\left[\dot{r}^2 + \frac{l^2}{m^2r^2}\right] - \frac{Km}{r} = \text{constant} \tag{8.19}$$

In Eq. (8.19) we recognize the kinetic energy term

$$T = \tfrac{1}{2}m(\mathbf{v})^2 = \tfrac{1}{2}m(\dot{r}^2 + r^2\dot{\phi}^2) \tag{8.20}$$

and the potential energy term

$$U = -\frac{Km}{r} \tag{8.21}$$

The physical interpretation of Eq. (8.19) is best understood by analogy to one-dimensional motion. In that case the kinetic energy is $T' = \frac{1}{2}m\dot{r}^2$ and we introduce an effective potential

$$U' = -\left[\frac{Km}{r} - \frac{l^2}{mr^2}\right] \tag{8.22}$$

so that

$$E = T' + U' = \tfrac{1}{2}m\dot{r}^2 - \left[\frac{Km}{r} - \frac{l^2}{2mr^2}\right] \tag{8.19'}$$

Thus, the Newtonian potential is modified by a 'centrifugal barrier' term leading to a one-dimensional potential well as shown in Fig. 8.5. We now see clearly that the total energy of the particle determines the allowed range of radial distances, and therefore the nature of the orbit.

Since in Eq. (8.19), $E$ and $l$ are constant we can solve for $\dot{r}$ and integrate to obtain $r(t)$ (it is given by an elliptic integral). Instead we are interested in a parametric equation of the orbit of the form $f(r, \phi) = 0$. This is facilitated by returning to the equation of motion and making the standard change of variable

$$u \equiv 1/r \tag{8.23}$$

Therefore, in terms of the new variable

$$\dot{\phi} = \frac{l}{m}\frac{1}{r^2} = \frac{l}{m}u^2 \tag{8.23'}$$

Fig. 8.5. Effective potential for motion under the influence of a central inverse square law force.

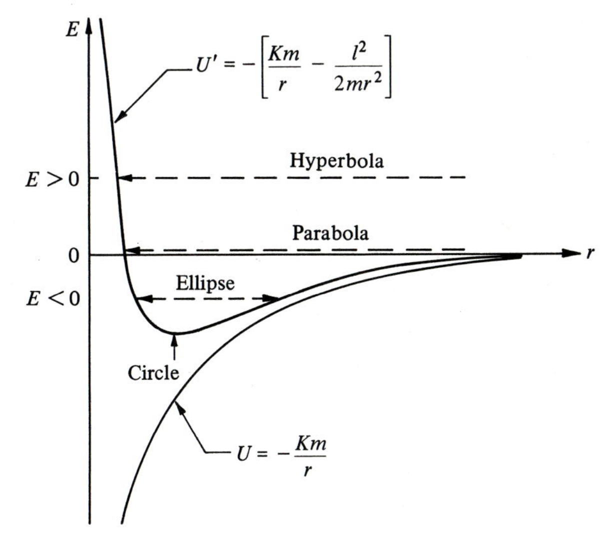

and

$$\dot{r} = \frac{\mathrm{d}r}{\mathrm{d}\phi}\dot{\phi} = \frac{\mathrm{d}}{\mathrm{d}\phi}\left(\frac{1}{u}\right)\dot{\phi} = -\frac{l}{m}\frac{\mathrm{d}u}{\mathrm{d}\phi}$$

$$\ddot{r} = -\frac{l}{m}\left[\frac{\mathrm{d}}{\mathrm{d}\phi}\left(\frac{\mathrm{d}u}{\mathrm{d}\phi}\right)\right]\dot{\phi} = -\left(\frac{l}{m}\right)^2\frac{\mathrm{d}^2u}{\mathrm{d}\phi^2}u^2 \qquad (8.23'')$$

Introducing these expressions for $\ddot{r}$, $r$ and $\dot{\phi}$ in Eq. (8.14) we obtain

$$\frac{\mathrm{d}^2u}{\mathrm{d}\phi^2} + u = \frac{K}{(l/m)^2} \qquad (8.24)$$

which can be solved by elementary methods. It is a harmonic oscillator equation with a constant driving term, and has the solution

$$u = k\cos(\phi - \phi_0) + \frac{K}{l/m^2} \qquad (8.25)$$

where $k$ and $\phi_0$ are arbitrary integration constants.

To interpret the orbit equation (Eq. (8.25)), we set the initial phase $\phi_0 = 0$ and express $k$ in terms of the total energy to obtain

$$\frac{1}{r} = \frac{K}{(l/m)^2}[1 \pm e\cos\phi] \qquad (8.26)$$

where

$$\frac{K}{(l/m)^2} = \pm\frac{1}{a(1-e^2)}, \qquad k = \frac{e}{a(1-e^2)} \qquad (8.27)$$

$e$ is the eccentricity and $a$ the semi-major axis of the orbit. Since $(K/(l/m)^2)$ must be positive, we use

+ sign when $e < 1$, i.e. an ellipse, or $E < 0$
− sign when $e > 1$, i.e. a hyperbola, or $E > 0$

The orbit equations are:

$$r = \frac{a(1-e^2)}{1 + e\cos\phi} \qquad E = -\frac{1}{2}\frac{Km}{a} \qquad e < 1 \qquad (8.28)$$

which represents an ellipse (see Eq. (8.9)), or

$$r = \frac{a(e^2-1)}{1 - e\cos\phi} \qquad E = \frac{1}{2}\frac{Km}{a} \qquad e > 1 \qquad (8.29)$$

which represents a hyperbola (see Eq. (8.10)), or

$$r = \frac{(l/m)^2}{K}\frac{1}{1-\cos\phi} \qquad E = 0 \qquad e = 1 \qquad (8.30)$$

which represents a parabola (see Eq. (8.11)). Finally we note that

$$e^2 = 1 + \frac{(l/a)^2}{2mE} = 1 + \frac{2mE(l/m)^2}{(Km)^2} \tag{8.31}$$

and recall that $E$ can be either positive or negative; $a$ can then be determined from the first of Eqs. 8.27.

*The velocity equation* is a first integral of the equations of motion also known by its latin name *vis-viva* or 'live force'. We start with the equation of motion (Eq. (8.12)) which is

$$\ddot{\mathbf{r}} = -\frac{K}{r^3}\mathbf{r} \tag{8.32}$$

Here we used $\mathbf{r} = r\hat{u}_r$, and the relevant vectors are shown again in Fig. 8.6. We form the scalar product of the vector equation, Eq. (8.32), with $\dot{\mathbf{r}}$ and obtain

$$2\mathbf{r}\cdot\ddot{\mathbf{r}} = -\frac{2K}{r^3}\dot{\mathbf{r}}\cdot\mathbf{r}$$

or

$$\frac{d}{dt}(\dot{\mathbf{r}}^2) = -\frac{2K}{r^2}\dot{r} \tag{8.33}$$

since $\dot{\mathbf{r}}\cdot\mathbf{r} = r\dot{r}$ (see Fig. 8.6).

Eq. (8.33) can be immediately integrated because $\dfrac{2K}{r^2}\dot{r} = -2K\dfrac{d}{dt}\left(\dfrac{1}{r}\right)$.

Noting that $\dot{\mathbf{r}}^2 = V^2$, with $V$ the velocity of the particle, we obtain

$$V^2 - [V(t=0)]^2 = 2K\left[\frac{1}{r} - \frac{1}{r(t=0)}\right] \tag{8.34}$$

Our goal is to simplify this result by a proper choice of $t = 0$. We consider an elliptical orbit and choose $t = 0$ to be the time when the particle is at the perifocus, the point of closest approach to the center of force. We indicate all the variables at that instant of time by a zero subscript

$$r_0 = a(1-e), \qquad \dot{r}_0 = 0$$

$$V_0^2 = r_0^2\dot{\phi}_0^2 = \left(\frac{l}{m}\right)^2 \frac{1}{r_0^2} = \left(\frac{l}{m}\right)^2 \frac{1}{a^2(1-e)^2}$$

Fig. 8.6. The position vector and its derivatives.

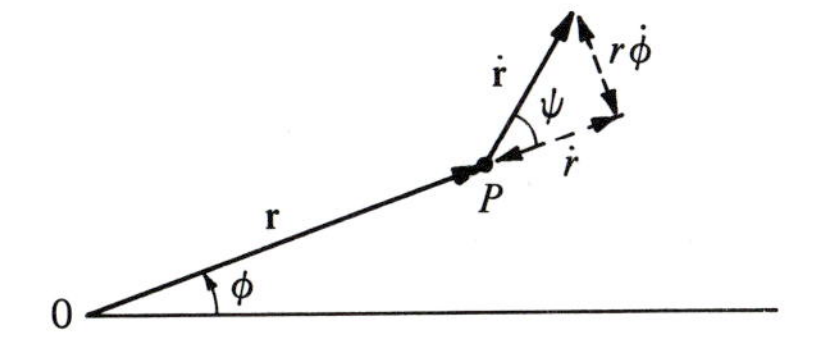

If we make use of Eq. (8.27) we can rewrite

$$V_0^2 = \frac{K}{a}\frac{(1+e)}{(1-e)}$$

and therefore Eq. (8.34) becomes

$$V^2 = \frac{2K}{r} - \frac{K}{a}\left[\frac{2}{1-e} - \frac{1+e}{1-e}\right]$$

or

$$V^2 = K\left[\frac{2}{r} \mp \frac{1}{a}\right] \tag{8.35}$$

where the *minus* sign is for elliptic orbits and the *plus* sign for hyperbolic orbits; $a$ is the semi-major axis of the orbit. For circular orbits $r = a$ and we obtain the familiar result $V^2 = K/a$.

We will refer to Eq. (8.35) as the *velocity equation*, because it relates the velocity of the particle to the distance from the center of force for different initial conditions. These conditions are expressed through the semi-major axis $a$. As a check on our result we calculate the total energy of the particle

$$E = T + U = \tfrac{1}{2}mV^2 - \frac{Km}{r}$$

Using Eq. (8.35) we find immediately

$$E = \mp\frac{Km}{2a} = \text{constant}$$

in agreement with Eqs. (8.28, 8.29). As another example, for a parabolic orbit $a \to \infty$, and therefore

$$V^2 = 2K/R \tag{8.36}$$

which yields the expression for the escape velocity from a planet of radius $R$ and gravitational attraction $K = GM$. For a low orbit around a planet we have $r = a = R$ and therefore $V^2 = K/R$.

## 8.3 Transfer orbits

During interplanetary travel the spacecraft executes certain well-defined maneuvers which we can classify as follows

(a) *Escape* from the local gravitational field of a planet.
(b) *Capture* in the local gravitational field of a planet.
(c) *Transfer* from the heliocentric orbit of one planet to that of another.

(d) *Encounter* with a planet, which affects both the direction and the magnitude of the heliocentric velocity of the craft.

The first three of these maneuvers are accomplished by *thrusting*, that is, by changing the velocity of the vehicle. The *velocity increment* $\Delta V$ is given by the integrated thrust,

$$\Delta V = \int \frac{F}{m} \, dt \tag{8.37}$$

and is a measure of the fuel that is needed for a particular maneuver. The change in the total energy of the vehicle is proportional to

$$\Delta(V^2) \simeq 2\mathbf{V} \cdot \Delta\mathbf{V}$$

and thus can be positive or negative or zero depending on the relative orientation of $\mathbf{V}$ and $\Delta\mathbf{V}$. Keep in mind that when a vehicle is in orbit around a planet its heliocentric velocity is the vector sum of the local vehicle velocity and of the planet's heliocentric velocity. In an encounter the velocity and orbit of the vehicle change due to the gravitational attraction of a planet or other massive body.

We will now consider the most economic trajectory for transferring a vehicle from an initial orbit $I$ to a final orbit $F$. For simplicity we will take the initial and final orbits to be circular with respective radii $a_{\mathrm{I}}$ and $a_{\mathrm{F}}$. These are indicated in Fig. 8.7 where we also show an elliptical trajectory *tangent* to the initial orbit at $A$ and to the final orbit at $B$. This trajectory is called the *Hohman ellipse* and we can calculate the necessary velocity increments at the points $A$ and $B$. Note that $B$ lies on the line joining $A$ with the force center.

We use the 'velocity equation' (Eq. (8.35)) to find the velocity of the

Fig. 8.7. Hohman transfer from one circular trajectory to another.

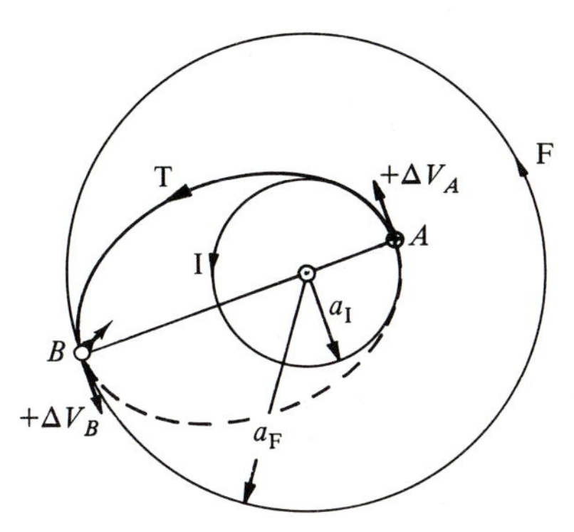

vehicle in the initial and final circular orbits

$$V_{\mathrm{I}} = \left[K\left(\frac{2}{r} - \frac{1}{a_{\mathrm{I}}}\right)\right]^{1/2} = \left(\frac{K}{a_{\mathrm{I}}}\right)^{1/2} \tag{8.38}$$

and

$$V_{\mathrm{F}} = \left[K\left(\frac{2}{r} - \frac{1}{a_{\mathrm{F}}}\right)\right]^{1/2} = \left(\frac{K}{a_{\mathrm{F}}}\right)^{1/2} \tag{8.38'}$$

The semi-major axis of the Hohman transfer ellipse is obviously

$$a_{\mathrm{T}} = (a_{\mathrm{I}} + a_{\mathrm{F}})/2 \tag{8.39}$$

and therefore the velocity of a vehicle following that ellipse is at the point $A$ (the perifocus),

$$V_{\mathrm{T}A} = \left[K\left(\frac{2}{a_{\mathrm{I}}} - \frac{2}{a_{\mathrm{I}} + a_{\mathrm{F}}}\right)\right]^{1/2} = \left[\frac{2K}{a_{\mathrm{I}}}\frac{a_{\mathrm{F}}}{(a_{\mathrm{I}} + a_{\mathrm{F}})}\right]^{1/2} \tag{8.40}$$

At the point $B$ (the apofocus) the vehicle velocity on the transfer ellipse is

$$V_{\mathrm{T}B} = \left[K\left(\frac{2}{a_{\mathrm{F}}} - \frac{2}{a_{\mathrm{I}} + a_{\mathrm{F}}}\right)\right]^{1/2} = \left[\frac{2K}{a_{\mathrm{F}}}\frac{a_{\mathrm{I}}}{(a_{\mathrm{I}} + a_{\mathrm{F}})}\right]^{1/2} \tag{8.40'}$$

By comparing Eqs. (8.38) and (8.40) we see that the velocity of the vehicle must be increased at the point $A$, in order to follow the Hohman ellipse; it must be again increased at the point $B$ in order to leave the Hohman ellipse. We speak of velocity increments $\Delta V$, where

$$\Delta V_A = V_{\mathrm{T}A} - V_{\mathrm{I}} = \left(\frac{K}{a_{\mathrm{I}}}\right)^{1/2}\left[\left(\frac{2a_{\mathrm{F}}}{a_{\mathrm{I}} + a_{\mathrm{F}}}\right)^{1/2} - 1\right] \tag{8.41}$$

$$\Delta V_B = V_{\mathrm{F}} - V_{\mathrm{T}B} = \left(\frac{K}{a_{\mathrm{F}}}\right)^{1/2}\left[1 - \left(\frac{2a_{\mathrm{I}}}{a_{\mathrm{I}} + a_{\mathrm{F}}}\right)^{1/2}\right] \tag{8.41'}$$

The total velocity increment $\Delta V_{\mathrm{H}} = \Delta V_A + \Delta V_B$ is often expressed in fractional terms

$$\frac{\Delta V_{\mathrm{H}}}{V_{\mathrm{I}}} = \left(1 - \frac{1}{R}\right)\left(\frac{2R}{1+R}\right)^{1/2} + \left(\frac{1}{R}\right)^{1/2} - 1 \tag{8.42}$$

where

$$R = a_{\mathrm{F}}/a_{\mathrm{I}}$$

As an example, let us calculate the velocity increment for a vehicle in a high earth orbit if it is to escape the solar system. In a high earth orbit the vehicle is free of the earth's attraction, but moves with the heliocentric velocity of the earth, which for a circular orbit is

$$V_{\mathrm{I}} = V_{\oplus} = \left(\frac{K_{\odot}}{a_{\oplus}}\right)^{1/2} = \left(\frac{1.325 \times 10^{20}\ \mathrm{m^3/s}}{1.485 \times 10^{11}\ \mathrm{m}}\right)^{1/2} = 29\,800\ \mathrm{m/s}$$

If the vehicle is to escape from the solar system, then $a_{\mathrm{F}} = \infty$, whereas

$a_I = a_\oplus$. Using this input, namely $R = a_F/a_I = \infty$ in Eq. (8.42) we find for the fractional velocity increment,

$$\Delta V_H/V_I = \sqrt{2} - 1 = 0.41$$

and therefore

$$\Delta V_H = 0.41 V_I = 12\,200 \text{ m/s} \tag{8.43}$$

This additional velocity must be imparted in a direction *parallel* to the vehicle's initial velocity, which is the same as the earth's heliocentric velocity.

In practice, if we wish to launch from the surface of the earth a vehicle that is to escape the solar system, we will have to provide not only the velocity increment shown in Eq. (8.43) but also the velocity increment necessary to place the vehicle in high earth orbit. The latter is given by

$$V^\oplus_{esc} = \left(\frac{2GM_\oplus}{R_\oplus}\right)^{1/2} = (2gR_\oplus)^{1/2} = 11\,200 \text{ m/s}$$

Thus the total velocity increment is

$$\Delta V_{total} = V^\oplus_{esc} + \Delta V_H = 23\,400 \text{ m/s} \tag{8.43$'$}$$

It is important, however, to appreciate that if the vehicle was launched directly from the earth's surface (in a direction *parallel* to the earth's heliocentric motion) without first going into a high orbit, the total velocity increment would be smaller even though the total energy gained by the probe is the same in both cases. To see this note that $V_I = V_\oplus$ as before, and the final velocity $V_t$ must be such that

$$\tfrac{1}{2}V_t^2 = \frac{K_\odot}{a_I} + \frac{GM_\oplus}{R_\oplus} = V_I^2 + \tfrac{1}{2}(V^\oplus_{esc})^2 \tag{8.44}$$

in order to overcome the sun's and the earth's gravitational fields. Thus

$$\Delta V = V_t - V_I = [2V_I^2 + (V^\oplus_{esc})^2]^{1/2} - V_I = 13\,800 \text{ m/s} \tag{8.44$'$}$$

which is significantly smaller than the result obtained in Eq. (8.43′). This is so because the escape velocity is added in 'quadrature' in Eq. (8.44′);* note that if $V^\oplus_{esc}$ was zero we would regain exactly the result of Eq. (8.43).

*Earth to Mars.* As a further application we consider the total velocity increment needed to launch a vehicle from earth and have it land on Mars. We will use (see Table 8.1)

$$M_♂ = 0.11\, M_\oplus \qquad \text{and} \qquad R = \frac{a_F}{a_I} = 1.524$$

* For a given velocity increment $\Delta V$, the gain in energy is $\Delta E = m\mathbf{V}\cdot\Delta\mathbf{V}$ and it is largest when $V$ is large.

From Eq. (8.42) we obtain

$$\frac{\Delta V_A}{V_I} = \left(\frac{2R}{1+R}\right)^{1/2} - 1 = 0.099; \qquad \Delta V_A = 2950 \text{ m/s}$$

$$\frac{\Delta V_B}{V_I} = \left(\frac{1}{R}\right)^{1/2} - \left[\frac{2}{R(1+R)}\right]^{1/2} = 0.089; \qquad \Delta V_B = 2650 \text{ m/s}$$

It is most economic to escape from a *low* earth orbit and to be captured into a low Mars orbit. The radius of Mars is $R_♂ = 0.52\, R_⊕$ and therefore the corresponding escape velocities are

$$V^{⊕}_{\text{esc}} = \left(\frac{2GM_⊕}{R_⊕}\right)^{1/2} = 11\,200 \text{ m/s}$$

$$V^{♂}_{\text{esc}} = \left(\frac{2GM}{R}\right)^{1/2} = 5150 \text{ m/s}$$

With *respect to the earth* the vehicle's kinetic energy must be such that after it has been decreased by the potential energy corresponding to the earth's attraction it still has the value $\Delta V_A$. Thus if we designate by $v_A$ the velocity that the vehicle must have in order to successfully leave the earth orbit and enter the transfer ellipse, $v_A$ must satisfy

$$\tfrac{1}{2}v_A^2 - \tfrac{1}{2}(V^{⊕}_{\text{esc}})^2 = \tfrac{1}{2}\Delta V_A^2$$

The required velocity increment is the difference between $v_A$ and the velocity in low earth orbit $v_0^{⊕}$, where $(v_0^{⊕})^2 = \tfrac{1}{2}(V^{⊕}_{\text{esc}})^2$. Thus

$$\Delta v_A = v_A - v_0^{⊕} = [(V_{\text{esc}})^2 + (\Delta V_A)^2]^{1/2} - (1/\sqrt{2})V^{⊕}_{\text{esc}} = 3660 \text{ m/s}$$

Fig. 8.8. Transfer trajectory from a parking orbit around the earth to a parking orbit around Mars; the wiggles are greatly exaggerated.

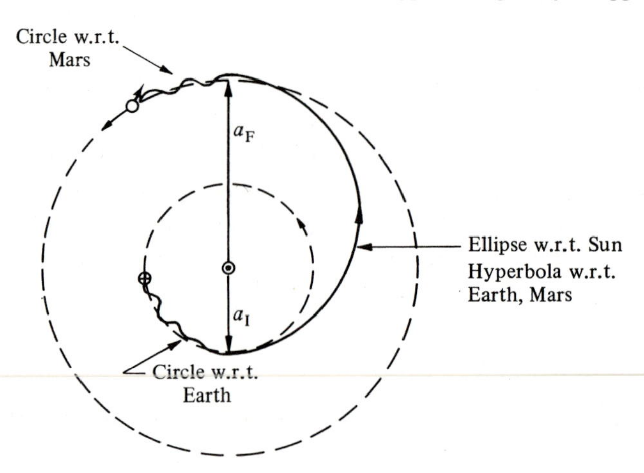

and correspondingly

$$\Delta v_B = v_B - v_0^{♂} = [(V_{\text{esc}})^2 + (\Delta V_B)^2]^{1/2} - (1/\sqrt{2})V_{\text{esc}}^{♂} = 2150 \text{ m/s}$$

In the total velocity balance we must, of course, add the velocity increments required to bring the vehicle into low earth orbit and to land it from low Mars orbit. Thus

$$\Delta V_{\text{T}} = v_0^{\oplus} + \Delta v_A + \Delta v_B + v_0^{♂} = 17\,370 \text{ m/s}$$

We stress that the thrust imparted to the vehicle must be parallel to its own motion and to its heliocentric motion for the above calculations to remain valid. The reader can easily show that if the transfer to the Hohman ellipse was executed from a high earth orbit (and into a high Mars orbit) the total velocity increment would have been $\Delta V'_{\text{T}} = 22\,000$ m/s. The trajectory of the vehicle as seen from the heliocentric system is shown in Fig. 8.8 where the effect of the planetary gravitational fields is included. Finally note that the velocity increments for the missions that we discussed are within the reach of present technology (see Chapter 7), especially for not too massive vehicles.

## 8.4 Encounters

When a vehicle approaches a planet from a large distance, the total energy of the vehicle with respect to the planet is positive, and in that reference frame the vehicle executes a hyperbolic trajectory with focus at the position of the planet. Thus, the asymptotic velocity *with respect to the planet* before and after the encounter will have the same magnitude but different direction. This change in direction of the relative velocity results in a change in the magnitude and direction of the heliocentric velocity of the vehicle. By properly chosen encounters the velocity of a vehicle can be boosted; this possibility is exploited in missions to the outer planets – such as carried out by the Pioneer and Voyager craft.

That the heliocentric velocity of a vehicle changes in an encounter with a planet can be easily understood from the sketches of Fig. 8.9. The velocity *relative* to the planet, at large distance (i.e. its asymptotic value) will be designated by $\mathbf{v}_\infty$ and its magnitude $v_\infty$ is referred to as the *hyperbolic excess velocity*. The heliocentric velocity of the planet is $\mathbf{u}$ and the heliocentric velocity of the vehicle before and after the encounter are labeled by $\mathbf{V}$ and $\mathbf{V}'$. By definition

$$\mathbf{V} = \mathbf{u} + \mathbf{v}_\infty \qquad \mathbf{V}' = \mathbf{u}' + \mathbf{v}'_\infty \tag{8.45}$$

For the duration of the encounter we can set $\mathbf{u} = \mathbf{u}'$ and even though $\mathbf{v}_\infty$ has changed direction, $|\mathbf{v}_\infty| = |\mathbf{v}'_\infty|$. The vector diagram before and after the encounter is shown in Figs. 8.9(*a*) and 8.9(*b*) where the new direction of $\mathbf{v}_\infty$ (i.e. $\mathbf{v}'_\infty$) is such that $|\mathbf{V}'| > |\mathbf{V}|$.

We can reach the same conclusion by a more detailed argument as follows: The total energy of the vehicle with respect to the sun is

$$E = T + U \qquad \text{or} \qquad h = \frac{2E}{m} = V^2 - \frac{2K}{R} \tag{8.46}$$

where $R$ is the heliocentric distance and $h$ is known as the energy constant. If we use the velocity equation (Eq. (8.35)) whereby $V^2 = (2K/R - K/a)$ we find immediately $h = -K/a$, in agreement with Eq. (8.28). The change in the energy of the vehicle during an encounter (where $R$ can be considered constant) is then

$$\frac{dh}{dt} = 2\mathbf{V}\cdot\frac{d\mathbf{V}}{dt} = 2(\mathbf{u} + \mathbf{v})\cdot\frac{d\mathbf{v}}{dt} \tag{8.47}$$

Here we expressed $\mathbf{V} = \mathbf{u} + \mathbf{v}$ with $\mathbf{v}$ the velocity relative to the planet; further $d\mathbf{V} = d\mathbf{v}$ because during the encounter $\mathbf{u}$ is practically constant.

The relative acceleration of the vehicle $d\mathbf{v}/dt$ depends only on the gravitational attraction of the planet $K_p = GM_p$ and on the distance $r$ from the planet

$$\frac{d\mathbf{v}}{dt} = -\left(\frac{K_p}{r^2}\right)\hat{u}_r$$

The change in the total energy of the vehicle is the integral of Eq. (8.47) over time

$$\Delta h = \Delta V^2 = \int_{-\infty}^{+\infty} \frac{dh}{dt}\, dt = -2K_p \int \frac{(\mathbf{u} + \mathbf{v})\cdot\hat{u}_r}{r^2} = -2K_p \int \frac{\mathbf{u}\cdot\hat{u}_r}{r^2}\, dt \tag{8.48}$$

The last step in Eq. (8.48) follows because by symmetry the integral of $\mathbf{v}\cdot(d\mathbf{v}/dt)$ vanishes. In Fig. 8.10 we show two trajectories of encounters with a planet that moves with heliocentric velocity $\mathbf{u}$. For case (*a*) the unit vector $\hat{u}_r$ is antiparallel to $\mathbf{u}$ (for small $r$) so that the integral of Eq.

Fig. 8.9. Vector diagrams for the velocity change during an encounter of a probe with a planet. The velocity of the planet is $\mathbf{u}$ and that of the probe $\mathbf{V}$, so that $\mathbf{v}_\infty$ is the relative velocity between probe and planet: (*a*) before the encounter, (*b*) after the encounter; even though the magnitude of $\mathbf{v}_\infty$ is not changed the magnitude of the heliocentric velocity of the probe $\mathbf{V}'$ has increased.

(8.48) is positive and the vehicle *gains* energy in the encounter. For case (*b*) the opposite is true and the vehicle *loses* energy. The angle by which the direction of $\mathbf{v}_\infty$ changes depends on $r_p$, the distance of closest approach to the planet.

In a more familiar context the change in the heliocentric velocity of the vehicle is analogous to the change in the speed of a tennis ball that is bounced off a moving backboard. If the backboard is moving towards the ball (case (*a*)) the ball will bounce off with higher speed; if the backboard is moving away from the ball (case (*b*)), the ball moves slower on its return path. Encounters are also referred to as 'swing-bys' or 'slingshot trajectories'. Their accuracy depends on the correct aiming of the vehicle when it is still far away from the planet. In what follows we will give a quantitative analysis of the expected velocity increment in an encounter and use it to discuss the Voyager-2 mission.

In Figs. 8.11(*a*, *b*) we show the orbit of a planet moving with heliocentric velocity $\mathbf{u}$ and the heliocentric trajectory of a vehicle before and after the encounter. The vehicle velocity is $\mathbf{V}_1$ before and $\mathbf{V}_2$ after the encounter and its corresponding velocity relative to the planet is designated by $\mathbf{v}_\infty$ and $\mathbf{v}'_\infty$. As shown in the figures

$$\mathbf{V}_1 = \mathbf{u} + \mathbf{v}_\infty \qquad \mathbf{V}_2 = \mathbf{u} + \mathbf{v}'_\infty \tag{8.49}$$

and we identify the angles

$$\left.\begin{aligned} \beta_1 &= \measuredangle(\mathbf{V}_1, \mathbf{u}) & \beta_2 &= \measuredangle(\mathbf{V}_2, \mathbf{u}) \\ \xi_1 &= \measuredangle(\mathbf{v}_\infty, \mathbf{u}) & \xi_2 &= \measuredangle(\mathbf{v}'_\infty, \mathbf{u}) \end{aligned}\right\} \tag{8.50}$$

and the angle $\Delta\xi$ by which the relative velocity has been rotated

$$\Delta\xi = \measuredangle(\mathbf{v}_\infty, \mathbf{v}'_\infty) \qquad |\mathbf{v}_\infty| = |\mathbf{v}'_\infty| \tag{8.50$'$}$$

Given $\mathbf{u}$ and $\mathbf{V}_1$ we can find $\mathbf{V}_2$ if we know $\Delta\xi$.

Fig. 8.10. Encounter of a probe with a planet moving with velocity $\mathbf{u}$; $\mathbf{v}_\infty$, $\mathbf{v}'_\infty$ are the relative velocities of the probe with respect to the planet before and after the encounter. In case (*a*) the probe gains velocity in the heliocentric system; in case (*b*) the heliocentric velocity of the probe decreases.

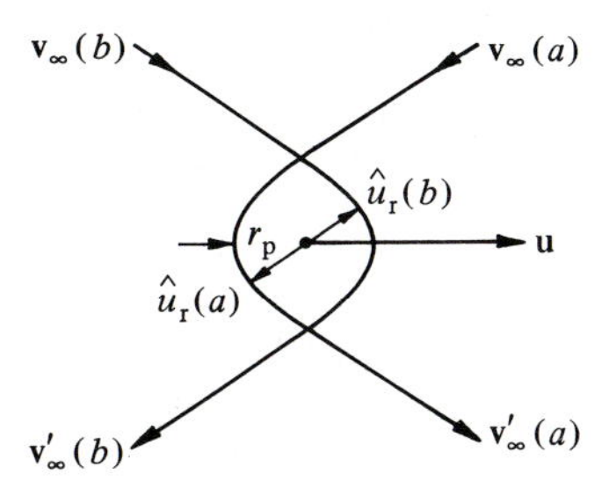

To calculate $\Delta\xi$ we consider the encounter in the reference frame of the planet as shown in Fig. 8.12. Here the trajectory of the vehicle is a hyperbola with semi-major axis $a$. The semi-major axis is related to the total energy (see Eq. (8.29)) $E = Km/2a$. At large distance from the planet the total energy is simply the kinetic energy $E = mv_\infty^2/2$. Therefore we obtain (setting $K = K_p$ for clarity)

$$a = \frac{K_p}{v_\infty^2} \tag{8.51}$$

Fig. 8.11. Vehicle trajectory and planet orbit during an encounter: (*a*) before the encounter, (*b*) after the encounter (in this case the vehicle gains velocity).

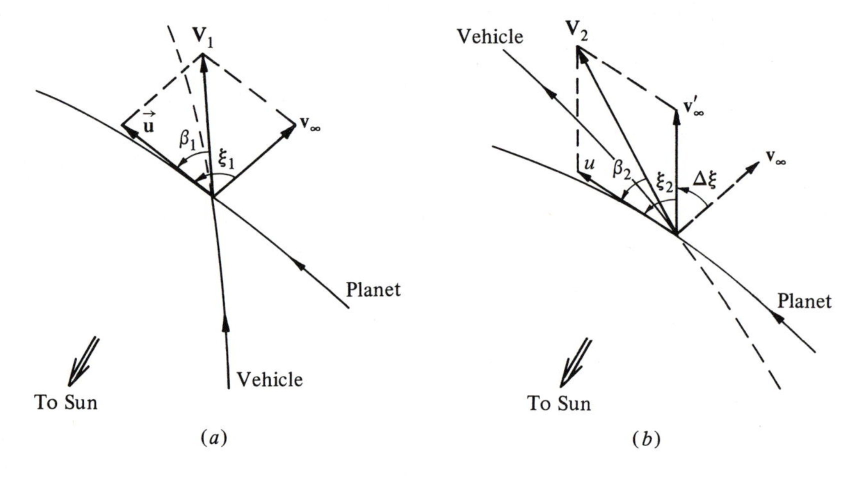


Fig. 8.12. Trajectory of a vehicle encountering a planet, as seen in the planet's rest frame (i.e. relative motion).

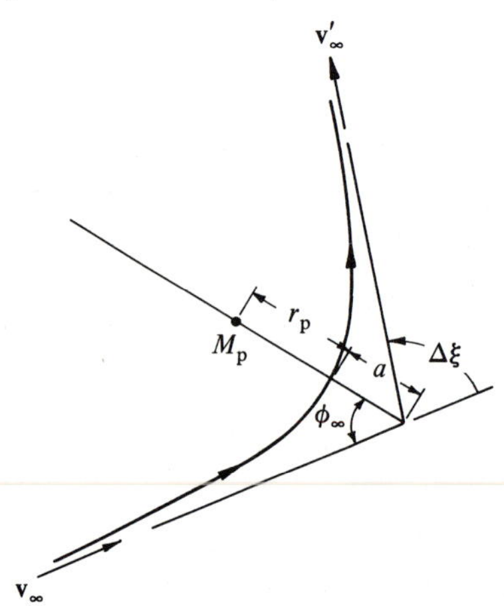

The distance of closest approach $r_p$ for the encounter is given and is related to the eccentricity (see Eq. (8.10)) through

$$r_p = a(e-1) \qquad \text{or} \qquad e = 1 + \frac{r_p}{a} \tag{8.52}$$

We also know that the angle of the asymptotes is related to the eccenticity (see Fig. 8.3(*b*) or Eq. (8.10) for $r \to \infty$)

$$\cos\phi_\infty = 1/e \tag{8.53}$$

where the angle $\phi_\infty$ is shown in Fig. 8.12 and defines $\Delta\xi$ since

$$1/e = \cos\phi_\infty = \sin\left(\frac{\pi}{2} - \phi_\infty\right) = \sin(\Delta\xi/2) \tag{8.54}$$

The calculation of $\mathbf{V}_2$ proceeds then as follows:

(a) Given $\mathbf{u}$ and $\mathbf{V}_1$ we find $\mathbf{v}_\infty = \mathbf{V}_1 - \mathbf{u}$.

(b) We find $\beta_1$ from Eq. (8.49) since

$$(v_\infty)^2 = u^2 + (V_1)^2 - 2uV_1\cos\beta_1 \tag{8.55}$$

and $\xi_1$, from

$$v_\infty \sin\xi_1 = V_1 \sin\beta_1 \tag{8.55'}$$

(c) Next we calculate the parameters of the hyperbolic trajectory from a knowledge of $K_p$ and $r_p$ (Eqs. (8.51) and (8.52))

$$a = K_p/v_\infty^2 \qquad e = 1 + r_p/a \tag{8.56}$$

and therefore we find $\Delta\xi$ from Eq. (8.54)

$$\Delta\xi = 2\sin^{-1}(1/e) \tag{8.57}$$

(d) Finally we form

$$\xi_2 = \xi_1 - \Delta\xi$$

and calculate $\mathbf{V}_2 = \mathbf{u} + \mathbf{v}'_\infty$, where

$$(V_2)^2 = u^2 + (v'_\infty)^2 + 2uv'_\infty \cos\xi_2 \tag{8.58}$$

and $\beta_2$ from

$$V_2 \sin\beta_2 = v'_\infty \sin\xi_2$$

We recall that $|\mathbf{v}'_\infty| = |\mathbf{v}_\infty|$. Furthermore as $r_p \to 0$, $e \to 1$ and therefore $\Delta\xi \to 180°$; of course, $r_p$ cannot be smaller than the radius of the planet. In practice, the desired value of $\Delta\xi$ fixes $r_p$; to achieve this $r_p$ the vehicle must be carefully aimed by suitable thrusting maneuvers before it comes under the influence of the gravitational attraction of the planet.

## 8.5 The Voyager-2 grand tour of the planets

The Voyager missions were designed to explore the outer planets by taking advantage of a very special conjunction of the planetary orbits.

The trajectories followed by the two craft are shown in Fig. 8.13. Voyager-2 was launched on August 20, 1977 and swung by Jupiter, Saturn and Uranus; it encountered Neptune in 1989 and will then leave the solar system. Communications have been maintained with the craft and the information acquired during the encounters has been received on earth in spite of the very large distance involved: the signal transit time from Uranus is 2 hr 40 min. A sketch of the Voyager craft is shown in Fig. 8.14.

From Fig. 8.13 we can obtain the approximate data for the mission collected in Table 8.2. We have given the average velocity between encounters even though in practice the velocity of the craft is not constant. One would expect that the velocity would decrease as the craft moved farther away from the sun. However, during each planetary encounter the craft receives a velocity boost and this is reflected by the tabulated mean velocities which are relatively constant. For convenience we note that a speed of 1 AU/year = 4750 m/s. The velocity of Voyager-2 when it leaves the solar system (at approximately 60 AU) will be $V_f = 16.1$ km/s.

It is instructive to analyse in some detail one of the encounters and we will consider the Saturn swing-by. This is shown in Figs. 8.15(*a*, *b*, *c*) which are drawn to indicate progressively more detail. The distance of closest approach to the satellites of Saturn is indicated by the arrows and is a concern, since too close an approach would alter the Voyager's trajectory. For instance, the craft approached to within 93 000 km of Tethys while it passed Phoebe at 2 076 000 km. The closest approach to Saturn was

$$r_p = 161\,094 \text{ km} \tag{8.59}$$

or approximately 2.67 Saturn radii. For the gravitational attraction of

Table 8.2. *Approximate data for the Voyager-2 mission*

| | | $a$ (AU) | $\Delta s$ (AU) | $\Delta t$ (years) | $\bar{V}$ (km/s) |
|---|---|---|---|---|---|
| Earth | 20/8/77 | 1.0 | | | |
| | | | 6.9 | 1.89 | 17.3 |
| Jupiter | 9/7/79 | 5.2 | | | |
| | | | 7.5 | 2.13 | 16.8 |
| Saturn | 26/8/81 | 9.5 | | | |
| | | | 18.0 | 4.41 | 19.3 |
| Uranus | 24/1/86 | 19.2 | | | |
| | | | 14.5 | 3.58 | 19.3 |
| Neptune | 24/8/89 | 30.1 | | | |
| Leave solar system | ~2000 AD | 60 | | | $V_f = 16.1$ |

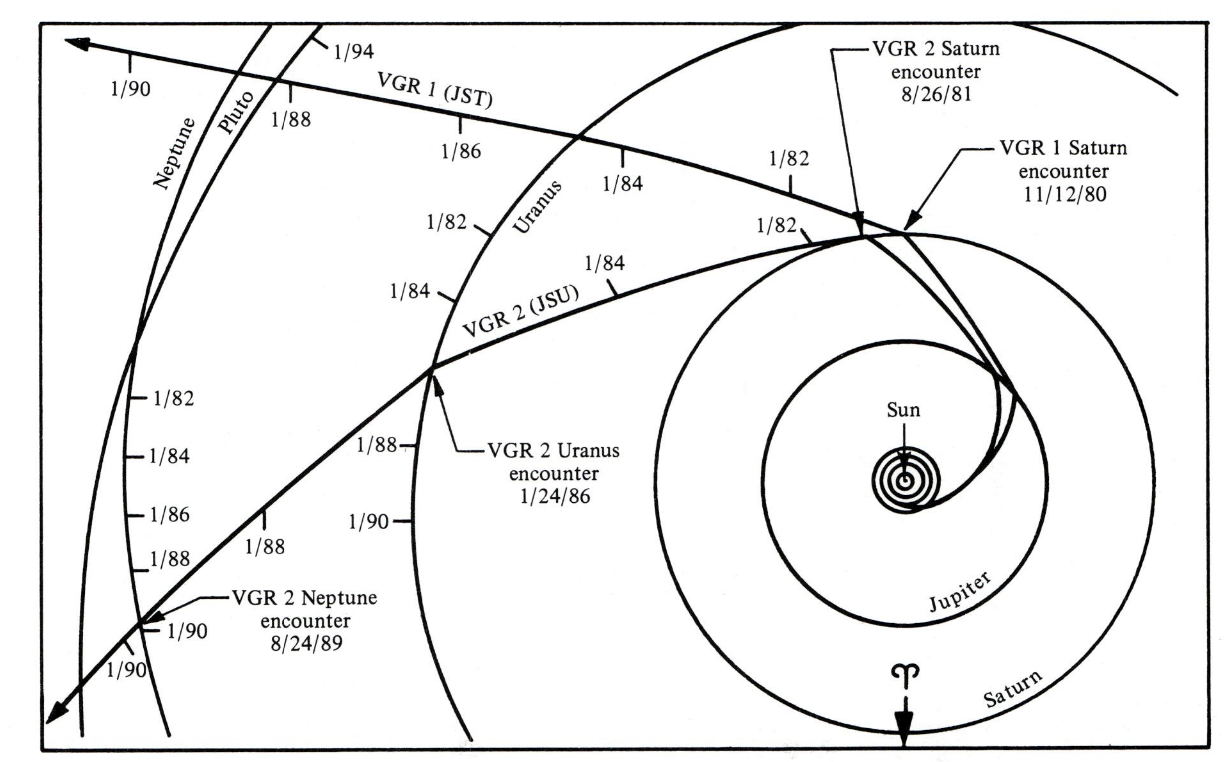


Fig. 8.13. Voyager-1 and Voyager-2 missions as seen in the plane of the elliptic. (Provided through the courtesy of the Jet Propulsion Laboratory, California Institute of Technology, Pasadena, California.)

Saturn we have

$$K_{\hbar} = 3.79 \times 10^7 \text{ km}^3/\text{s}^2 \tag{8.60}$$

and the heliocentric velocity of Saturn is

$$u_{\hbar} \simeq \frac{2\pi a}{T} = \frac{2\pi \times 9.54}{29.46} (\text{AU/yr}) = 9660 \text{ m/s} \tag{8.61}$$

where we used data from Table 8.1.

We are given that the excess hyperbolic velocity of the craft with respect to Saturn is

$$v_{\infty} = 10\,680 \text{ km/s} \tag{8.62}$$

and from Fig. 8.13 we infer that $\beta_1 = 53°$. We can then calculate $V_1$, by solving the quadratic equation* (see Eq. (8.55))

$$V_1^2 = v_{\infty}^2 - u^2 + 2uV_1 \cos\beta_1$$

to find

$$V_1 = 13.3 \text{ km/s} \tag{8.63}$$

This is a reasonable value, since the mean velocity between Jupiter and Saturn was $\bar{V} = 16.8$ km/s.

Also we find the eccentricity of the hyperbolic orbit using Eq. (8.56)

$$a = \frac{K_{\hbar}}{v_{\infty}^2} = \frac{3.79 \times 10^7 \text{ km}^3/\text{s}^2}{114 \text{ km}^2/\text{s}^2} = 3.32 \times 10^5 \text{ km}$$

$$e = 1 + r_p/a = 1.484$$

Fig. 8.14. The Voyager-2 spacecraft.

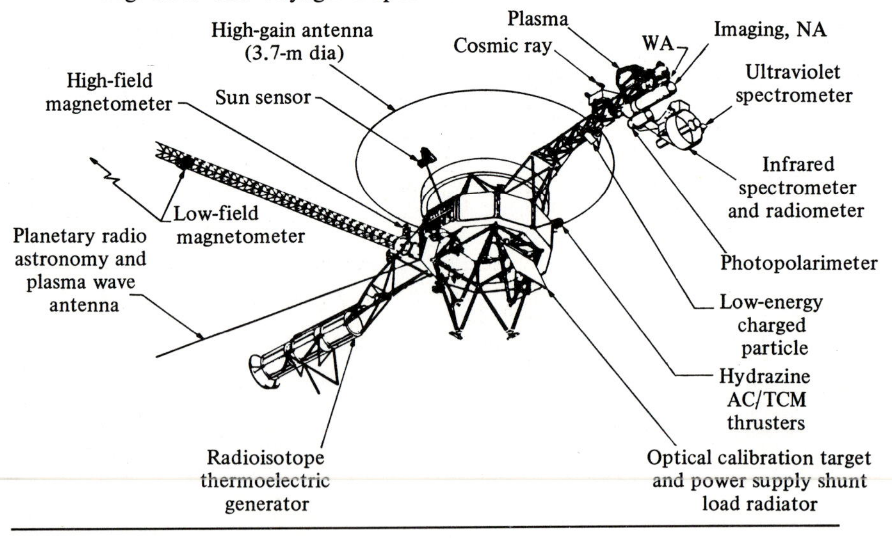


* One can also use a graphical method unless high precision is desired.

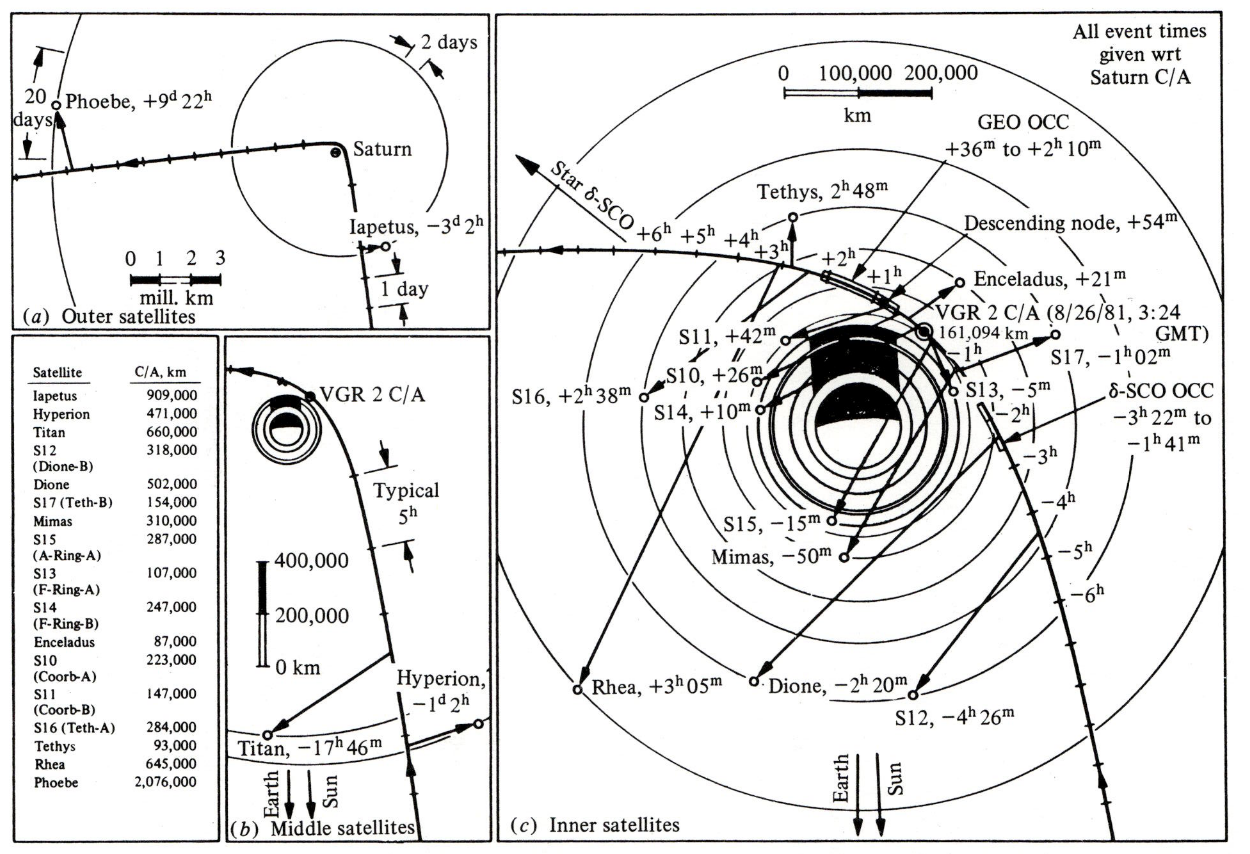


| Satellite | C/A, km |
|---|---|
| Iapetus | 909,000 |
| Hyperion | 471,000 |
| Titan | 660,000 |
| S12 (Dione-B) | 318,000 |
| Dione | 502,000 |
| S17 (Teth-B) | 154,000 |
| Mimas | 310,000 |
| S15 (A-Ring-A) | 287,000 |
| S13 (F-Ring-A) | 107,000 |
| S14 (F-Ring-B) | 247,000 |
| Enceladus | 87,000 |
| S10 (Coorb-A) | 223,000 |
| S11 (Coorb-B) | 147,000 |
| S16 (Teth-A) | 284,000 |
| Tethys | 93,000 |
| Rhea | 645,000 |
| Phoebe | 2,076,000 |

Fig. 8.15. Encounter of Voyager-2 with Saturn shown in the equator plane for three different scales. (Provided through the courtesy of the Jet Propulsion Laboratory, California Institute of Technology, Pasadena, California.)

Thus

$$\Delta\xi = 2\sin^{-1}(1/e) = 85^\circ \tag{8.64}$$

in agreement with the trajectory indicated in Fig. 8.15(*a*).

To calculate $V_2$, the departure velocity from Saturn, we first evaluate (see Eq. (8.55′))

$$\xi_1 = \sin^{-1}\left[\frac{V_1}{v_\infty}\sin\beta_1\right] = 86^\circ$$

Therefore

$$\xi_2 = \xi_1 - \Delta\xi = 1^\circ$$

and finally from Eq. (8.58) we obtain

$$V_2^2 = u^2 + (v_\infty)^2 + 2uv_\infty\cos\xi_2 = 20.2 \text{ km/s} \tag{8.65}$$

Thus Voyager-2 gained a velocity increment $\Delta V \simeq 7$ km/s on swinging by Saturn. This is to be compared with the velocity increments in thrusting maneuvers which are typically $\Delta V \simeq 15$ m/s.

Having found $V_2$ let us estimate the arrival velocity at Uranus. From energy conservation we have

$$V_{♅}^2 - \frac{2K_\odot}{R_{♅}} = V_{♄}^2 - \frac{2K_\odot}{R_{♄}} \tag{8.66}$$

where for simplicity we assumed circular orbits for Uranus and Saturn, with radii $R_{♅}$ and $R_{♄}$. Setting $V_{♄} = V_2 = 20.2$ km/s as found in Eq. (8.65) we obtain

$$V_{♅} = (V_1)_{♅} = 17.9 \text{ km/s}$$

Fig. 8.16. Trajectories of the Pioneer and Voyager spacecraft after leaving the solar system; the positions are shown for the year 2000 AD.

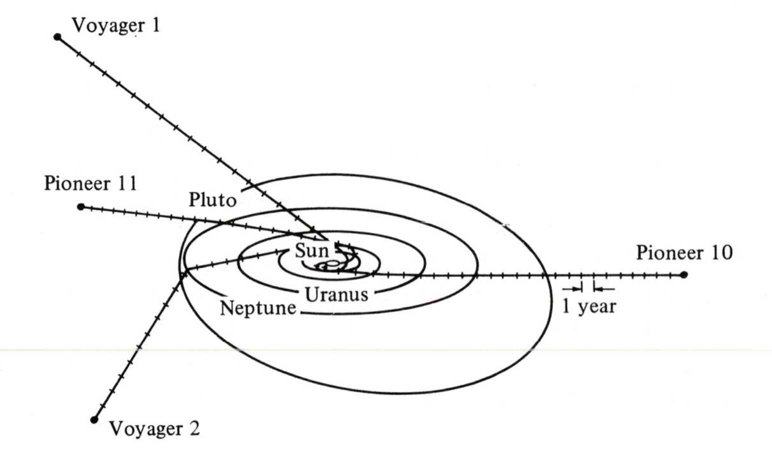

Therefore we predict an average velocity between Saturn and Uranus of $\bar{V} = 19.05$ km/s in good agreement with the value $\bar{V} = 19.3$ km/s listed in Table 8.2. Given that the final velocity after escape from the solar system is $V_f = 16.1$ km/s, the departure velocity from Uranus must be $(V_2)_{♅} \gtrsim 19$km/s. Thus Voyager-2 received a boost at the Uranus swing-by as well.

We can also work backwards from the data to estimate the initial launch velocity relative to the earth. A Hohman transfer from earth to Jupiter ($R_{♃} = 5.2$ AU) requires a fractional velocity increment

$$\frac{\Delta V_H}{V_I} = \left(1 - \frac{1}{5.2}\right)\left(\frac{10.4}{6.4}\right)^{1/2} + \left(\frac{1}{5.2}\right)^{1/2} - 1 = 0.48$$

Therefore $\Delta V_H = 14.3$ km/s where we used $V_I = V_{\oplus} = 29.8$ km/s. If we include the escape velocity from the earth, we find for the launch velocity

$$v = [(14.3)^2 + (11.2)^2]^{1/2} = 18.2 \text{ km/s}$$

This is a slight underestimate because the orbit to Jupiter is hyperbolic, but it shows that the vehicle velocities are within the capabilities of available launching systems.

In closing, we show in Fig. 8.16 an oblique view of the trajectories of the Pioneers and Voyagers at their expected position in the year 2000 AD. These are the first probes that have been sent from earth to explore the region outside the solar system. They have also provided us with a wealth of information about the planets of our own solar system. The missions were feasible only by taking advantage of the gravitational field of the planets and required a highly refined tracking and command network. No doubt there will be much more exploration of our solar system, but we already have a fair understanding of the conditions prevailing on the planets as well as of the properties of the sun.

## 8.6 Interstellar travel

Interstellar travel is often chronicled in fascinating detail by science fiction writers. Yet, travel to another solar system is not practical today and there are fundamental limitations on the type of travel, the length of the journey and the energy requirements for such a venture. We will consider these aspects of space travel and discuss the methods of propulsion that have been proposed. These methods are based on correct scientific fact but their practical realization is far from proven, and the energy necessary for a single mission dwarfs the present energy production capabilities on earth.

The presumption motivating interstellar travel is that at least some of the stars that we see in the sky have planetary systems similar to our own, and may harbor planets hospitable to human life. While this may be plausible, it is not proven, because planets are practically impossible to observe at astronomical distances. Furthermore, in our part of the galaxy the density of stars is low, and the nearest stars are several light years away. We give below a list of the four nearest stars and show their relative positions in Fig. 8.17.

| Star | Distance | Properties |
|---|---|---|
| Alpha-Centauri | 4.3 LY | Triple star |
| Barnard's star | 6.0 LY | Red dwarf |
| Sirius | 8.2 LY | Large bright star |
| Epsilon-Eridani | 10.8 LY | Similar to sun |

Note that distances are now measured in light years (LY) where

$$1\,\mathrm{AU} = 1.495 \times 10^{11}\ \mathrm{m}$$
$$1\,\mathrm{LY} = 9.458 \times 10^{15}\ \mathrm{m}$$
$$1\,\mathrm{pc} = 3.086 \times 10^{16}\ \mathrm{m} = 3.26\ \mathrm{LY}$$

The last unit is the 'parsec' and is defined as the distance at which 1 AU subtends a parallax of one second of arc. It is the unit commonly used to describe extended astronomical objects and distances. It follows that the parallax subtended by the above stars over one earth revolution is of the order of 1 arcsecond and therefore can be measured directly. Still, as compared to the dimensions of the solar system (50 AU) the projected travel represents an extrapolation in distance by a factor of $10^4$.

*Kinematics.* As our first task we want to explore the time necessary to complete a mission. Supposing that an *unmanned* craft is sent to $\alpha$-Centauri we would have to wait for 4.3 years after its arrival to receive the signal on earth. To complete the mission in a reasonable time (as measured by the span of human life) the craft would have to reach a velocity which is a fraction of the speed of light. We can imagine that the

Fig. 8.17. The nearest stars to our solar system.

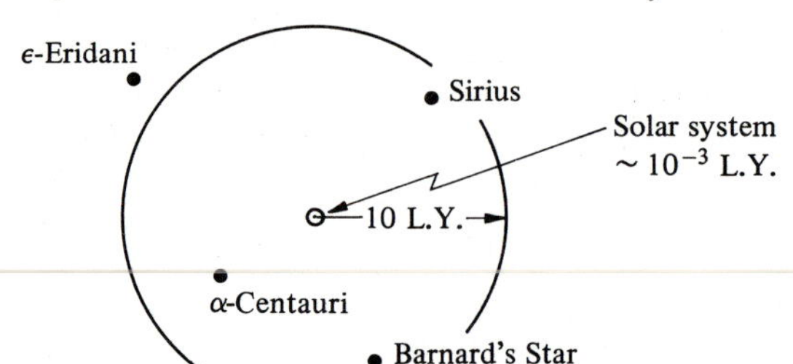

acceleration imparted to the craft is small (low thrust acceleration) but sustained for a very long time.

As an example we choose an acceleration

$$a = 0.1\,g \simeq 1\,\mathrm{m/s^2}$$

and calculate the time needed to complete a mission to $\alpha$-Centauri for different final velocities $V_f$ of the craft. The results are given in Table 8.3 and it is evident that little is gained for $V_f/c \gtrsim 0.3$. The table has been calculated using non-relativistic kinematics since they introduce only very small differences. In a realistic mission the craft should decelerate when it approaches its target and this would further add to the mission time.

Next we make an estimate of the energy requirements. For a probe of mass $M = 100$ kg (which is $\frac{1}{10}$th of the Voyager mass, and $\frac{1}{1000}$th of the shuttle orbiter) at velocity $V_f/c = 0.1$, the kinetic energy is

$$T = \tfrac{1}{2}V_f^2 = 4.5 \times 10^{16}\,\mathrm{J} \sim 10^{10}\,\mathrm{kW\text{-}hours} \tag{8.67}$$

This corresponds to one day's energy production in the U.S. Therefore the propulsion engine should be efficient, but it must also be highly compact. These conditions are best met by nuclear fuels. Furthermore the recoil velocities of the nuclear fragments are in the range $V_f/c \sim 0.1$ and are therefore well matched to the speed of the craft.

Apart from self-propelled vehicles it has also been proposed to use an earth based energy source, such as a laser or beamed microwaves to 'push' the craft along. A variant is the direct use of the solar radiation impinging on a solar sail extended by the spaceship. In analogy to the planetary swing-by one could consider a craft that 'rides a comet' out of the solar system. These methods yield rather low velocities as compared to the desired $V_f/c \sim 0.1$ and can accelerate only small payloads. We will return to these ideas after first discussing a proposal for a large ship propelled by nuclear explosions.

*Dyson's ship.* In the 1950s, F. Dyson and S. Ulam proposed the

Table 8.3. *Total mission time to α-Centauri for $a = 1\,m/s^2$*

| Final velocity ($V_f/c$) | Acceleration time (years) | Coasting time (years) | Data return (years) | Mission time (years) |
|---|---|---|---|---|
| ~1.0 | 10 | 0 | 4.3 | 14 |
| 0.5 | 5 | 6 | 4.3 | 15 |
| 0.3 | 3 | 13 | 4.3 | 20 |
| 0.1 | 1 | 43 | 4.3 | 48 |
| 0.05 | 0.5 | 85 | 4.3 | 90 |

construction of a spaceship based on the concepts sketched in Fig. 8.18. The ship carries a large complement of nuclear bombs which are dropped one by one behind the ship and exploded; forward going particles are caught by a shield so that about half of the explosion's momentum is transferred to the ship. An ablation shield is used to absorb the generated heat and a shock absorber to buffer the ship from the impact of the explosion. Note that the payload is large, of the order of an old fashioned ocean liner, and that the explosions must take place well behind the craft in order to assure the radiation protection of the crew.

The shock absorber limits the velocity increment that we can impart to the vehicle in each explosion. We choose $\Delta V = 30$ m/s per explosion which is reasonable. Further, we take the average mass of the ship to be $\bar{M} = 150\,000$ tonnes so that $\Delta V = 30$ m/s corresponds to a momentum change

$$\Delta p = \bar{M}\Delta V = 4.5 \times 10^9 \text{ kg-m/s}$$

A rough calculation shows that a 1 megatonne (Mt) nuclear explosive can provide the thrust, as follows: Let the mass of the fragments be $dm_f = 10^3$ kg and their velocity $v_f = 0.1c$. If $\frac{1}{3}$ of the total momentum is transferred to the shield the thrust is

$$F_t = \tfrac{1}{3} dm_f V_f \simeq 10^{10} \text{ kg-m/s}$$

However, to sustain an average acceleration $a = 0.1\,g = 1$ m/s$^2$ by velocity increments of $\Delta V = 30$ m/s, we will have to explode one thermonuclear device every 30 s!

Next we calculate the total fuel required, for a final velocity of the ship, $V_f/c = 0.03$. From the simple non-relativistic expression $V_f = a\Delta t$ the acceleration time is $\Delta t \simeq 10^7$ s $\simeq 0.3$ years. Therefore we will need

Fig. 8.18. Schematic of a very large spaceship propelled by nuclear explosions as proposed by F. Dyson.

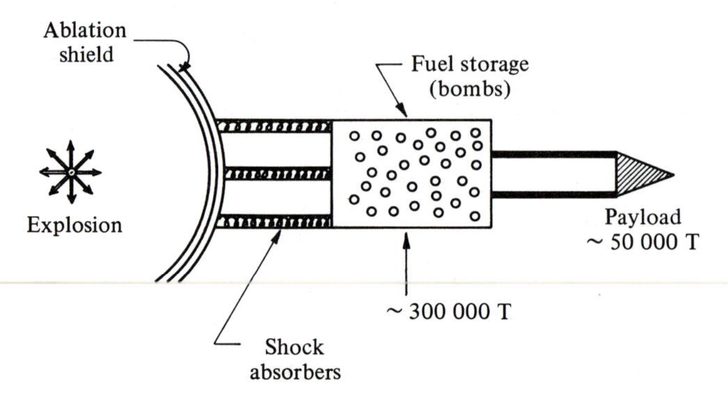

approximately

$$3 \times 10^5 \quad 1\,\text{Mt bombs} \tag{8.68}$$

each of a mass of 1 tonne. That amount of fuel is approximately 30 times the present nuclear arsenal of the U.S. In terms of cost, if the price of deuterium is \$200/kg and even if we ignore all other components of the bombs, we have an expenditure of $\$6 \times 10^{10}$ which is approximately 60 times the (yearly) U.S. federal budget. Thus a mission of that type is well outside our present capabilities.

*Beamed power*. Here we consider a payload attached to a reflector as shown in Fig. 8.19. The reflector should be as large as possible and also as light as possible. It could be constructed out of a thin mesh in the case of a microwave beam, or out of thin aluminum in the case of optical radiation. We will assume a thickness $t = 16$ nm (160 Å – 5% of the laser light passes through); the density of the reflector, or sail, is then $\rho = 0.4\ \text{g/m}^2$. This density is sufficient to sustain the thermal heating. If the diameter of the reflector is $d = 3.6$ km, the total area and total mass (divided approximately equally between the sail, the structure and the payload) would be

$$A = 10^7\ \text{m}^2 \qquad M = 10^3\ \text{kg}$$

Next we calculate the necessary power to impart an acceleration $a = 0.03\,g = 0.3\ \text{m/s}^2$ to the craft. For a perfect reflector $\mathrm{d}p/\mathrm{d}t = 2P/c$ where $P$ is the electromagnetic power (in Watts) incident on the reflector, and $\mathrm{d}p/\mathrm{d}t$ is the momentum transfer. Since $p = Mv$ and using $M = 10^3$ kg. we obtain

$$P = \frac{c}{2}\frac{\mathrm{d}p}{\mathrm{d}t} = \frac{Mc}{2}\frac{\mathrm{d}v}{\mathrm{d}t} = \frac{Mc}{2}a \simeq 0.5 \times 10^{11}\ \text{W} \tag{8.69}$$

Namely, 50 GW of laser power must be focussed to within a narrow cone. For reference we recall that a large U.S. city with a population of half a million people has a power consumption of the order of 1 GW.

If the angular divergence of the beam is set by the diffraction limit and we use 10 km optics at $\lambda = 300$ nm we could achieve $\theta_{\mathrm{d}} \simeq 3.6 \times 10^{-11}$ rad.

Fig. 8.19. Principle of propulsion by beamed power.

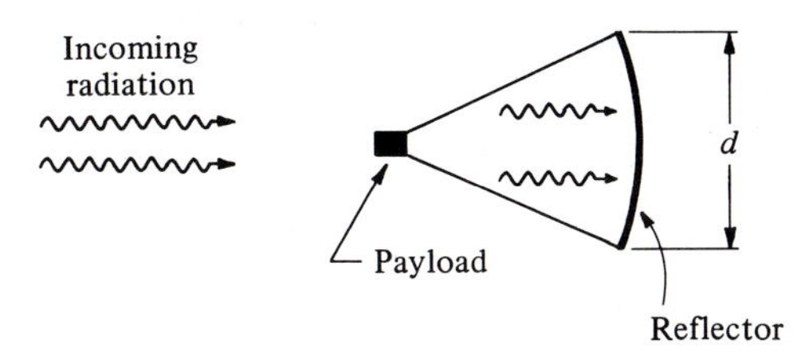

Such an angle would preserve efficient collection of the beamed power for a distance of 350 AU giving a final velocity to the craft $V_f \sim 0.02c$. We could envisage even larger lenses constructed in space by robots; to achieve this a significant infrastructure in space technology will be needed.

*Solar sails.* The idea here is to use the momentum of solar radiation either by absorbing or reflecting the radiation as shown in Fig. 8.20. At the earth's orbit the solar flux is $S_0 = 1.36 \times 10^3$ W/m$^2$ and therefore at a distance $R$ (expressed in AU)

$$S = \frac{S_0}{R^2} \quad \text{and} \quad \frac{dV}{dt} = \frac{2S_0 A}{Mc} \frac{1}{R^2} \tag{8.70}$$

where $M$ is the mass of the craft, and $A$ the area of the reflector. If we use the parameters of the previous example $A = 10^7$ m$^2$, $M = 10^3$ kg, we find

$$a = \frac{0.1 \text{ m/s}^2}{[R(\text{AU})]^2} \tag{8.70'}$$

Thus, in the vicinity of the earth a solar sailing craft could be effective, but the acceleration decreases rapidly as the craft distances itself from the sun. For a craft with the above parameters, starting from earth radius and escaping the solar system, the final velocity would be $V_f \sim 0.005c$ which, even though slow for interstellar travel purposes, still equals 100 times the escape velocity of the Voyager craft.

*Antimatter propulsion.* The annihilation of antimatter with matter is well studied in the laboratory and is the most efficient nuclear reaction in the sense that over half of the initial mass is converted into energy. Typically, antiproton–proton annihilations lead to final states with $\pi$-mesons

$$\bar{p}p \rightarrow n_+\pi^+ + n_-\pi^- + n_0\pi^0 \tag{8.71}$$

where $n_+$, $n_-$, $n_0$ are the number of positive, negative and neutral $\pi$-mesons (or pions) produced. On the average five pions are produced so that the rest mass of the annihilation products is $5m_\pi \sim 700$ MeV, as compared to

Fig. 8.20. Solar sailing using: (*a*) an absorbing or (*b*) reflecting sail.

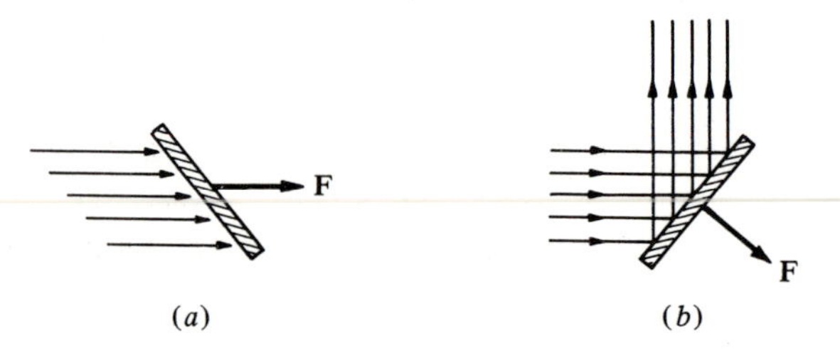

the initial mass $2m_p = 1880$ MeV; the balance is converted into the kinetic energy of the decay products. The $\pi^0$ decays to two $\gamma$-rays with a lifetime $\tau = 10^{-17}$ s; charged pions also decay but with longer lifetimes.

The annihilation products are emitted isotropically and therefore it will be necessary to focus them to the rear of the ship in order to gain thrust. This can be accomplished with magnetic fields. Typical exhaust velocities will be very near the speed of light, $v_e \sim 0.8c$. Containing and handling antimatter in macroscopic quantities is a problem of unprecedented challenge and may not be solvable in practice. We will nevertheless assume that our craft carries a fuel of antiprotons of mass $M_f = 9$ kg and can use the antiprotons, intelligently, for propulsion.

For the parameters of the craft we choose a payload of $M_L = 1$ tonne and a mass of matter (to be exhausted at high velocity) $M_e = 4$ tonnes. We assume that all of the $\bar{p}$ rest-mass is converted to energy and that this energy is transferred with 100% efficiency to the craft. Then the total energy gained by the payload will be

$$E_{\bar{p}p} = 2(M_f c^2) = 2 \times 9 \times (3 \times 10^8)^2 \simeq 1.6 \times 10^{18}\ \text{J} \qquad (8.71')$$

and, the final velocity of the payload

$$V = (2E_{\bar{p}p}/M_L)^{1/2} = 6 \times 10^7\ \text{m/s} \simeq 0.2c \qquad (8.71'')$$

The above estimates are based on optimal efficiency and on a very large amount of antimatter fuel.

At present antimatter is produced at high energy accelerators, where the antiprotons are stored at high velocity in a magnetically confining evacuated ring. The storage rate is typically $2 \times 10^{11}\bar{p}$/day, which translates to $10^{-10}$ g/year. While this rate could be increased by several orders of magnitude, we still must find alternate ways of producing antimatter, if we want to use it as fuel for an interstellar mission.

It is generally accepted that chemical fuels are not adequate for interstellar travel, because of their low specific efficiency and the relatively low exhaust velocities. Ion propulsion can produce higher exhaust velocities but does not resolve the energy problem. Nuclear fuels are better suited to interstellar travel but the propulsion systems are highly complex and expensive. This will not stop humans from thinking about new techniques for interstellar travel and hopefully, future generations may succeed in venturing beyond our own solar system.

## 8.7 Inertial guidance

The flight of missiles and spacecraft is controlled by instruments which can sense the rotation of the craft's axis and the acceleration of the craft with respect to a given reference frame. In certain space flights the reference frame may be fixed to the stars, whereas in terrestrial flights, the earth is used as reference. Rotation is sensed by comparing the orientation of the vehicle to the axis of one or more free gyroscopes; a gyroscope on which no torques act, maintains its axis of rotation in a fixed direction. Special relativity precludes the possibility of measuring the uniform velocity of a vehicle by purely internal instruments. Contrary to that, the acceleration of the vehicle can be measured due to the *inertial* force $\mathbf{F}_I = -m\mathbf{a}$ which is experienced by any massive body in accelerated motion. Integrating the measured acceleration yields the velocity, and a further integration of the velocity gives the position of the vehicle.

The above statements are true only in an inertial frame, a frame where Newton's laws are strictly valid. The earth is not an inertial frame because of its daily rotation; however the acceleration of the earth due to its motion around the sun is a much smaller effect and can be ignored. Furthermore, in the vicinity of the earth all bodies are acted upon by the force of gravity and this must be known in order to deduce the correct acceleration. Therefore, accelerometers must be mounted on a stable platform with respect to the earth. It can be said that one navigates in reference to the direction of the earth's gravitational field.

The principle of operation of an accelerometer can be understood by the analogy to 'Einstein's' elevator. We suspend a mass from a vertical spring which in turn is fastened to the roof of the elevator. When the elevator is at rest, as in Fig. 8.21(*a*) the spring is stretched by an amount $\Delta x$ where $k\Delta x = mg$ with $k$ the restoring constant of the spring and $m$ the

Fig. 8.21. Principle of operation of an accelerometer demonstrated with the help of 'Einstein's' elevator: (*a*) elevator at rest (the force of gravity acts on the mass and extends the spring), (*b*) elevator in free fall (no force acts on the mass), (*c*) elevator accelerates upwards with $a = -g$ (the force on the mass is twice that of gravity and extends the spring twice as much as in (*a*)).

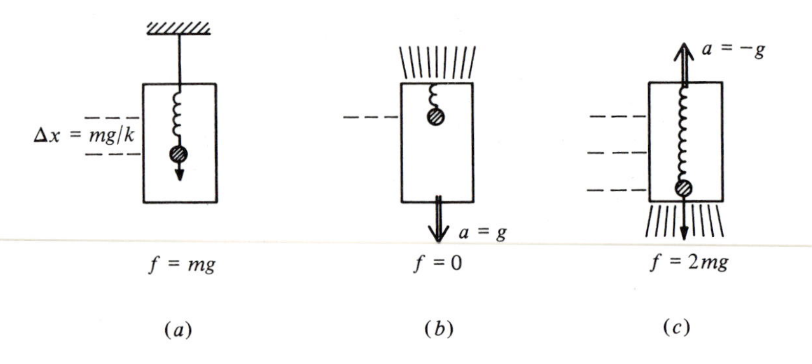

mass. If the elevator is in free fall as in (*b*) of the figure, the spring will be in its free, or equilibrium position ($\Delta x = 0$) because no force acts on it; note that in this case the elevator accelerates downwards with acceleration $a = g$. Finally, if the elevator accelerates upwards with acceleration $a = -g$, the spring will indicate twice the displacement of case (*a*) because $k\Delta x = m(g + a)$; this is sketched in (*c*) of the figure.

In a frame of reference which has acceleration **a** with respect to the earth, a body of mass *m* feels a *specific force*

$$\mathbf{f} = \mathbf{F}_g + \mathbf{F}_I = m(\mathbf{g} - \mathbf{a}) \tag{8.72}$$

where $\mathbf{F}_g$ is the force of gravity and $\mathbf{F}_I$ the inertial force. Accelerometers measure $(\mathbf{g} - \mathbf{a})$ along three orthogonal axes and if the direction and magnitude of **g** are precisely known, one obtains **a**. The simplest type of accelerometer is a pendulum with very low friction, one realization being sketched in Fig. 8.22(*a*). The effects of friction can be canceled by using a suitable feedback system; when the shaft of the pendulum turns, a signal is produced which is used to drive a torque motor to restore the pendulum to its equilibrium position. The current in the servo loop is a direct measure of the specific force. Another design is based on the mass–spring idea but instead of a spring, the mass is supported by a 'force generator' driven by an electric current as shown in Fig. 8.22(*b*). The moving mass is kept at the equilibrium position by a servo system where, as before, the signal current is a measure of the specific acceleration along the axis of the instrument.

Fig. 8.22. Practical accelerometers always involve a feedback mechanism: (*a*) the sensing element is a pendulous mass, (*b*) the motion of the sensing element is linear.

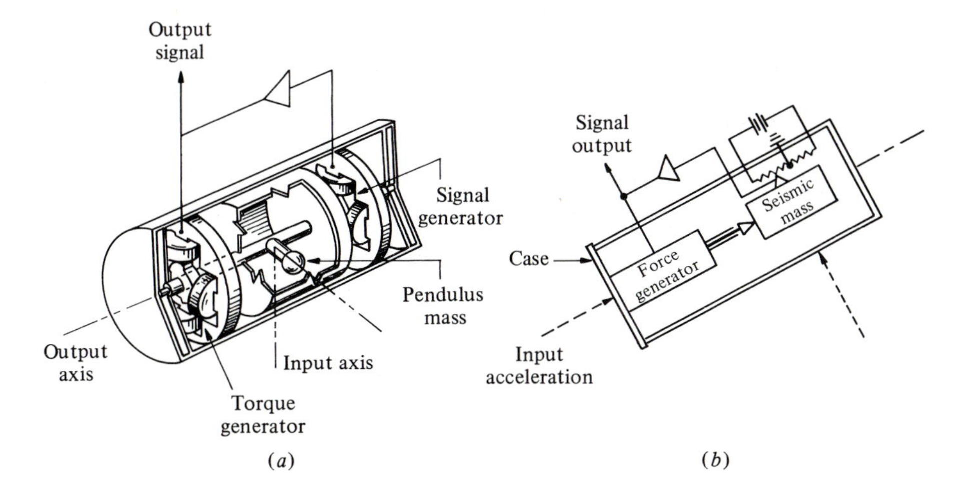

To maintain a fixed direction in space whether in airplanes, missiles, or ships, gyroscopes are always used. Mechanical gyroscopes have been developed to great perfection but recently laser ring gyros are being introduced in many commercial applications. Fundamentally, a gyroscope is a wheel spinning fast about its principal axis of inertia as shown in Fig. 8.23. The suspension of the shaft must be as free as possible so that no external torques are applied; in the absence of torques the angular momentum is conserved and the axis of the gyro will remain fixed in the inertial frame. To show this explicitly we will present an elementary analysis of gyroscopic motion.

We use the coordinate system shown in Fig. 8.23 and designate the moment of inertia with respect to the $x$-axis by $I$, whereas for the transverse directions we use $I_t$. The gyro is constructed so that $I \gg I_t$. The angular momentum is

$$\mathbf{L} = I\omega_x\hat{i} + I_t\omega_y\hat{j} + I_t\omega_z\hat{k} \tag{8.73}$$

where the unit vectors $\hat{i}, \hat{j}, \hat{k}$ are fixed to the *body of the gyro*. It is clear from Eq. (8.73) that the angular momentum vector does *not* necessarily lie along the symmetry axis (the $x$-axis in this case); however, since $I \gg I_t$ *and* $\omega_x \gg \omega_y, \omega_z$, in a good gyro $\mathbf{L}$ is always near the principal axis. The equation of motion for $\mathbf{L}$ in vector notation is

$$\frac{d\mathbf{L}}{dt} = \boldsymbol{\tau} \tag{8.74}$$

and as $\mathbf{L}$ changes, the gyro rotates so as to maintain its principal axis aligned with the angular momentum vector.

The characteristic property of gyros is that when a torque $\boldsymbol{\tau}$ is applied, they usually rotate in a direction normal to $\boldsymbol{\tau}$. Such response is contrary to the experience with non-spinning bodies but is a simple consequence of Eq. (8.74). In Fig. 8.24($a$) we show a torque $\boldsymbol{\tau}$ acting on a gyro of

Fig. 8.23. A gyroscope is a wheel spinning fast about its principal axis of inertia; **L** is the angular momentum vector.

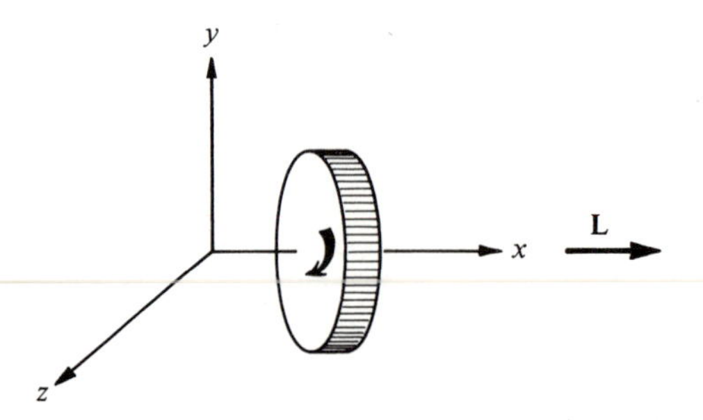

angular momentum **L**, where $\boldsymbol{\tau}$ is orthogonal to **L**. In a time interval d$t$ the change in **L** is $\Delta\mathbf{L} = \int \boldsymbol{\tau}\,\mathrm{d}t$ and is directed along $\boldsymbol{\tau}$ as shown in the figure. Thus **L** changes its direction to **L**′ but not its magnitude; **L** is rotated by an angle $\Delta\theta$ where

$$\Delta\theta = \frac{|\Delta\mathbf{L}|}{|\mathbf{L}|} = \frac{1}{L}\int \tau\,\mathrm{d}t$$

As long as $\boldsymbol{\tau}$ continues to act on the gyro in a direction normal to the gyro axis, the gyro rotates about the direction normal to **L** and $\boldsymbol{\tau}$ at a rate

$$\Omega = \frac{\mathrm{d}\theta}{\mathrm{d}t} = \lim \frac{\Delta\theta}{\Delta t} = \frac{|\boldsymbol{\tau}|}{|\mathbf{L}|}$$

This result can be expressed vectorially by the equation

$$\boldsymbol{\tau} = \boldsymbol{\Omega} \times \mathbf{L} \tag{8.75}$$

where $\boldsymbol{\Omega}$ specifies both the axis and magnitude of the slow rotation; such motion is called *precession.*

In the absence of torques, **L** must remain fixed in space, but this does not imply that the principal axis of the gyroscope will also necessarily remain fixed. In general, the principal axis will slowly rotate around the direction of **L**, as shown in Fig. 8.24(*b*); this motion is called *nutation* and will be superimposed on the precession that results from the presence of a torque. To find the half-angle of the nutation cone, we consider the case of no external torques; thus **L** and $L^2$ are conserved

$$L^2 = I^2\omega_x^2 + I_t^2(\omega_y^2 + \omega_z^2) = \text{constant} \tag{8.76}$$

The kinetic energy of the gyro is also conserved

$$E = \tfrac{1}{2}\mathbf{L}\cdot\boldsymbol{\omega} = \tfrac{1}{2}I\omega_x^2 + \tfrac{1}{2}I_t(\omega_y^2 + \omega_z^2) = \text{constant} \tag{8.76$'$}$$

Fig. 8.24. (*a*) Precession of a gyroscope under the influence of a torque $\boldsymbol{\tau}$. (*b*) Nutation of a gyroscope in the absence of external torques; the principal rotation axis describes a cone around the angular momentum vector which is fixed in space.

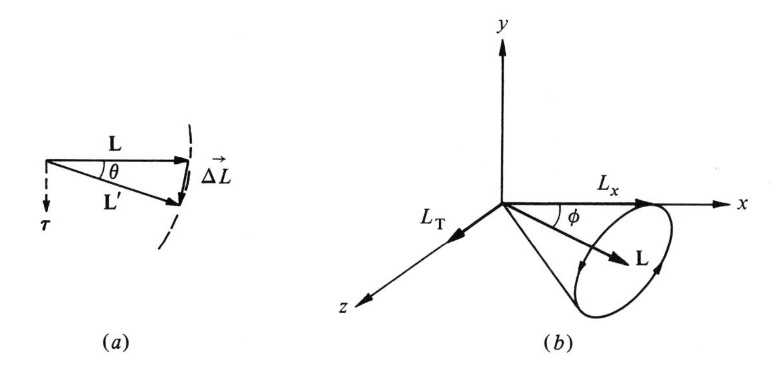

Since $I$ and $I_t$ are independent, the two Eqs. (8.76) can be satisfied simultaneously only if $\omega_x^2$ and $(\omega_y^2 + \omega_z^2)$ are independently constant throughout the motion. Therefore the principal axis lies on a cone centered on **L** and with half-angle $\phi$, where

$$\phi = \tan^{-1}\left\{\frac{I_t[(\omega_y^2 + \omega_z^2)]^{1/2}}{I\omega_x}\right\} = \text{constant} \tag{8.76''}$$

In terms of the sketch of Fig. 8.24(*b*), the transverse angular momentum is free to rotate in the (gyro-fixed) *y*–*z* plane, and its motion is determined by the initial conditions.*

Gyros are built with freedom to rotate about one or both axes, in addition to the principal axis. As an example we will discuss the single-degree-of-freedom gyro shown in Fig. 8.25. This type of gyro was developed by C. S. Draper at M.I.T. and is widely used; the spinning wheel assembly is contained in an enclosure which floats at neutral buoyancy in a high viscosity fluid to reduce gravitational forces on the bearings. When the case rotates about the 'input axis' the spin reference axis tends to precess about the free axis – labeled in the figure as the 'output axis' – in the direction shown. Such rotation induces in the generator a signal which after amplification drives the torque motor. However, exerting a torque on the gyro (along the output axis) forces the gyro, and therefore its case, to precess around the input axis; the precession

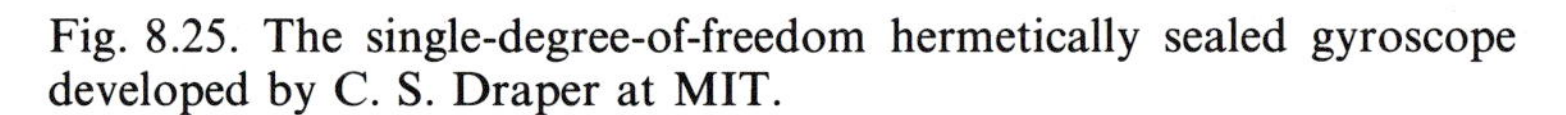

Fig. 8.25. The single-degree-of-freedom hermetically sealed gyroscope developed by C. S. Draper at MIT.

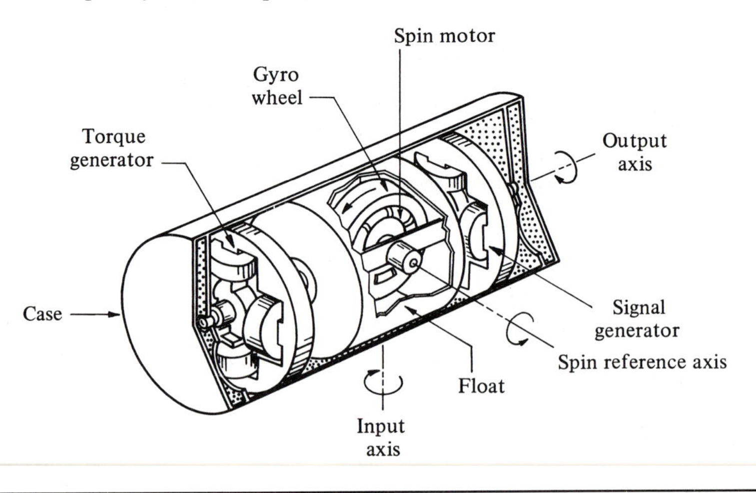


* The exact motion of the spin axis is obtained from solving Eq. (8.74) for $\tau = 0$ after introducing **L** from Eq. (8.73).

is such that the reference axis is maintained exactly at its original position in space. Different arrangements in the servomechanism make it possible to measure either the rate of precession or the total angle of rotation.

The measurement of the very small rate of rotation that is necessary for precise intertial guidance can also be achieved with optical means. If two em waves, for instance light beams, travel in a closed path in opposite directions, the transit time will be different for the two waves if the whole assembly rotates with respect to an inertial frame. This can be understood with the help of Fig. 8.26(*a*) if it is argued that because of the relative motion between the sending and receiving end, the path in one direction is effectively shorter than in the other direction.

We can obtain the correct answer for the path difference by an elementary argument as follows: the path difference will be $\Delta L = T \times \Delta v$ where $\Delta v$ is the difference in the velocity of propagation of the two beams, and $T$ is the time of propagation around the loop. If we take a circular path of radius $R$, rotating with angular velocity $\Omega$

$$\Delta v = 2R\Omega \qquad T = 2\pi R/c$$

and therefore

$$\Delta L = \frac{4\pi R^2\Omega}{c} = \frac{4A\Omega}{c} \tag{8.77}$$

We expressed the path difference in terms of the area $A$ enclosed by the loop, and the result of Eq. (8.77) is exact even through the precise derivation involves considerations on the propagation of light in an

Fig. 8.26. Principle of operation of an optical (ring laser) gyro. (*a*) The effective path length for light propagating in an optical fiber loop along, or counter to the direction of rotation is different and can be detected by interference techniques. (*b*) The open ring gyro is formed using mirrors. (*c*) An active medium is introduced in the path so that the system can lase, eliminating the need to inject and extract the light.

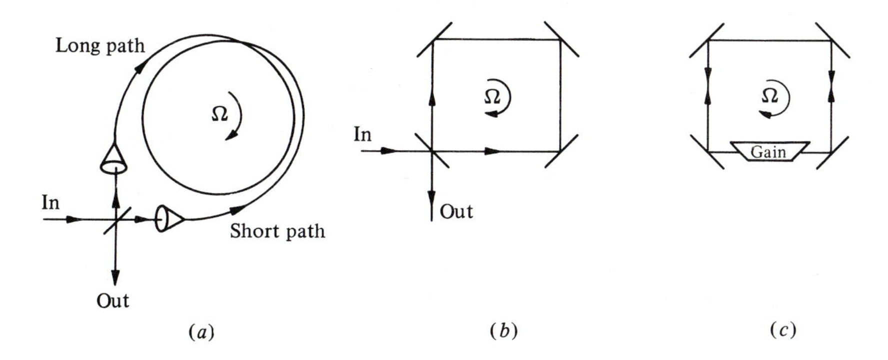

accelerated frame of reference. This phenomenon is known as the *Sagnac* effect, first discussed in 1913.

In spite of the smallness of $\Delta L$, one can use interferometric techniques to measure the phase difference between the two beams, and amplify the effect by letting the beams make many, say $N$, revolutions around the ring. Then

$$\Delta\phi = \frac{8\pi AN}{\lambda c}\Omega \tag{8.78}$$

To appreciate the orders of magnitude involved we choose $A = 100\ \text{cm}^2$, $N = 1000$, $\lambda = 500$ nm and a rotation rate $\Omega \simeq 7 \times 10^{-8}$ rad/s; this corresponds to $\Omega \simeq 10^{-3}\Omega_E$ where $\Omega_E$ is the rotation of the earth. For these parameters $\Delta\phi \simeq 10^{-5}$ radians, namely $3 \times 10^{-6}$ of an interference fringe. Ring laser gyros have reached this accuracy which in more familiar units corresponds to $\delta\Omega \sim 0.01°$/hour.

Ring laser gyros can be configured as shown in Fig. 8.26(*a*) by using optical fibers and injecting/extracting the two beams. Other possibilities are to use open mirrors arranged in a ring pattern as shown in (*b*) of the figure and again inject and extract the beams. More interestingly, a lasing (gain) medium can be included in the ring path as in (*c*). In this case the two beams have different frequency where $\Delta f = 4A\Omega/l\lambda$, with $l$ the length of the ring. The development of laser gyros is being pursued because of their compactness and relatively simpler construction as compared to mechanical gyros, rather than because of superior precision.

## Exercises

### *Exercise 8.1*

The term 'low orbit' implies a circular orbit around a planet with $r = R_{\text{planet}}$.

(a) Calculate the speed of a satellite in low orbit around the earth and around Mercury.

(b) Find the time for one revolution.

### *Exercise 8.2*

Consider a Hohman transfer between two circular orbits of radius $a_I$ and $a_F$ ($a_F > a_I$). Show that the energy increments corresponding to the velocity increments $\Delta V_A$ and $\Delta V_B$ add up, as they must, to

$$\Delta E_A + \Delta E_B = \frac{Km}{2}\left[\frac{1}{a_I} - \frac{1}{a_F}\right]$$

Here $K = GM_\odot$ and $m$ is the mass of the vehicle.

### *Exercise 8.3*

Analyse the Voyager-2 swing by Uranus that took place in January of 1986. The trajectory in the heliocentric system is shown in Fig. 8.13 and the hyperbolic excess velocity is

$$v_\infty(\text{Uranus}) = 14\,730 \text{ m/s}$$

The aim point is at a distance of closest approach

$$\text{C.A.} = 4.21(\text{Uranus radii}) = 107\,080 \text{ km}$$

(a) Find the heliocentric velocities of approach and departure from Uranus.
(b) Find the heliocentric velocity of approach to Neptune and the time of flight.
(c) Assuming that the Neptune encounter does not alter the trajectory significantly, find the velocity of Voyager-2 after escape from the solar system.

### *Exercise 8.4*

(a) If you were to use a 100% reflecting surface as a solar sail so as to navigate away from the earth, how would you deploy it?
(b) For a sail area of 1 $\text{km}^2$ what would be the thrust?
(c) What is the velocity change for a 1000 kg vehicle after a year of sailing at the earth's orbit?

The solar constant at the radius of the earth is $S = 0.136 \text{ W/cm}^2$.

### *Exercise 8.5*

Consider a rocket accelerated by the ejection of ions. Use reasonable values to calculate the rocket velocity that can be achieved for a payload of 10 tons.

### *Exercise 8.6*

Show that a coordinate system fixed to the *center* of the earth can be considered as an inertial coordinate system to an accuracy of a few parts in $10^8$.

# Appendix 1

## THE FOURIER TRANSFORM

Consider a periodic function of time $V(t)$ with period $T$

$$f_0 = \frac{1}{T} \qquad \omega_0 = 2\pi f_0 = \frac{2\pi}{T} \tag{A1.1}$$

$V(t)$ can always be decomposed into an even and an odd part

$$V(t) = V_e(t) + V_o(t)$$

$$V_e(t) = V_e(-t) \qquad V_o(t) = -V_o(t) \tag{A1.2}$$

The Fourier theorem assures us that we can express $V_e(t)$ and $V_o(t)$ as a series of harmonics of $\omega_0$

$$V_e(t) = A_0 + \sum_{n=1}^{\infty} A_n \cos(n\omega_0 t) \tag{A1.3}$$

$$V_o(t) = \sum_{n=1}^{\infty} B_n \sin(n\omega_0 t) \tag{A1.4}$$

The coefficients $A_n$ are found by multiplying both sides of Eq. (A1.3) by $\cos(m\omega_0 t)$ and integrating over $\mathrm{d}t$ from $-T/2$ to $+T/2$

$$\int_{-T/2}^{T/2} V_e(t)\cos(m\omega_0 t)\,\mathrm{d}t = \int_{-T/2}^{T/2} A_0 \cos(m\omega_0 t)\,\mathrm{d}t + \sum_{n=1}^{\infty} \int_{-T/2}^{+T/2} A_n \cos(n\omega_0 t)\cos(m\omega_0 t)\,\mathrm{d}t \tag{A1.5}$$

The cosine functions of the harmonics of $\omega_0$ form an orthogonal set in the interval $-T/2$ to $+T/2$

$$\int_{-T/2}^{T/2} \cos(n\omega_0 t)\cos(m\omega_0 t)\,\mathrm{d}t = \frac{T}{2}\delta_{mn} \qquad m, n \neq 0$$

$$= T \qquad m = n = 0$$

Thus we obtain from Eq. (A1.5)

$$A_0 = \frac{1}{T}\int_{-T/2}^{T/2} V(t)\,\mathrm{d}t \tag{A1.6}$$

$$A_n = \frac{2}{T}\int_{-T/2}^{T/2} V(t)\cos(n_0\omega_0 t)\,\mathrm{d}t \tag{A1.7}$$

By a similar argument and using the orthogonality of the sine functions we obtain for the $B_n$ coefficients

$$B_n = \frac{2}{T}\int_{-T/2}^{T/2} V(t)\sin(n\omega_0 t)\,\mathrm{d}t \tag{A1.8}$$

For a continuous, non-periodic function we must use a Fourier integral. We will use complex notation, so that even if the function $V(t)$ is real, the Fourier transform $A(\omega)$ may be complex. Complex notation greatly simplifies calculations and is widely used; note that a complex function contains both amplitude and phase information. We express the function $V(t)$ through the integral

$$V(t) = \frac{1}{(2\pi)^{1/2}}\int_{-\infty}^{\infty} A(\omega)\mathrm{e}^{-\mathrm{i}\omega t}\,\mathrm{d}\omega \tag{A1.9}$$

To determine $A(\omega)$ we multiply by $\mathrm{e}^{\mathrm{i}\omega' t}$ and integrate over time

$$\int_{-\infty}^{\infty} V(t)\mathrm{e}^{\mathrm{i}\omega' t}\,\mathrm{d}t = \frac{1}{(2\pi)^{1/2}}\int_{-\infty}^{\infty} A(\omega)\,\mathrm{d}\omega \int_{-\infty}^{\infty} \mathrm{e}^{-i(\omega-\omega')t}\,\mathrm{d}t$$

$$= \frac{1}{(2\pi)^{1/2}}\int_{-\infty}^{\infty} A(\omega)2\pi\,\delta(\omega-\omega')\,\mathrm{d}\omega$$

$$= (2\pi)^{1/2}A(\omega')$$

Thus

$$A(\omega) = \frac{1}{(2\pi)^{1/2}}\int_{-\infty}^{\infty} V(t)\mathrm{e}^{\mathrm{i}\omega t}\,\mathrm{d}t \tag{A1.10}$$

If $V(t)$ is real, $A(\omega)$ can be expressed in terms of trigonometric functions

$$\left.\begin{aligned} \mathrm{Re}[A(\omega)] &= \frac{1}{(2\pi)^{1/2}}\int_{-\infty}^{\infty} V(t)\cos(\omega t)\,\mathrm{d}t \\ \mathrm{Im}[A(\omega)] &= \frac{1}{(2\pi)^{1/2}}\int_{-\infty}^{\infty} V(t)\sin(\omega t)\,\mathrm{d}t \end{aligned}\right\} \tag{A1.11}$$

Furthermore when $V(t)$ is real, negative and positive frequencies are related through

$$A(-\omega) = A^*(\omega) \tag{A1.12}$$

as follows immediately from Eqs. (A1.11).

If $V(t)$ is real and an even function of $t$, $A(\omega)$ is real and even in $\omega$; then

$$V_e(t) = \frac{2}{(2\pi)^{1/2}} \int_0^\infty A(\omega)\cos(\omega t)\, d\omega \tag{A1.13}$$

If $V(t)$ is real and an odd function of $t$, $A(\omega)$ is imaginary and odd in $\omega$; then

$$V_o(t) = \frac{-2i}{(2\pi)^{1/2}} \int_0^\infty A(\omega)\sin(\omega t)\, d\omega \tag{A1.14}$$

Note that in the expressions (A1.13, 14) the integration is only over positive frequencies. If the symmetry properties of $V(t)$ are specified, the integration over $dt$ in Eqs. (A1.11) can be restricted to the interval $0 < t < \infty$.

In deriving Eq. (A1.10) we made use of the fundamental relation

$$\int_{-\infty}^{\infty} e^{i(\omega - \omega')t}\, dt = 2\pi\, \delta(\omega - \omega') \tag{A1.15}$$

Eq. (A1.15) is a representation of the Dirac delta function which can be defined through

$$\int_{-\infty}^{\infty} f(x)\, \delta(x - x_0)\, dx = f(x_0) \tag{A1.16}$$

One way of arriving at the $\delta$-function is based on a limiting procedure that can be obtained from Eq. (A1.15). We let $\omega - \omega' = \lambda$, then

$$\delta(\lambda) = \lim_{T\to\infty} \frac{1}{2\pi} \int_{-T}^{T} e^{i\lambda T}\, dt = \lim_{T\to\infty} \frac{1}{2\pi i\lambda}\left(e^{i\lambda T} - e^{-i\lambda T}\right)$$

$$= \lim_{T\to\infty} \left[\frac{T}{\pi} \frac{\sin \lambda T}{\lambda T}\right]$$

As $T \to \infty$, $\quad \dfrac{\sin(\lambda T)}{\lambda T} \to 1 \quad$ if $\lambda = 0$

$\qquad\qquad \to 0 \quad \lambda \neq 0$

Thus $\quad \delta(\lambda) \to (T/\pi) \to \infty \quad \lambda = 0$

$\qquad\qquad \to 0 \quad \lambda \neq 0$

But

$$\int_{-\infty}^{\infty} \delta(\lambda)\, d\lambda = \lim_{T\to\infty} \int_{-\infty}^{\infty} \frac{T}{\pi} \frac{\sin \lambda T}{\lambda T}\, d\lambda = \lim \frac{2}{\pi} \int_0^\infty \frac{\sin x}{x}\, dx = 1$$

# Appendix 2

## THE POWER SPECTRUM

We define the *correlation* $C(\tau)$ between two functions of time $f_1(t)$ and $f_2(t)$ as the average value of their shifted overlap*

$$C(\tau) = \lim_{T\to\infty} \frac{1}{T}\int_{-T/2}^{T/2} f_1(t) f_2(t-\tau)\,\mathrm{d}t \tag{A2.1}$$

The *autocorrelation* $R(\tau)$ is defined through

$$R(\tau) = \lim_{T\to\infty} \frac{1}{T}\int_{-T/2}^{T/2} f(t) f(t-\tau)\,\mathrm{d}t \tag{A2.2}$$

$R(\tau)$ measures how the function $f(t)$ remains self-similar over the time interval $\tau$. For instance if $f(t)$ is constant, then

$$R(\tau) = \frac{1}{T}\int_{-T/2}^{T/2} A^2\,\mathrm{d}t = A^2$$

If $f(t) = A\cos\omega t$, then

$$\begin{aligned} R(\tau) &= \frac{A^2}{T}\int_{-T/2}^{T/2} \cos\omega t\cos(\omega t-\omega\tau)\,\mathrm{d}t \\ &= \frac{A^2}{T}\int_{-T/2}^{T/2} [\cos^2\omega t\cos\omega\tau + \cos\omega t\sin\omega t\sin\omega\tau]\,\mathrm{d}t \\ &= \frac{A^2}{2}\cos\omega\tau \end{aligned}$$

as expected.

Next we evaluate the Fourier transform of the autocorrelation of $y(t)$

$$\int_{-\infty}^{\infty} R(\tau)\mathrm{e}^{\mathrm{i}\omega\tau}\,\mathrm{d}\tau = \int_{-\infty}^{\infty} \lim_{T\to\infty}\frac{1}{T}\int_{-T/2}^{T/2} y(t)y(t-\tau)\mathrm{e}^{\mathrm{i}\omega\tau}\,\mathrm{d}t\,\mathrm{d}\tau \tag{A2.3}$$

* In more advanced treatments the ensemble average is used instead of an integral over time.

We can insert $e^{i\omega t}e^{-i\omega t} \equiv 1$ into the integrand and obtain for the r.h.s. of Eq. (A2.3)

$$\lim_{T\to\infty} \frac{2\pi}{T}\frac{1}{(2\pi)^{1/2}}\int_{-T/2}^{T/2} y(t)e^{i\omega t}\,dt\,\frac{1}{(2\pi)^{1/2}}\int_{-\infty}^{\infty} y(t-\tau)e^{-i\omega(t-\tau)}\,d\tau$$

This is the product of two terms, each of which is the Fourier transform of $y(t)$, the second one being complex conjugated. Thus the r.h.s. of Eq. (A2.3) becomes

$$\lim_{T\to\infty} \frac{2\pi}{T}[g_T(\omega)g_T^*(\omega)] = G(\omega)$$

and therefore we have shown that the power spectrum is the Fourier transform of the autocorrelation function (Eq. (3.34)).

$$G(\omega) = \int_{-\infty}^{\infty} R(\tau)e^{i\omega\tau}\,d\tau \tag{A2.4}$$

The inverse relation is obviously (Eq. (3.35))

$$R(\tau) = \frac{1}{2\pi}\int_{-\infty}^{\infty} G(\omega)e^{-i\omega\tau}\,d\omega \tag{A2.5}$$

Therefore it follows immediately that

$$R(0) = \langle y(t)^2\rangle = \int_{-\infty}^{\infty} G(\omega)\,df \tag{A2.6}$$

This result is an expression of *Parseval's theorem* which states that the same total power is obtained either in the time or frequency domain

$$\int_{-\infty}^{\infty} |y(t)|^2\,dt = \int_{-\infty}^{\infty} |g(\omega)|^2\,d\omega \tag{A2.7}$$

where $g(\omega)$ is the Fourier transform of $y(t)$.

# Appendix 3

## THE EQUATIONS OF FLUID MECHANICS

The dynamics of a non-viscous fluid are completely determined by two equations: the continuity equation, and Euler's equation which expresses Newton's equations for a fluid. We have made use of these equations in the text but always in a simplified form. For completeness we give here their full expressions using the notation of vector calculus, but without seeking or suggesting solutions.

(a) *The continuity equation*

$$\nabla \cdot (\rho \mathbf{v}) + \frac{\partial \rho}{\partial t} = 0 \tag{A3.1}$$

For an incompressible fluid, $\rho$ is constant in space and time so that

$$\frac{\partial \rho}{\partial t} = \frac{\partial \rho}{\partial x} = \frac{\partial \rho}{\partial y} = \frac{\partial \rho}{\partial z} = 0 \tag{A3.2}$$

and the continuity equation takes the form

$$\nabla \cdot \mathbf{v} = 0 \tag{A3.3}$$

For stationary (or steady) flow

$$\frac{\partial \rho}{\partial t} = \frac{\partial \mathbf{v}}{\partial t} = \frac{\partial P}{\partial t} = 0 \tag{A3.4}$$

and the continuity equation takes the form

$$\nabla \cdot (\rho \mathbf{v}) = \rho(\nabla \cdot \mathbf{v}) + \mathbf{v} \cdot (\nabla \rho) = 0 \tag{A3.5}$$

(b) *Euler's equation*

$$\frac{\mathrm{d}\mathbf{v}}{\mathrm{d}t} = \mathbf{f} - \frac{1}{\rho} \nabla P \tag{A3.6}$$

Here **f** represents the external body forces *per unit mass* and **v** is the velocity of a fluid element; $P$ is the pressure. Note that $\mathrm{d}\mathbf{v}/\mathrm{d}t$ is a total derivative so that

$$\frac{\mathrm{d}\mathbf{v}}{\mathrm{d}t}=\frac{\partial\mathbf{v}}{\partial x}\frac{\partial x}{\partial t}+\frac{\partial\mathbf{v}}{\partial y}\frac{\partial y}{\partial t}+\frac{\partial\mathbf{v}}{\partial z}\frac{\partial z}{\partial t}+\frac{\partial\mathbf{v}}{\partial t} \tag{A3.7}$$

where $x$, $y$, $z$ and $t$ are treated as independent variables. Eq. (A3.7) can be written in vector notation as

$$\frac{\mathrm{d}\mathbf{v}}{\mathrm{d}t}=(\mathbf{v}\cdot\nabla)\mathbf{v}+\frac{\partial\mathbf{v}}{\partial t}=\left[\mathbf{v}\cdot\nabla+\frac{\partial}{\partial t}\right]\mathbf{v} \tag{A3.7$'$}$$

where $(\mathbf{v}\cdot\nabla)$ has the special meaning implied by Eqs. (A3.7). Thus we introduce the concept of *convective derivative*

$$\frac{\mathrm{d}}{\mathrm{d}t}=\frac{\partial}{\partial t}+\mathbf{v}\cdot\nabla \tag{A3.8}$$

(c) *Special cases: low velocity*

When $v \ll v_s$ we can neglect terms proportional to $v$, but we keep the derivatives of $v$. Then the continuity equation becomes

$$\rho(\nabla\cdot\mathbf{v})=-\frac{\partial\rho}{\partial t} \tag{A3.9}$$

Euler's equation, with $\mathbf{f}=0$, yields

$$\frac{\mathrm{d}\mathbf{v}}{\mathrm{d}t}=(\mathbf{v}\cdot\nabla)\mathbf{v}+\frac{\partial\mathbf{v}}{\partial t}=-\frac{1}{\rho}\nabla P$$

and ignoring the non-linear term

$$\frac{\partial\mathbf{v}}{\partial t}=-\frac{1}{\rho}\nabla P \tag{A3.10}$$

Next we take the divergence of Eq. (A3.10) and the partial time derivative of Eq. (A3.9) (where we neglect a term of order $\rho(\nabla\cdot\mathbf{v})^2$), to eliminate the velocity term. We then find

$$\nabla^2 P=\frac{\partial^2\rho}{\partial t^2} \tag{A3.11}$$

The equation of state of the gas or liquid provides a relation between $\rho$ and $P$ for the particular process that takes place. This allows us to express $P$ as a function of $\rho$ (or vice versa) so as to obtain a wave equation for the pressure or density. This is the subject of Appendix 4 but we wanted to stress here the approximations that go into deriving the linear

Eq. (A3.11). For an adiabatic process in an ideal gas we find

$$\nabla^2 \rho - \frac{1}{c^2}\frac{\partial^2 \rho}{\partial t^2} = 0 \tag{A3.12}$$

with

$$c^2 = \gamma P/\rho \tag{A3.12$'$}$$

and $\gamma = c_P/c_V$. The speed of sound is then $v_s = \sqrt{c^2}$.

(d) *Special cases: supersonic flow*

From the equation of state (see Eqs. (A3.11 and 12)) we have the relation

$$\nabla P = c^2 \nabla \rho \tag{A3.13}$$

so that Euler's equation with $\mathbf{f} = 0$ gives

$$\frac{d\mathbf{v}}{dt} = -\frac{c^2}{\rho}\nabla \rho \tag{A3.14}$$

We take the dot product of Eq. (A3.14) with $\mathbf{v}$, to obtain

$$\mathbf{v}\cdot\frac{d\mathbf{v}}{dt} = -\frac{c^2}{\rho}\mathbf{v}\cdot(\nabla \rho) \tag{A3.14$'$}$$

We now expand the continuity equation

$$\mathbf{v}\cdot(\nabla \rho) + \rho(\nabla\cdot\mathbf{v}) = -\frac{\partial \rho}{\partial t} \tag{A3.15}$$

Comparing Eqs. (A3.14′ and 15) we have the equation for supersonic flow

$$-\frac{\rho}{c^2}\mathbf{v}\cdot\frac{d\mathbf{v}}{dt} + \rho(\nabla\cdot\mathbf{v}) = -\frac{\partial \rho}{\partial t} \tag{A3.16}$$

This is a highly non-linear equation because $(d\mathbf{v}/dt)$ is already a non-linear term (see Eq. (A3.7′)) and it is further multiplied by $\mathbf{v}$.

(e) *Viscous fluid*

In the presence of viscosity the Euler equation is replaced by the Navier–Stokes equation

$$\frac{d\mathbf{v}}{dt} = \mathbf{f} - \frac{1}{\rho}\nabla P + \frac{1}{3}\frac{\eta}{\rho}\nabla(\nabla\cdot\mathbf{v}) + \frac{\eta}{\rho}(\nabla^2\mathbf{v}) \tag{A3.17}$$

Here $\eta$ is the coefficient of viscosity and we note again that the equation is highly non-linear. It is for this reason that the solutions to fluid flow problems are highly complex and difficult to obtain analytically. The Navier–Stokes equation is valid for small Reynolds numbers.

# Appendix 4

## THE SPEED OF SOUND

Sound is due to *small* local variations in the pressure and density of a liquid, a gas or a solid. These variations obey a wave equation and therefore can propagate through the medium. In Appendix 3 we have shown that small variations in density and pressure are related through Eq. (A3.11).

$$\nabla^2 P = \frac{\partial^2 \rho}{\partial t^2} \tag{A4.1}$$

For a gas the relation between pressure and density depends on the thermodynamic process that takes place. Sound is an *adiabatic* process because the variations in density and pressure are too fast to permit the transfer of heat. Thus

$$PV^{\gamma} = \text{constant} \tag{A4.2}$$

Since the density $\rho$ is given by $\rho = nM/V$ ($n$ = number of moles, $M$ = molecular weight) we have

$$P = K\rho^{\gamma} \tag{A4.3}$$

where $K$ is an arbitrary constant and $\gamma = c_P/c_v$.

We take the gradient of Eq. (A4.3)

$$\nabla P = \gamma K \rho^{\gamma-1} \nabla \rho = \frac{\gamma}{\rho} K \rho^{\gamma} \nabla \rho \tag{A4.4}$$

and if we take the divergence of Eq. (A4.4) then we obtain to *first order*

$$\nabla \cdot (\nabla P) = \nabla^2 P = \frac{\gamma}{\rho} K \rho^{\gamma} \nabla^2 \rho = \left( \frac{\gamma P}{\rho} \right) \nabla^2 \rho \tag{A4.4$'$}$$

Thus Eq. (A4.1) becomes a wave equation

$$\nabla^2 \rho - \frac{1}{c^2} \frac{\partial^2 \rho}{\partial t^2} = 0 \tag{A4.5}$$

with

$$c^2 = (\gamma P/\rho) \tag{A4.6}$$

For an ideal gas

$$PV = nRT$$

with $n$ the number of moles and $R$ the gas constant. Thus

$$\frac{P}{\rho} = \frac{RT}{M}$$

and the speed of sound in a gas is given by

$$v_s = (\gamma RT/M)^{1/2} \tag{A4.7}$$

Eq. (A4.7) was used in the main text to evaluate $v_s$ for air at s.t.p.

The result of Eq. (A4.4′) can be written in general in the form

$$\nabla^2 P = \left(\frac{dP}{d\rho}\right)\nabla^2 \rho \tag{A4.8}$$

Therefore the speed of sound is given by

$$v_s = (dP/d\rho)^{1/2} \tag{A4.9}$$

For gases we found $dP/d\rho$ from Eq. (A4.3); for liquids we can use the compressibility $\kappa$, which is defined through

$$\kappa = -\frac{1}{\rho}\frac{\partial\rho}{\partial P} \tag{A4.10}$$

For instance for water $\kappa = -50 \times 10^{-6}$ (atm$^{-1}$). Then

$$\frac{dP}{d\rho} = \frac{1}{\kappa\rho} = \frac{1}{(10^3\ \mathrm{kg/m^3})50 \times 10^{-11}(\mathrm{m^2/N})} = 2 \times 10^6\ \frac{\mathrm{m^2}}{\mathrm{s^2}}$$

and therefore

$$v_s = (2 \times 10^6)^{1/2} = 1.4 \times 10^3\ \mathrm{m/s}$$

For solids $(dP/d\rho)$ is equivalent to $(Y/\rho)$ where $Y$ is Young's modulus (defined as the ratio of normal stress $S_n = F/A$ to strain $\varepsilon = \delta l/l$). Then

$$Y = \frac{F/A}{\delta l/l} \quad \text{and} \quad \frac{Y}{\rho} = \frac{Fl}{\rho A\delta l} = \frac{Fl}{\rho\delta V} = \frac{Fl/V}{\rho\delta V/V} = -\frac{\delta P}{\delta\rho} = v_s^2 \tag{A4.11}$$

For steel, $Y = 2 \times 10^{11}\ \mathrm{N/m^2}$, $\rho \simeq 8 \times 10^3\ \mathrm{kg/m^3}$, and we find for the speed of sound

$$v_s = \left(\frac{Y}{\rho}\right)^{1/2} = \left(\frac{2 \times 10^{11}}{8 \times 10^3}\right)^{1/2} = 1.6 \times 10^4\ \mathrm{m/s}$$

# REFERENCES AND SUGGESTIONS FOR FURTHER READING


CHAPTER 1

A. S. Grove, *Physics and Technology of Semiconductor Devices*, J. Wiley, New York, 1967.

C. Mead and L. Conway, *Introduction to VLSI Systems*, Addison-Wesley, Reading, MA, 1980.

R. F. Pierret and G. W. Neudeck, *Modular Series on Solid State Devices*, Addison-Wesley, Reading, MA, 1983.

CHAPTER 2

W. C. Holton, The large scale integration of microelectronic circuits, *Scientific American*, September 1977.

P. Horowitz and W. Hill, *The Art of Electronics*, Cambridge University Press, Cambridge, U.K., 1980.

M. M. Mano, *Computer System Architecture*, Prentice-Hall, Englewood Cliffs, N.J., 1980.

CHAPTER 3

J. Brown and E. V. D. Glazier, *Telecommunications*, Chapman and Hall, London, 1974.

R. M. Gagliardi, *Satellite Communications*, Wadsworth, London, 1984.

CHAPTER 4

J. B. Marion, *Classical Electromagnetic Radiation*, Academic Press, New York, 1965.

Y. Suematsu and K.-I. Iga, *Optical Fiber Communications*, J. Wiley, New York, 1976.

O. Svelto, *Principles of Lasers*, translated by D. C. Hanna, Plenum Press, New York, 1982.

CHAPTER 5

D. R. Inglis, *Nuclear Energy*, Addison-Wesley, Reading, MA, 1973.

J. D. McGervey, *Introduction to Modern Physics*, Academic Press, New York, 1983.

E. H. Thorndike, *Energy and Environment*, Addison-Wesley, Reading, MA, 1976.

CHAPTER 6

P. Craig and J. Jungerman, *Nuclear Arms Race*, McGraw-Hill, New York, 1985.

S. Glasstone and P. J. Dolan, *The Effects of Nuclear Weapons*, U.S. Government Printing Office, Washington, 1977.

D. Schroeer, *Science Technology and the Nuclear Arms Race*, J. Wiley, New York, 1984.

The science and technology of directed energy weapons, *Reviews of Modern Physics*, Vol. 59, July 1987.

CHAPTER 7

J. D. Anderson, *Introduction to Flight*, McGraw-Hill, New York, 1985.
A. H. Shapiro, *Shape and Flow*, Doubleday, Anchor Books, Garden City, N.Y., 1961.
G. P. Sutton, *Rocket Propulsion Elements*, J. Wiley, New York, 1963.
Th. von Karman, *Aerodynamics*, McGraw-Hill, New York, 1954.

CHAPTER 8

A. I. Berman, *The Physical Principles of Astronautics*, J. Wiley, New York, 1961.
C. S. Draper, W. Wrigley and J. Hovorka, *Inertial Guidance*, Pergamon, New York, 1960.
H. Seifert (editor), *Space Technology*, J. Wiley, New York, 1961.
A. B. Sergeyevsky, Voyager-2: A grand tour of the giant planets, *AAS/AIAA Astrodynamics Conference, Lake Tahoe, Nevada*, August 1981.

# INDEX